T0135362

Lecture Notes in Information Systems and Organisation

Volume 58

Lecture Notes in Information Systems and Organization—LNISO—is a series of scientific books that explore the current scenario of information systems, in particular IS and organization. The focus on the relationship between IT, IS and organization is the common thread of this collection, which aspires to provide scholars across the world with a point of reference and comparison in the study and research of information systems and organization. LNISO is the publication forum for the community of scholars investigating behavioral and design aspects of IS and organization. The series offers an integrated publication platform for high-quality conferences, symposia and workshops in this field. Materials are published upon a strictly controlled double blind peer review evaluation made by selected reviewers.

LNISO is abstracted/indexed in Scopus

Fred D. Davis · René Riedl · Jan vom Brocke ·
Pierre-Majorique Léger · Adriane B. Randolph ·
Gernot R. Müller-Putz

Editors

Information Systems and Neuroscience

NeuroIS Retreat 2022

Editors
Fred D. Davis
Information Technology Division
Texas Tech University
Lubbock, TX, USA

Jan vom Brocke
Department of Information Systems
University of Liechtenstein
Vaduz, Liechtenstein

Adriane B. Randolph
Department of Information Systems
and Security
Kennesaw State University
Kennesaw, GA, USA

René Riedl 🆔
Digital Business Institute
University of Applied Sciences Upper
Austria
Steyr, Austria

Institute of Business
Informatics—Information Engineering
Johannes Kepler University Linz
Linz, Austria

Pierre-Majorique Léger
Department of Information Technologies
HEC Montreal
Montréal, QC, Canada

Gernot R. Müller-Putz
Institute of Neural Engineering
Graz University of Technology
Graz, Austria

ISSN 2195-4968   ISSN 2195-4976 (electronic)
Lecture Notes in Information Systems and Organisation
ISBN 978-3-031-13063-2   ISBN 978-3-031-13064-9 (eBook)
https://doi.org/10.1007/978-3-031-13064-9

Organization

Conference Co-chairs

Fred D. Davis, Texas Tech University, Lubbock, USA
René Riedl, University of Applied Sciences Upper Austria, Steyr, Austria and Johannes Kepler University Linz, Linz, Austria

Programme Co-chairs

Jan vom Brocke, University of Liechtenstein, Vaduz, Liechtenstein
Pierre-Majorique Léger, HEC Montréal, Montreal, Canada
Adriane B. Randolph, Kennesaw State University, Kennesaw, USA
Gernot R. Müller-Putz, Graz University of Technology, Graz, Austria

Programme Committee

Marc T. P. Adam, University of Newcastle, Callaghan, Australia
Bonnie B. Anderson, Brigham Young University, Utah, USA
Ricardo Buettner, University of Bayreuth, Bayreuth, Germany
Colin Conrad, Dalhousie University, Halifax, Canada
Constantinos Coursaris, HEC Montréal, Montreal, Canada
Alan R. Dennis, Indiana University, Indiana, USA
Thomas M. Fischer, Johannes Kepler University Linz, Linz, Austria
Rob Gleasure, Copenhagen Business School, Frederiksberg, Denmark
Jacek Gwizdka, University of Texas at Austin, Austin, Texas
Alan R. Hevner, Muma College of Business, Florida, USA
Marco Hubert, University of Aarhus, Aarhus, Denmark

Organization Support

Preface

The proceedings contain papers presented at the 14th annual NeuroIS Retreat held June 14–16 2022. NeuroIS is a field in Information Systems (IS) that uses neuroscience and neurophysiological tools and knowledge to better understand the development, adoption, and impact of information and communication technologies (www.neurois.org).

The NeuroIS Retreat is a leading academic conference for presenting research and development projects at the nexus of IS and neurobiology. This annual conference promotes the development of the NeuroIS field with activities primarily delivered by and for academics, though works often have a professional orientation.

In 2009, the inaugural NeuroIS Retreat was held in Gmunden, Austria. Since then, the NeuroIS community has grown steadily, with subsequent annual Retreats in Gmunden from 2010 to 2017. Beginning in 2018, the conference is taking place in Vienna, Austria. Due to the Corona crisis, the organizers decided to host the NeuroIS Retreat virtually in 2020 and 2021. This year, the NeuroIS Retreat took place again in a physical face-to-face-format in Vienna.

The NeuroIS Retreat provides a platform for scholars to discuss their studies and exchange ideas. A major goal is to provide feedback for scholars to advance their research papers toward high-quality journal publications. The organizing committee welcomes not only completed research but also work in progress. The NeuroIS Retreat is known for its informal and constructive workshop atmosphere. Many NeuroIS presentations have evolved into publications in highly regarded academic journals.

This year is the 8th time that we publish the proceedings in the form of an edited volume. A total of 35 research papers were accepted and are published in this volume, and we observe diversity in topics, theories, methods, and tools of the contributions in this book. The 2022 keynote presentation entitled "The Neurobiology of Trust: Benefits and Challenges for NeuroIS" is given by Frank Krueger, professor of systems social neuroscience at the School of Systems Biology at George Mason University (GMU), USA. Moreover, Jan vom Brocke, professor of information systems at the University of Liechtenstein, gives a hot topic talk entitled "From Neuro-adaptive

Systems to Neuro-adaptive Processes: Opportunities of NeuroIS to Contribute to the Emerging Field of Process Science".

Altogether, we are happy to see the ongoing progress in the NeuroIS field. Also, we can report that the NeuroIS Society, established in 2018 as a non-profit organization, has been developing well. We foresee a prosperous development of NeuroIS.

Lubbock, USA   Fred D. Davis
Steyr, Austria   René Riedl
Vaduz, Liechtenstein   Jan vom Brocke
Montréal, Canada   Pierre-Majorique Léger
Kennesaw, USA   Adriane B. Randolph
Graz, Austria   Gernot R. Müller-Putz
June 2022

Sponsors

We thank the sponsors of the NeuroIS Retreat 2022:

Main Sponsors

Technology Sponsors

Noldus

Information Technology

The Neurobiology of Trust: Benefits and Challenges for NeuroIS (Keynote)

Frank Krueger

Trust pervades nearly every aspect of our daily lives; it penetrates not only our human social interactions but also our interactions with information and communication technologies (ICTs). The talk provides an overarching neurobiological framework of trust—focusing on empirical, methodological, and theoretical aspects—that serves as a common basis for the broad and transdisciplinary community of trust research. The integration into a unified conceptual framework of trust can guide future investigations to better understand both fundamental and applied NeuroIS research in developing new theories and designing innovative ICT artifacts that positively affect practical outcomes for individuals, groups, organizations, and society.

From Neuro-adaptive Systems to Neuro-adaptive Processes: Opportunities of NeuroIS to Contribute to the Emerging Field of Process Science (Hot Topic Talk)

Jan vom Brocke

A hugely fascinating aspect of NeuroIS is the prospect of developing neuro-adaptive systems—in simple terms, information systems (IS) that are sensitive to emotions and thoughts. A recent research agendum published in the European Journal of Information Systems presents four areas to advance NeuroIS research towards societal contributions: (1) IS design, (2) IS use, (3) emotion research, and (4) neuro-adaptive systems. All four areas contribute to an intriguing new field called Process Science, which can further leverage the emission sensitivity of systems to processes, that way making important contributions of value to society. This Hot Topic Talk further outlines this idea and makes a call for NeuroIS contributions to Process Science.

Contents

Our Brain Reads, While We Can't: EEG Reveals Word-Specific Brain Activity in the Absence of Word Recognition 1
Peter Walla, Robin Leybourne, and Samuil Pavlevchev

Information Overload and Argumentation Changes in Product Reviews: Evidence from NeuroIS 9
Florian Popp, Bernhard Lutz, and Dirk Neumann

Flow in Knowledge Work: An Initial Evaluation of Flow Psychophysiology Across Three Cognitive Tasks 23
Karen Bartholomeyczik, Michael Thomas Knierim, Petra Nieken, Julia Seitz, Fabio Stano, and Christof Weinhardt

Biosignal-Based Recognition of Cognitive Load: A Systematic Review of Public Datasets and Classifiers 35
Julia Seitz and Alexander Maedche

Enhancing Wireless Non-invasive Brain-Computer Interfaces with an Encoder/Decoder Machine Learning Model Pair 53
Ernst R. Fanfan, Joe Blankenship, Sumit Chakravarty, and Adriane B. Randolph

How Does the Content of Crowdfunding Campaign Pictures Impact Donations for Cancer Treatment 61
Andreas Blicher, Rob Gleasure, Ioanna Constantiou, and Jesper Clement

All Eyes on Misinformation and Social Media Consumption: A Pupil Dilation Study 73
Mahdi Mirhoseini, Spencer Early, and Khaled Hassanein

Increased Audiovisual Immersion Associated with Mirror Neuron System Enhancement Following High Fidelity Vibrokinetic Stimulation 81
Kajamathy Subramaniam, Jared Boasen, Félix Giroux,
Sylvain Sénécal, Pierre-Majorique Léger, and Michel Paquette

Recognizing Polychronic-Monochronic Tendency of Individuals Using Eye Tracking and Machine Learning 89
Simon Barth, Moritz Langner, Peyman Toreini, and Alexander Maedche

The Effect of SVO Category on Theta/Alpha Ratio Distribution in Resource Allocation Tasks 97
Dor Mizrahi, Ilan Laufer, and Inon Zuckerman

Is Our Ability to Detect Errors an Indicator of Mind Wandering? An Experiment Proposal 105
Colin Conrad, Michael Klesel, Kydra Mayhew, Kiera O'Neil,
Frederike Marie Oschinsky, and Francesco Usai

Resolving the Paradoxical Effect of Human-Like Typing Errors by Conversational Agents 113
R. Stefan Greulich and Alfred Benedikt Brendel

The View of Participants on the Potential of Conducting NeuroIS Studies in the Wild 123
Anke Greif-Winzrieth, Christian Peukert, Peyman Toreini,
and Christof Weinhardt

Emoticons Elicit Similar Patterns of Brain Activity to Those Elicited by Faces: An EEG Study 133
Alessandra Flöck and Marc Mehu

Facilitating NeuroIS Research Using Natural Language Processing: Towards Automated Recommendations 147
Nevena Nikolajevic, Michael T. Knierim, and Christof Weinhardt

CareCam: An Intelligent, Camera-Based Health Companion at the Workplace 155
Dimitri Kraft, Angelina Schmidt, Frederike Marie Oschinsky,
Lea Büttner, Fabienne Lambusch, Kristof Van Laerhoven,
Gerald Bieber, and Michael Fellmann

Decision Delegation and Intelligent Agents in the Context of Human Resources Management: The Influence of Agency and Trust. A Research Proposal 163
Marion Korosec-Serfaty, Sylvain Sénécal, and Pierre-Majorique Léger

A Stress-Based Smart Retail Service in Shopping Environments: An Adoption Study 171
Nurten Öksüz

**Collecting Longitudinal Psychophysiological Data in Remote
Settings: A Feasibility Study** 179
Sara-Maude Poirier, Félix Giroux, Pierre-Majorique Léger,
Frédérique Bouvier, David Brieugne, Shang-Lin Chen,
and Sylvain Sénécal

**Eye-Gaze and Mouse-Movements on Web Search as Indicators
of Cognitive Impairment** 187
Jacek Gwizdka, Rachel Tessmer, Yao-Cheng Chan,
Kavita Radhakrishnan, and Maya L. Henry

**Mixed Emotions: Evaluating Reactions to Dynamic Technology
Feedback with NeuroIS** 201
Sophia Mannina and Shamel Addas

**Smart Production and Manufacturing: A Research Field with High
Potential for the Application of Neurophysiological Tools** 211
Josef Wolfartsberger and René Riedl

New Measurement Analysis for Emotion Detection Using ECG Data ... 219
Verena Dorner and Cesar Enrique Uribe Ortiz

**Caption and Observation Based on the Algorithm for Triangulation
(COBALT): Preliminary Results from a Beta Trial** 229
Pierre-Majorique Léger, Alexander J. Karran, Francois Courtemanche,
Marc Fredette, Salima Tazi, Mariko Dupuis, Zeyad Hamza,
Juan Fernández-Shaw, Myriam Côté, Laurène Del Aguila,
Chantel Chandler, Pascal Snow, Domenico Vilone, and Sylvain Sénécal

**Leveraging Affective Friction to Improve Online Creative
Collaboration: An Experimental Design** 237
Maylis Saigot

A Look Behind the Curtain: Exploring the Limits of Gaze Typing 251
Marius Schenkluhn, Christian Peukert, and Christof Weinhardt

**Towards Mind Wandering Adaptive Online Learning and Virtual
Work Experiences** 261
Colin Conrad and Aaron J. Newman

**Measurement of Heart Rate and Heart Rate Variability: A Review
of NeuroIS Research with a Focus on Applied Methods** 269
Fabian J. Stangl and René Riedl

**Measurement of Heart Rate and Heart Rate Variability in NeuroIS
Research: Review of Empirical Results** 285
Fabian J. Stangl and René Riedl

Investigating Mind-Wandering Episodes While Using
Digital Technologies: An Experimental Approach Based
on Mixed-Methods 301
Caroline Reßing, Frederike M. Oschinsky, Michael Klesel,
Björn Niehaves, René Riedl, Patrick Suwandjieff,
Selina C. Wriessnegger, and Gernot R. Müller-Putz

The Effects of Artificial Intelligence (AI) Enabled Personality
Assessments During Team Formation on Team Cohesion 311
Nicolette Gordon and Kimberly Weston Moore

Age-Related Differences on Mind Wandering While Using
Technology: A Proposal for an Experimental Study 319
Anna Zeuge, Frederike Marie Oschinsky, Michael Klesel,
Caroline Reßing, and Bjoern Niehaves

A Brief Review of Information Security and Privacy Risks
of NeuroIS Tools 329
Rosemary Tufon and Adriane B. Randolph

Picture Classification into Different Levels of Narrativity Using
Subconscious Processes and Behavioral Data: An EEG Study 339
Leonhard Schreiner, Hossein Dini, Harald Pretl, and Luis Emilio Bruni

Usability Evaluation of Assistive Technology for ICT Accessibility:
Lessons Learned with Stroke Patients and Able-Bodied
Participants Experiencing a Motor Dysfunction Simulation 349
Félix Giroux, Loic Couture, Camille Lasbareille, Jared Boasen,
Charlotte J. Stagg, Melanie K. Fleming, Sylvain Sénécal,
and Pierre-Majorique Léger

Correction to: Information Systems and Neuroscience C1
Fred D. Davis, René Riedl, Jan vom Brocke, Pierre-Majorique Léger,
Adriane B. Randolph, and Gernot R. Müller-Putz

Author Index 361

Our Brain Reads, While We Can't: EEG Reveals Word-Specific Brain Activity in the Absence of Word Recognition

Peter Walla, Robin Leybourne, and Samuil Pavlevchev

Abstract This electroencephalography (EEG) study provides significant neuro-physiological evidence for the processing of words outside conscious awareness. Brain potentials were recorded while 25 participants were presented with words and shapes, each of them with 17, 34, and 67 ms presentation duration times ("no presentation" was included as control condition). Participants were instructed to report whether they have seen anything and if yes, what it was (shape or word). If they saw a word and were able to read it, they were asked to report having read it. Crucially, even though only 2.9% of the 17 ms word presentations were classified as readable, these presentations elicited significant brain activity near the Wernicke area that was missing in the case of 17 ms shape presentations. The respective brain activity difference lasted for about 150 ms being statistically significant between 349 and 409 ms after stimulus onset. It is suggested that this neurophysiological difference reflects non-conscious (i.e., subliminal) word processing. Some aspects of the NeuroIS discipline include text messages and the current findings demonstrate that those text messages can be processed by the nervous system even in the absence of their conscious recognition.

Keywords Subliminal words · EEG · ERPs · Non-conscious processing

P. Walla (✉)
Faculty of Psychology, Freud CanBeLab, Sigmund Freud University, Freudplatz 1, 1020 Vienna, Austria
e-mail: peter.walla@sfu.ac.at

Faculty of Medicine, Sigmund Freud University, Freudplatz 3, 1020 Vienna, Austria

School of Psychology, Centre for Translational Neuroscience and Mental Health Research, University of Newcastle, University Drive, Callaghan, NSW, Australia

R. Leybourne
Faculty of Psychology, Webster Vienna Private University, Praterstraße 23, 1020 Vienna, Austria

S. Pavlevchev
Center for Research in Modern European Philosophy (CREMP), Kingston University, Penrhyn Road Campus, Kingston upon Thames, UK

1 Introduction

Non-conscious brain activities, especially in the context of verbal information processing have long been an interesting focus [1]. For instance, it has been shown that superficially (i.e., alphabetically) encoded words lead to measurable brain activity during their subsequent, repeated presentation reflecting detection of their repeated nature even in the absence of conscious recognition [2, 3]. This has been interpreted as non-conscious verbal memory traces. The time window, during which event-related potentials (ERPs) differed between words that failed to be identified as repeated (participant response: "I have not seen this word before"), but were actually presented before, and words that were correctly identified as new (participant response: "I have not seen this word before") was only 200 ms long. While this might seem short, it is long enough to be detected via electroencephalography (EEG), which is known for its excellent temporal resolution due to its direct sensitivity to neurophysiological responses in the brain.

The processing of text (words) in general is of high relevance to the NeuroIS discipline. Verbal stimuli have been used for various investigations [4–6]. Some NeuroIS research successfully used brain imaging tools to study potential effects on trustworthiness perceptions through using text messages like they appear on eBay websites [7].

While in the Rugg et al. study [2] lack of recognition was rather a consequence of low-level attention (due to only alphabetical (letter-based) word encoding during first word exposures), other conditions can cause detectable non-conscious word processing. Such conditions can be related to weak stimulation features like low contrast or short presentation durations. Several early studies demonstrated that short or weak word presentations can still lead to measurable behavioral effects in the absence of awareness [e.g. 8], but only a few studies reported distinct neurophysiological correlates of subliminal word processing. One such study found evidence for neurophysiological traces in response to short word stimulations as short as 1 ms [9]. However, it remains unclear if those traces are indeed specific to subliminal word processing or if they would look similar for other types of stimuli. A very recent study comparing short word presentations with short shape presentations in one experiment was able to show that EEG is indeed capable of detecting word-specific subliminal brain processes [10]. In addition to comparing word with shape processing, this experiment also included varying presentation durations. Words and shapes were visually presented for 17, 34, 67, and 100 ms. Participants had to respond to all stimulus presentations by reporting whether they saw "nothing", a "blur" (something but not sure whether it was word or shapes), a "word", or a "shape" (a string of simple symbols). Crucially, even though recognition for words presented for 17 ms was only 6% (a very insignificant recognition rate), these words elicited brain electrical amplitudes significantly different from those elicited by 17 ms long shape presentations. For their ERP analysis, the authors chose an electrode position located around the well-known Wernicke area, which is commonly understood as a cortical area involved in comprehension, i.e., understanding semantic content [e.g. 11].

The present study was meant to replicate the findings of Pavlevchev et al. [10], while also adding one further condition during which nothing was presented. Besides replicating their results, the hypothesis was that the "nothing" condition should not elicit any processing-related brain electrical amplitudes and could serve as a control condition to compare with both words and shapes. If a physiological difference is detected between the newly introduced control condition and the short 17 ms presentations then this will go to show that despite their insignificant recognition rate of 6%, they are stimuli transduced into neural signals that enter the brain subliminally via the visual system. Finally, the present study asked participants whether they had read the presented word or merely thought it was a word but could not read it.

2 Materials and Methods

2.1 Participants

A total of 25 young adults (10 females and 15 males between the ages of 19 and 27) participated in this study. The mean age of the participants was 22.04 years (SD = 1.99). All participants reported being right-handed, having normal or corrected-to-normal vision, and not having any neuropathological history.

2.2 Stimuli

For this study, the same 30 low-frequency words (6-letters; neutral object nouns) from two databases [12, 13] together with 30 shape stimuli, which were self-created variations of sequences of 6 different simple symbols (same font and contrast as the words) were used as in the Pavlevchev study [10]. Figure 1 shows examples for both stimulus types (taken from [10]).

Fig. 1 Examples for word and shape stimulus types (taken from [10])

2.3   Electroencephalography (EEG)

A 64 channel actiCHamp Plus System from Brain Products was used for recording brain electrical signals (in the Freud CanBeLab at Sigmund Freud University in Vienna). The active electrodes were all embedded in an actiCAP connected to the amplifier. The free software PsychoPy 2021.2.3 for Windows was used to design the experiment and to control stimuli delivery to study participants. Brain potential changes were recorded with a sampling rate of 1.000 Hz (filtered: DC to 100 Hz). Offline, all EEG data were down-sampled to 250 Hz and a bandpass filter from 0.1 to 30 Hz was applied.

2.4   Procedure

All stimuli controlled by the software package PsychoPy3 (v2021.1.0) were presented on a computer monitor in random order (stimulus type and duration were randomly varied across presentations). Each presentation (one trial) consisted of a 2 s long "+" symbol (fixation), a 500 ms blank screen, a stimulus (17, 33, or 67 ms), and a 1 s blank screen. The 100 ms duration condition from the Pavlevchev et al. study [10] was left out and replaced by a "nothing" condition for both words and shapes. After each trial, participants were instructed to indicate via a button press, whether nothing, a blur, a shape, a word they could not read, or a word they could read was observed.

2.5   Data Analyses

EEG signal processing was carried out with the EEGDISPLAY 6.4.9 software [14]. Epochs were generated from 100 ms before stimulus onset to 1 s after stimulus onset. The duration of 100 ms prior to stimulus onset was used as a baseline. All epochs with artifacts were automatically excluded. Event-related potentials (ERPs) were calculated for all conditions and re-referenced to the common average across all electrode sites. Finally, only data collected from one electrode location (near the Wernicke area; P7; see Fig. 1) were further processed for this paper.

Behavioral data (button presses) were counted as correct recognition of the stimuli for each condition and then averaged across all participants. Finally, percentages of correct recognitions were calculated.

3 Results

3.1 Behavior

Of all 30 shapes that were presented for 17 ms 22.8% were identified as "I saw a shape". Of all 30 words that were presented for 17 ms 25.9% were classified as "I saw a word". However, across all participants, only 2.9% of all words were classified as "I could read the word".

42.5% of all words that were presented for 34 ms were classified as "I could read the word", and words that were presented for 67 ms were classified as readable at a rate of 71.1%. Of all shapes presented for 34 ms, 68.4% were correctly identified and the correct identification rate for shapes presented for 67 ms was 86%.

The low reading rate for words presented for only 17 ms forms the behavioral basis for this study and is interpreted as more or less a lack of ability to consciously read the word.

3.2 Electroencephalography (EEG)

Visual inspection and direct comparison to the results of the Pavlevchev et al. study [10] reveals a remarkably similar pattern of ERP modifications. The present data, however, displays slightly delayed neural activity that starts from around 350 ms and lasts until 410 ms post-stimulus onset. Whether the delay results from different neural processing or represents an unknown technical issue remains unclear. This means that the temporal aspects (time delay) of the current study need to be dealt with a certain amount of caution. However, the above-mentioned period was further analyzed. A repeated measures ANOVA revealed a significant "duration" (0, 17, 34, 67 ms) times "type" (nothing, word, shape) interaction ($p = 0.040$; Greenhouse–Geisser corrected) for this time window. A following t-test comparing the mean amplitude of the "17 ms word" condition with the mean amplitude of the "17 ms shape" condition revealed a highly significant difference ($p < 0.001$). See Fig. 2 showing all generated and overlaid event-related potentials (ERPs). The most important ERP difference (between both 17 ms presentations) is marked in grey color. It is noteworthy to mention that this effect can also be seen at various surrounding electrode locations (around P7), even in the right hemisphere, but slightly reduced. However, no further statistical analysis was carried out for those other electrode locations.

4 Discussion

First, the findings of this study largely replicate prior results published by Pavlevchev et al. [10], while also showing a clear difference between "no presentation" and all

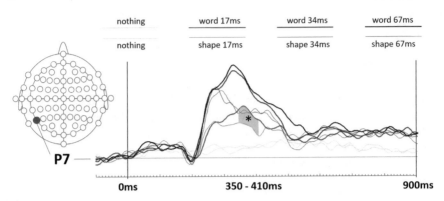

Fig. 2 Event-related potentials (ERPs) overlaid for all eight conditions. Note, that the "nothing" conditions indeed resulted in ERPs similar to the baseline and that the "17 ms word" condition (thick red curve) elicited an ERP significantly different from the ERP elicited by the "17 ms shape" condition (thin red curve)

other presentations. Most importantly, this study also replicates how ERPs elicited by 17 ms shape and word presentations differed significantly in the vicinity of the well-known Wernicke area (comprehension center), which is interpreted as neurophysiological evidence for subliminal word processing. However, it must be noted that this effect in the current study appears slightly later than in the Pavlevchev et al. study [10]. At this point it remains unclear if this delay is of any neurophysiological nature or simply a technical issue, maybe related to some unknown trigger problem. Further, while these authors [10] only asked their participants to report if they have seen a word, in this study, participants were also asked to report if they could read what they believed to have been a word. The 2.9% rate at which participants reported to have read words presented for 17 ms is sufficiently low to assume an overall lack of reading for this short presentation condition. Nevertheless, these shortly presented words elicited brain activity significantly different from that elicited by shapes.

As a next step, non-words should be introduced to this experimental design to test if the found effect is a result of semantic non-conscious (i.e., subliminal) word processing and not a result of only non-conscious (i.e., subliminal) lexical (alphabetical) processing.

Much of the existing literature on subliminal processing reports about respective spatial, physiological correlates related to it [11–14]. We have mainly focused on its temporal aspect and together with existing knowledge about subliminal processing in general, we herewith introduce this fascinating topic to the NeuroIS discipline. NeuroIS research includes constructs that can benefit from insight into how short exposures to some material (e.g. text messages) may enter into or influence information processing in the context of decision making, technology use etc., even at a level below conscious awareness.

References

1. Williams, A. C. (1938). Perception of subliminal visual stimuli. *The Journal of Psychology, 6*(1), 187–199.
2. Rugg, M. D., Mark, R. E., Walla, P., Schloerscheidt, A. M., Birch, C. S., & Allan, K. (1998, April 9). Dissociation of the neural correlates of implicit and explicit memory. *Nature, 392,* 595–598.
3. Walla, P., Endl, W., Lindinger, G., Deecke, L., & Lang, W. (1999). Implicit memory within a word recognition task: An event-related potential study in human subjects. *Neuroscience Letters, 269*(3), 129–132.
4. Kim, D., & Benbasat, I. (2006). The effects of trust-assuring arguments on consumer trust in internet stores: Application of Toulmin's model of argumentation. *Information Systems Research, 17*(3), 286–300.
5. Pavlou, P. A., & Dimoka, A. (2006). The nature and role of feedback text comments in online marketplaces: Implications for trust building, price premiums, and seller differentiation. *Information Systems Research, 17*(4), 392–414.
6. Gefen, D., Benbasat, I., & Pavlou, P. A. (2008). A research agenda for trust in online environments. *Journal of Management Information Systems, 24*(4), 275–286.
7. Riedl, R., Hubert, M., & Kenning, P. (2010). Are there neural gender differences in online trust? An fMRI study on the perceived trustworthiness of eBay offers. *MIS Quarterly, 34*(2), 397–428.
8. Bornstein, R. F., Leone, D. R., & Galley, D. J. (1987). The generalizability of subliminal mere exposure effects: Influence of stimuli perceived without awareness on social behavior. *Journal of Personality and Social Psychology, 53*(6), 1070–1079.
9. Bernat, E., Bunce, S., & Shevrin, H. (2001). Event-related brain potentials differentiate positive and negative mood adjectives during both supraliminal and subliminal visual processing. *International Journal of Psychophysiology, 42*(1), 11–34.
10. Pavlevchev, S., Chang, M., Flöck, A. N., & Walla, P. (2022). Subliminal word processing: EEG detects word processing below conscious awareness. *Brain Sciences, 12,* 464. https://doi.org/10.3390/brainsci12040464
11. Walla, P., Hufnagl, B., Lindinger, G., Imhof, H., Deecke, L., & Lang, W. (2001). Left temporal and temporo-parietal brain activity depends on depth of word encoding: A magnetoencephalographic (MEG) study in healthy young subjects. *NeuroImage, 13,* 402–409.
12. Thorndike, E. L., & Lorge, I. (1944). *The teacher's word book of 30,000 words.* Bureau of Publications.
13. Brysbaert, M., & New, B. (2009). Moving beyond Kučera and Francis: A critical evaluation of current word frequency norms and the introduction of a new and improved word frequency measure for American English. *Behavior Research Methods, 41,* 977–990. https://doi.org/10.3758/BRM.41.4.977
14. Fulham, W. R. (2015). *EEG display* (Version 6.4.9) [Computer software]. Functional Neuroimaging Laboratory, University of Newcastle.

Information Overload and Argumentation Changes in Product Reviews: Evidence from NeuroIS

Florian Popp, Bernhard Lutz, and Dirk Neumann

Abstract Information overload theory suggests that consumers can only process a certain amount and complexity of information. In this study, we focus on product reviews with different complexity in terms of argumentation changes, i.e., alternations between positive and negative arguments. We present the results of a NeuroIS experiment, where participants processed product reviews with low or high rates of argumentation changes. Participants were asked to state their perceived helpfulness of the product review, their purchase intention for the product, and self-reported information overload. During the experiment, we measure cognitive activity based on eye-tracking and electroencephalography (EEG). Our preliminary results suggest that a higher rate of argumentation changes is linked to greater self-reported information overload, and greater cognitive activity as measured by EEG. In addition, we find that greater self-reported overload is linked to lower perceived review helpfulness, and lower purchase intention.

Keywords Product reviews · Information overload · EEG · Eye-tracking

1 Introduction

On modern retailer platforms, product reviews assist customers in their purchase decision-making process [1, 2]. Previous work has demonstrated that reviews that are perceived as more helpful also have a stronger influence on sales numbers [3]. More helpful reviews can further translate to a more positive attitude towards a

F. Popp (✉) · B. Lutz · D. Neumann
University of Freiburg, Freiburg, Germany
e-mail: florian.popp@is.uni-freiburg.de

B. Lutz
e-mail: bernhard.lutz@is.uni-freiburg.de

D. Neumann
e-mail: dirk.neumann@is.uni-freiburg.de

© The Author(s), under exclusive license to Springer Nature Switzerland AG 2022
F. D. Davis et al. (eds.), *Information Systems and Neuroscience*,
Lecture Notes in Information Systems and Organisation 58,
https://doi.org/10.1007/978-3-031-13064-9_2

product [4, 5]. As such, product reviews serve as focal point for the study of human purchase decision-making in information systems (IS) research [1, 6–9].

The questions of what makes helpful product reviews and how certain review characteristics are linked to sales numbers are subject to a large number of publications. Extensive literature reviews are provided by Hong et al. [10] and Zheng [11]. Product reviews consist of a star rating and a textual description [12]. The length of the text and the star rating are among the most studied determinants of review helpfulness [e.g., 6, 13–16]. While existing studies largely agree that longer reviews are perceived as more helpful [e.g., 6, 16], there is less agreement in regard to whether positive or negative reviews are perceived as more helpful. Sen and Lerman [15] find that negative reviews are perceived as more helpful, while the study by Pan and Zhang [17] finds the opposite. Other studies found that consumers perceive reviews of different polarity as more or less helpful depending on factors such as product characteristics [6], review quality [18], review extremity [13], product type [17], and goal orientations [19].

Customers are generally provided with a large number of product reviews [20]. The studies by Park and Lee [21] and Zinko et al. [22] find that, if users are confronted with too many reviews, they can experience information overload. The concept of information overload suggests that a medium amount of information is more helpful for customers' purchase decision-making than little or too much information [23–26]. However, information overload is not only limited to high quantities of information. Instead, information overload can also occur if the presented information exhibits high complexity [27]. While it seems intuitive that information overload can be caused by a large number of product reviews, little is known about whether information overload can also be caused by individual reviews.

In this study, we focus on the line of argumentation in product reviews in regard to how positive and negative aspects are discussed. The line of argumentation in product reviews can be structured in multiple ways. A review can be one-sided, so that it mentions positive or negative aspects only, or two-sided with a mixture of positive and negative arguments. Two-sided reviews can start with positive arguments followed by negative arguments, or vice versa [28]. Such clear cut reviews exhibit only a single argumentation change. As an alternative, the review can alternate between positive and negative arguments, which implies a higher rate of argumentation changes and greater review complexity. We hypothesize that product reviews with a higher rate of argumentation changes (i.e., frequent alternations between positive and negative arguments) are harder to comprehend and that they require greater cognitive effort for being processed. As a consequence, reviews with a high rate of argumentation changes should, in particular, be more likely to cause information overload. In our prior empirical study Lutz et al. [29], we analyzed the link between argumentation changes and review helpfulness solely based on secondary data. In particular, we did not specifically measure (proxies) of information overload as it was considered as a latent construct. The goal of this study is to validate our previous findings in a controlled setting using tools from neuroscience.

We conducted a NeuroIS experiment with eye-tracking and EEG measurements to analyze the effects of a high rate of argumentation changes.[1] Participants were shown product reviews of different products with either low (control group) or high (treatment group) rates of argumentation changes. Subsequently, they were asked to state self-reported information overload, review helpfulness, and purchase intention. Concordant with information overload theory, we find that a higher rate of argumentation changes is linked to greater self-reported information overload and greater cognitive activity as measured by EEG.

2 Research Hypotheses

The term *"information overload"* was coined by Gross [30]. One of the most influential studies on information overload was written by Jacoby et al. [25]. The authors found an inverted U-shape relation between information load and decision outcomes. Within that, consumers make the best choices when being provided with a medium amount of information, and the worst choices when being provided with too little or too much information. The information overload theory is based on the fact that consumers have limited cognitive processing capabilities [31]. Although Jacoby et al. [25], Jacoby [26] and Malhotra [31, 32] produced mixed results regarding the question of whether consumers can truly be overloaded in practice, they all agree that information overload can at least occur in experimental setups. In the area of e-commerce, information overload theory suggests that consumers can process a certain amount and complexity of information during online shopping, and that information which exceeds these capacities leads to poorer purchase decisions [33]. Nowadays, IS researchers largely agree that information overload is a significant issue in e-commerce which needs to be mitigated [34–37].

Interestingly, information overload is not limited to high amounts of information; instead it can also occur when information complexity is high [38, 39]. Otondo et al. [38] and Schneider [39] argue that among others, complexity and ambiguity are possible causes for information overload. For instance, Hiltz and Turoff [27] argue that information might not be recognized as important if it is not sufficiently organized by topic or content. Lurie [40] finds that information should not just be measured by its quantity; instead, it should also account for the information structure. Product reviews are provided in the form of a start rating and a textual description, which describes prior experiences and pros and cons of a product [41, 42]. Reviews can be written in a one-sided, or two-sided way [43, 44]. A two-sided review can, again, be written in several ways, by enumerating pros followed by cons, cons followed by pros, or by interweaving pros and cons [45]. As a motivating example, consider the

---

[1] An earlier version of this paper was presented at the 17th International Conference on Wirtschaftsinformatik 2022. However, this version only described the experimental design without the results.

following two hypothetical reviews, both evaluating pros (highlighted in light gray) and cons (highlighted in dark gray) of a coffee.

> Review A: "This coffee tastes great. However, I don't like the design of its packaging. The smell when opening the bag is awesome. I am a bit disappointed with the strength of it."
>
> Review B: "This coffee tastes great. The smell when opening the bag is awesome. However, I don't like the design of its packaging. I am a bit disappointed with the strength of it."

Both reviews present the same set of pro and contra arguments. Keeping everything else equal, the existing literature on review helpfulness suggests that both reviews are approximately equally helpful as both are of the same length (three sentences), language, and words. However, the rate of argumentation changes in both reviews differs: Review A alternates between positive and negative arguments, whereas Review B presents the content in a more organized manner by discussing positive arguments first and then negative arguments. We argue that a higher rate of argumentation changes increases information complexity [27], which requires more cognitive effort to comprehend the review. Concordant with information overload theory, we expect that a higher rate of argumentation changes increases the cognitive effort that is required to process the review. We therefore propose.

> **Hypothesis 1a (H1a)**. A higher rate of argumentation changes in a product review is linked to increased cognitive activity.
>
> **Hypothesis 1b (H1b)**. A higher rate of argumentation changes in a product review is linked to higher self-reported information overload.

Assuming that a higher rate of argumentation changes in a review is likely to cause information overload, we argue that a review with a high rate of argumentation changes is also perceived as less helpful. Additionally, previous work has linked information overload to a reduced consumer experience [46]. For instance, Gross [47] found that information overload could have a damaging effect on the way users view the merchant, and on their commitment to learn about the product's specifications. Another study found that the helpfulness of the information within reviews depends on the readability of the text [48]. Furthermore, previous work has demonstrated a link between helpfulness and sales numbers [3]. Hence, we propose.

> **Hypothesis 2a (H2a)**. Greater self-reported information overload is linked to lower perceived helpfulness.
>
> **Hypothesis 2a (H2b)**. Greater perceived helpfulness is linked to higher purchase intention.

Product types can be distinguished between low- and high-involvement products. While low-involvement products feature a lower perceived risk of poor purchase decisions (due to a lower price and less durability), high-involvement products feature a higher price and greater durability, and therefore a higher perceived risk [49]. Consequently, consumers have an incentive to invest more cognitive effort into collecting information (including product reviews) for high involvement products than for low-involvement products. Therefore, one could argue that customers seeking information for high-involvement products are already geared towards investing greater cognitive efforts, so that the (negative) effect of a higher rate of argumentation changes is stronger for low-involvement products. Conversely, one could argue that users prefer well structured high quality reviews when deciding upon buying high-involvement products, which implies a stronger effect of argumentation changes for high-involvement products. Accordingly, we propose two alternative hypotheses.

Hypothesis 3 (H3a). The effect of the rate of argumentation changes on self-reported information overload and cognitive activity is stronger for low involvement products than for high-involvement products.

Hypothesis 3 (H3b). The effect of the rate of argumentation changes on self-reported information overload and cognitive activity is stronger for highinvolvement products than for low-involvement products.

3 Method

3.1 Materials and Treatment

We collect product reviews based on an existing dataset of product reviews from Amazon [50] as this is the prevalent choice in the related literature when studying product reviews (e.g., [6, 12, 16, 51]). We select 16 online reviews for one low-involvement product (Arabica coffee) and 16 reviews for one high-involvement product (a digital camera), which yields a total of 32 reviews. We then manipulate these reviews to exhibit either low or high rates of argumentation changes by changing the order in which positive and negative arguments are presented. For this purpose, we need to remove words and phrases like "Therefore" or "And this is why", while ensuring that the manipulated reviews remain grammatically sound. Product names are removed to account for potential biases against a given brand among the participants.

3.2 Participants, Procedure, and Measures

We recruited 60 subjects via student mailing lists and announcements during lectures. Ethics approval was granted by the University of Freiburg. Each participant receives a fixed compensation of 12 Euros. All students are undergraduates from economics, computer science, or engineering. 28 out of the 60 participants identified as female (46.66%), and 32 as male. The mean age is 24.8 years. The participants are randomly assigned to the treatment (high rate of argumentation changes) or control group (low rate of argumentation changes). Each participant is presented a total of 32 reviews in randomized order. The experimental procedure starts with calibration of the eye-tracking and EEG devices. Before the actual experiment, we perform a test run with a dummy review to make the participants familiar with the general procedure. All product reviews are displayed without any images or product description to avoid potential confounding effects. Participants are given unlimited time for reading a review as we do not intent to induce time pressure. Subsequently, they are asked to state their perceived review helpfulness, the purchase intention, and self-reported information overload on 7-point Likert scales from 1 = low to 7 = high. Finally, participants fill out a survey, where they provide their age, gender, experience with online purchases, and propensity to trust [52].

We measure cognitive activity with EEG and eye-tracking. We use the EMOTIV EPOC X, a 14-channel wireless EEG device, which has been used by a range of previous work [e.g., 53]. As illustrated in Fig. 1, the 14 electrodes are positioned across the scalp. We process the EEG raw data with the following steps using EEGLAB [54]. First, we calibrate the channel location, clean the data by rejecting any unwanted artifacts and noise. Second, we decompose the data by rejecting any unwanted artifacts and noise. Second, we decompose the data by ICA, re-reference the channels to the device's reference channels, filter for alpha wave frequency (8–13 Hz), and define our epochs. Hereby, the epochs were defined as the first 10 s after the review as shown. Third, we determine the alpha wave desynchronization per epoch and participant as a proxy for cognitive load [53, 55–57]. Since a desynchronization of alpha waves results in an overall lower alpha value, lower alpha values indicate greater cognitive activity. For 54 out of 1,888 (ca. 2.9%) observations the signal quality was not sufficient. Thus, we rejected those observations resulting in a dataset comprising 1,834 observations. For the purpose of our cognitive load analysis, we focused on the recordings of the electrode F3, which is positioned on the left prefrontal cortex [53]. In addition, we use the Tobii Pro Fusion eye-tracker to measure the mean fixation duration as a proxy for cognitive load [58]. Hereby, longer fixation durations indicate greater cognitive load. We extract the mean fixation duration with the velocity-based algorithm proposed by Engbert and Kliegl [59].

4 Results

We now present the results from our experiment. We estimate OLS regression models with different dependent variables according to the hypothesis to be tested. We always control for the sequence number of the review in the experiment, the product type, the length of a review in words, the reading time in seconds, and the treatment variable, which equals 1 for reviews with a high rate of argumentation changes (treatment) and 0 for reviews with a low rate of argumentation changes (control).

4.1 Measured Cognitive Activity

We first test hypothesis H1a in regard to the measured cognitive activity. For this purpose, we consider the regression results presented in Table 1. Here, columns (a) and (b) present the results with the mean fixation duration as dependent variable,

Fig. 1 Positioning of
electrode of the EMOTIV
EPOC X device across the
scalp [60]

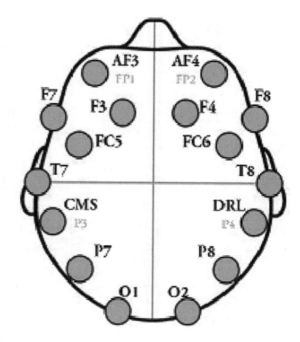

while columns (c) and (d) present the results with EEG alpha wave desynchronization
as dependent variable. The coefficient of the treatment variable is always negative
and significant with $p < 0.001$. However, the coefficient of the treatment variable for
the models explaining mean fixation duration (columns (a) and (b)) contradicts our
expectation that a higher rate of argumentation changes should be linked to longer
fixations [61, 62].

One possible explanation is that the participants read the reviews less carefully
due to greater review complexity. The results from columns (c) and (d) are in line with
our expectation that alpha wave desynchronization is a proxy for increased cognitive
load [53, 55–57]. Hence, we find support for H1a when measuring cognitive activity
using EEG.

Next, we test H3 in regard to the measured cognitive activity. For this purpose, we
consider the coefficient of Product type high × Treatment in Table 1. Evidently, the
coefficient of the interaction between product type and treatment is not statistically
significant. This indicates that the product type does not moderate the effect between
argumentation changes and cognitive activity.

4.2 Self-reported Information Overload

Next, we test hypothesis H1b, which links argumentation changes to self-reported
information overload. The regression results are shown in Table 2. The coefficient

Table 1 Regression results for measured cognitive activity

	Mean fixation duration		EEG Alpha desynchronization	
	(a)	(b)	(c)	(d)
Treatment	−2.183*** (0.171)	−2.235*** (0.173)	−2.382*** (0.238)	−2.384*** (0.242)
Sequence number	0.007 (0.016)	0.008 (0.016)	0.026 (0.022)	0.026 (0.022)
Length (words)	−0.042 (0.029)	−0.042 (0.029)	0.022 (0.040)	0.022 (0.040)
Product type high	−0.045 (0.031)	−0.094* (0.043)	0.001 (0.043)	−0.001 (0.060)
Reading time (s)	0.040 (0.034)	0.041 (0.034)	−0.029 (0.047)	−0.029 (0.047)
Product type high × Treatment		0.104 (0.062)		0.003 (0.087)
Intercept	1.388*** (0.122)	1.412*** (0.123)	2.317*** (0.171)	2.317*** (0.172)
Subject fixed-effects	✓	✓	✓	✓
R^2	0.543	0.544	0.195	0.195
Observations	1888	1888	1834	1834

***$p < 0.001$; **$p < 0.01$; *$p < 0.05$

of the treatment variable is always positive and significant with $p < 0.001$, which indicates that reviews with a high rate of argumentation changes are linked to higher self-reported information overload. Therefore, H1b is supported.

To test hypothesis H3 for self-reported information overload, we consider in the coefficient of the interaction between product type and treatment as shown in column (d) of Table 2. The coefficient of product type × treatment is not statistically significant, which indicates that the product type does not moderate the effect between argumentation changes and self-reported information overload. Given this finding and our previous finding in regard to measured cognitive activity, we reject hypotheses H3a and H3b.

4.3 Perceived Helpfulness and Purchase Intention

Finally, we test hypotheses H2a and H2b based on the regression results shown in Table 3. H2a and H2b cover known effects from the literature, so that the hypotheses tests serve as a validity check of our experiment. Columns (a) and (b) show the results that explain the perceived helpfulness, while columns (c) and (d) show the results that explain purchase intention.

We first consider the coefficient of self-reported information overload in columns (a) and (b). The coefficient is negative and significant with $p < 0.001$, which indicates that higher self-reported information overload is linked to lower perceived review helpfulness. Accordingly, H2a is supported. We also find that the review length has a

Table 2 Regression results for self-reported information overload

	(a)	(b)	(c)	(d)
Treatment	0.406*** (0.065)	0.361*** (0.066)	0.379*** (0.067)	0.439*** (0.094)
Sequence number	−0.038 (0.033)	−0.035 (0.033)	−0.032 (0.033)	−0.033 (0.033)
Length (words)	0.871*** (0.046)	0.865*** (0.046)	0.877*** (0.047)	0.877*** (0.047)
Product type high	−0.077 (0.065)	−0.079 (0.064)	−0.081 (0.065)	−0.024 (0.090)
Reading time (s)	0.058 (0.047)	0.068 (0.047)	0.058 (0.047)	0.058 (0.047)
Mean fixation duration (s)		−0.038 (0.033)	−0.044 (0.033)	−0.044 (0.033)
Mean fixation duration × Treatment		−0.483** (0.172)	−0.483** (0.173)	−0.484** (0.173)
Alpha desynchronization			0.172*** (0.047)	0.174*** (0.047)
Alpha desynchronization × treatment			−0.179** (0.068)	−0.180** (0.068)
Product type high × treatment				−0.120 (0.131)
Intercept	1.810*** (0.055)	1.814*** (0.055)	1.806*** (0.056)	1.778*** (0.064)
R^2	0.310	0.314	0.320	0.320
Observations	1888	1888	1834	1834

***$p < 0.001$; **$p < 0.01$; *$p < 0.05$

significantly positive effect on review helpfulness with $p < 0.001$. This is in line with previous findings that review length is a key determinant of the perceived helpfulness [6, 16, 17, 63–65]. Second, we consider the coefficient of the perceived helpfulness in columns (c) and (d). Evidently, the coefficient is positive and significant with $p < 0.001$, which indicates that greater review helpfulness is linked to higher purchase intention. Hence, H2b is supported. Furthermore, the results do not suggest that reviews of high-involvement products are perceived as more helpful, or a higher purchase intention for highinvolvement products. We also found evidence for habituation as the coefficient of the sequence number is negative and significant for both dependent variables.

5   Discussion and Contribution

We expect to contribute to the IS literature in the following ways. To the best of our knowledge, our study is the first to use tools from NeuroIS to study human

Table 3 Regression results for perceived helpfulness and purchase intention

	(a)	(b)	(c)	(d)
Treatment	−0.030 (0.069)	−0.025 (0.097)	0.149* (0.069)	0.119 (0.098)
Sequence number	−0.181* (0.035)	−0.181* (0.035)	−0.094** (0.035)	−0.094** (0.035)
Length (words)	0.340*** (0.053)	0.340*** (0.053)	−0.063 (0.054)	−0.063 (0.054)
Product type high	0.399*** (0.068)	0.404*** (0.094)	−0.105 (0.069)	−0.134 (0.095)
Reading time (s)	−0.006 (0.049)	−0.006 (0.049)	0.110* (0.049)	0.110* (0.049)
Self-reported information overload	−0.228*** (0.024)	−0.228*** (0.024)	0.033 (0.025)	0.033 (0.025)
Perceived helpfulness			0.262*** (0.023)	0.262*** (0.023)
Product type high × treatment		−0.010 (0.136)		0.061 (0.137)
Intercept	3.807*** (0.073)	3.805*** (0.079)	0.954*** (0.115)	0.968*** (0.119)
Observations	1888	1888	1834	1834

***$p < 0.001$; **$p < 0.01$; *$p < 0.05$

process of product reviews in regard to information overload for which research is still scant [20]. In that sense, the method and findings of this study are different from our prior study Lutz et al. [29], where we employed a data science approach to explain review helpfulness with the rate of argumentation changes. In particular, we did not specifically measure information overload. Instead, the rate of argumentation changes was directly used to explain review helpfulness, while information overload was considered a latent construct. Other studies found that information overload can occur at the product-level if users are provided with too many reviews [21], and that users can experience information overload if a review contains too much information [35].

From a practical perspective, our findings allow online retailers to present and promote reviews that can be read with less cognitive effort. Importantly, retailers may improve their sales numbers by providing more comprehensible reviews with a low rate of argumentation changes which foster an increase in customers' purchase intention. In addition, our findings can be used by online review system designers as writing guidelines to encourage consumers to write less complex reviews. The next steps are the computation of robustness checks, and the collection of valuable feedback from the NeuroIS Retreat to extend the study for publication to a journal.

References

1. Dellarocas, C. (2003). The digitization of word of mouth: Promise and challenges of online feedback mechanisms. *Management Science, 49*(10), 1407–1424.
2. Chevalier, J. A., & Mayzlin, D. (2006). The effect of word of mouth on sales: Online book reviews. *Journal of Marketing Research, 43*(3), 345–354.
3. Dhanasobhon, S., Chen, P. Y., Smith, M. (2007). An analysis of the differential impact of reviews and reviewers at Amazon.com. In *Proceedings of the 28th International Conference on Information Systems (ICIS)*.
4. Kim, S. J., Maslowska, E., & Malthouse, E. C. (2018). Understanding the effects of different review features on purchase probability. *International Journal of Advertising, 37*(1), 29–53.
5. Walther, J. B., Liang, Y., Ganster, T., Wohn, D. Y., & Emington, J. (2012). Online reviews, helpfulness ratings, and consumer attitudes: An extension of congruity theory to multiple sources in web 2.0. *Journal of Computer-Mediated Communication, 18*(1), 97–112.
6. Mudambi, S. M., & Schuff, D. (2010). What makes a helpful online review? A study of customer reviews on amazon.com. *MIS Quarterly, 34*(1), 185–200.
7. Archak, N., Ghose, A., & Ipeirotis, P. G. (2011). Deriving the pricing power of product features by mining consumer reviews. *Management Science, 57*(8), 1485–1509.
8. Branco, F., Sun, M., & Villas-Boas, J. M. (2012). Optimal search for product information. *Management Science, 58*(11), 2037–2056.
9. Godes, D., & Mayzlin, D. (2004). Using online conversations to study word-of-mouth communication. *Marketing Science, 23*(4), 545–560.
10. Hong, H., Xu, D., Wang, G. A., & Fan, W. (2017). Understanding the determinants of online review helpfulness: A meta-analytic investigation. *Decision Support Systems, 102*, 1–11.
11. Zheng, L. (2021). The classification of online consumer reviews: A systematic literature review and integrative framework. *Journal of Business Research, 135*, 226–251.
12. Ghose, A., & Ipeirotis, P. G. (2011). Estimating the helpfulness and economic impact of product reviews: Mining text and reviewer characteristics. *IEEE Transactions on Knowledge and Data Engineering, 23*(10), 1498–1512.
13. Cao, Q., Duan, W., & Gan, Q. (2011). Exploring determinants of voting for the "helpfulness" of online user reviews: A text mining approach. *Decision Support Systems, 50*(2), 511–521.
14. Chen, C. C., & Tseng, Y. D. (2011). Quality evaluation of product reviews using an information quality framework. *Decision Support Systems, 50*(4), 755–768.
15. Sen, S., & Lerman, D. (2007). Why are you telling me this? An examination into negative consumer reviews on the web. *Journal of Interactive Marketing, 21*(4), 76–94.
16. Yin, D., Mitra, S., & Zhang, H. (2016). Research note—When do consumers value positive vs. negative reviews? An empirical investigation of confirmation bias in online word of mouth. *Information Systems Research, 27*(1), 131–144.
17. Pan, Y., & Zhang, J. Q. (2011). Born unequal: A study of the helpfulness of user generated product reviews. *Journal of Retailing, 87*(4), 598–612.
18. Wu, J. (2017). Review popularity and review helpfulness: A model for user review effectiveness. *Decision Support Systems, 97*, 92–103.
19. Zhang, J. Q., Craciun, G., & Shin, D. (2010). When does electronic word-of-mouth matter? A study of consumer product reviews. *Journal of Business Research, 63*(12), 1336–1341.
20. Gottschalk, S. A., & Mafael, A. (2017). Cutting through the online review jungle—Investigating selective ewom processing. *Journal of Interactive Marketing, 37*, 89–104.
21. Park, D. H., & Lee, J. (2008). eWOM overload and its effect on consumer behavioral intention depending on consumer involvement. *Electronic Commerce Research and Applications, 7*(4), 386–398.
22. Zinko, R., Stolk, P., Furner, Z., & Almond, B. (2020). A picture is worth a thousand words: How images influence information quality and information load in online reviews. *Electronic Markets, 30*(4), 775–789.

23. Edmunds, A., & Morris, A. (2000). The problem of information overload in business organisations: A review of the literature. *International Journal of Information Management, 20*(1), 17–28.
24. Eppler, M. J., & Mengis, J. (2004). The concept of information overload: A review of literature from organization science, accounting, marketing, MIS, and related disciplines. *The Information Society, 20*(5), 325–344.
25. Jacoby, J., Speller, D. E., & Kohn, C. A. (1974). Brand choice behavior as a function of information load. *Journal of Marketing Research, 11*(1), 63–69.
26. Jacoby, J. (1984). Perspectives on information overload. *Journal of Consumer Research, 10*(4), 432–435.
27. Hiltz, S. R., & Turoff, M. (1985). Structuring computer-mediated communication systems to avoid information overload. *Communications of the ACM, 28*(7), 680–689.
28. Crowley, A. E., & Hoyer, W. D. (1994). An integrative framework for understanding two-sided persuasion. *Journal of Consumer Research, 20*(4), 561–574.
29. Lutz, B., Pröllochs, N., & Neumann, D. (2022). Are longer reviews always more helpful? Disentangling the interplay between review length and line of argumentation. *Journal of Business Research, 144*, 888–901.
30. Gross, B. M. (1964). *The managing of organizations: The administrative struggle.* Free Press of Glencoe.
31. Malhotra, N. K. (1984). Reflections on the information overload paradigm in consumer decision making. *Journal of Consumer Research, 10*(4), 436–440.
32. Malhotra, N. K. (1982). Information load and consumer decision making. *Journal of Consumer Research, 8*(4), 419–430.
33. Speier, C., Valacich, J. S., & Vessey, I. (1999). The influence of task interruption on individual decision making: An information overload perspective. *Decision Sciences, 30*(2), 337–360.
34. Chen, Y. C., Shang, R. A., & Kao, C. Y. (2009). The effects of information overload on consumers' subjective state towards buying decision in the internet shopping environment. *Electronic Commerce Research and Applications, 8*(1), 48–58.
35. Furner, C. P., & Zinko, R. A. (2017). The influence of information overload on the development of trust and purchase intention based on online product reviews in a mobile vs. web environment: An empirical investigation. *Electronic Markets, 27*(3), 211–224.
36. Gao, J., Zhang, C., Wang, K., & Ba, S. (2012). Understanding online purchase decision making: The effects of unconscious thought, information quality, and information quantity. *Decision Support Systems, 53*(4), 772–781.
37. Xiao, B., & Benbasat, I. (2007). E-commerce product recommendation agents: Use, characteristics, and impact. *MIS Quarterly, 31*(1), 137–209.
38. Otondo, R. F., van Scotter, J. R., Allen, D. G., & Palvia, P. (2008). The complexity of richness: Media, message, and communication outcomes. *Information & Management, 45*(1), 21–30.
39. Schneider, S. C. (1987). Information overload: Causes and consequences. *Human Systems Management, 7*(2), 143–153.
40. Lurie, N. H. (2004). Decision making in information-rich environments: The role of information structure. *Journal of Consumer Research, 30*(4), 473–486.
41. Gutt, D., Neumann, J., Zimmermann, S., Kundisch, D., & Chen, J. (2019). Design of review systems–A strategic instrument to shape online reviewing behavior and economic outcomes. *The Journal of Strategic Information Systems, 28*(2), 104–117.
42. Zimmermann, S., Herrmann, P., Kundisch, D., & Nault, B. R. (2018). Decomposing the variance of consumer ratings and the impact on price and demand. *Information Systems Research, 29*(4), 984–1002.
43. Lutz, B., Pröllochs, N., & Neumann, D. (2018). Understanding the role of two-sided argumentation in online consumer reviews: A language-based perspective. In *Proceedings of the 39th International Conference on Information Systems (ICIS)*.
44. Willemsen, L. M., Neijens, P. C., Bronner, F., & de Ridder, J. A. (2011). "Highly recommended!": The content characteristics and perceived usefulness of online consumer reviews. *Journal of Computer-Mediated Communication, 17*(1), 19–38.

45. Jackson, S., & Allen, M. (1987). Meta-analysis of the effectiveness of one-sided and two-sided argumentation. In *Annual Meeting of the International Communication Association* (Vol. 196, pp. 78–92).
46. Furner, C. P., Zinko, R., & Zhu, Z. (2016). Electronic word-of-mouth and information overload in an experiential service industry. *Journal of Service Theory and Practice*.
47. Gross, R. (2014). A theoretical consumer decision making model: The influence of interactivity and information overload on consumers intent to purchase online. *International Journal of Business Management & Economic Research, 5*(4), 64–70.
48. Scammon, D. L. (1977). "information load" and consumers. *Journal of Consumer Research, 4*(3), 148–155.
49. Gu, B., Park, J., & Konana, P. (2012). Research note—The impact of external word of-mouth sources on retailer sales of high-involvement products. *Information Systems Research, 23*(1), 182–196.
50. He, R., & McAuley, J. (2016). Ups and downs: Modelling the visual evolution of fashion trends with one-class collaborative filtering. In *Proceedings of the 25th International Conference on World Wide Web* (pp. 507–517).
51. Forman, C., Ghose, A., & Wiesenfeld, B. (2008). Examining the relationship between reviews and sales: The role of reviewer identity disclosure in electronic markets. *Information Systems Research, 19*(3), 291–313.
52. Gefen, D., & Straub, D. W. (2004). Consumer trust in B2C e-commerce and the importance of social presence: Experiments in e-products and e-services. *Omega, 32*(6), 407–424.
53. Moravec, P., Kim, A., Dennis, A., & Minas, R. (2019). Fake news on social media: People believe what they want to believe when it makes no sense at all. *MIS Quarterly, 43*(4), 1343–1360.
54. Delorme, A., & Makeig, S. (2004). EEGLAB: An open-source toolbox for analysis of single-trial EEG dynamics including independent component analysis. *Journal of neuroscience methods, 134*(1), 9–21.
55. Klimesch, W. (2012). Alpha-band oscillations, attention, and controlled access to stored information. *Trends in Cognitive Sciences, 16*(12), 606–617.
56. Makeig, S., Westerfield, M., Jung, T. P., Enghoff, S., Townsend, J., Courchesne, E., & Sejnowski, T. J. (2002). Dynamic brain sources of visual evoked responses. *Science, 295*(5555), 690–694.
57. Kelly, S. P., Lalor, E. C., Reilly, R. B., & Foxe, J. J. (2006). Increases in alpha oscillatory power reflect an active retinotopic mechanism for distracter suppression during sustained visuospatial attention. *Journal of Neurophysiology, 95*(6), 3844–3851.
58. Rayner, K. (1998). Eye movements in reading and information processing: 20 years of research. *Psychological Bulletin, 124*(3), 372.
59. Engbert, R., & Kliegl, R. (2003). Microsaccades uncover the orientation of covert attention. *Vision Research, 43*(9), 1035–1045.
60. Lang, M. (2012). Investigating the EMOTIV Epoc for cognitive control in limited training time.
61. Zagermann, J., Pfeil, U., & Reiterer, H. (2018). Studying eye movements as a basis for measuring cognitive load. In *Extended Abstracts of the 2018 CHI Conference on Human Factors in Computing Systems* (pp. 1–6).
62. Findlay, J. M., & Kapoula, Z. (1992). Scrutinization, spatial attention, and the spatial programming of saccadic eye movements. *The Quarterly Journal of Experimental Psychology Section A, 45*(4), 633–647.
63. Baek, H., Ahn, J., & Choi, Y. (2014). Helpfulness of online consumer reviews: Readers' objectives and review cues. *International Journal of Electronic Commerce, 17*(2), 99–126.
64. Korfiatis, N., García-Bariocanal, E., Sánchez-Alonso, S.: Evaluating content quality and helpfulness of online product reviews: The interplay of review helpfulness vs. review content. *Electronic Commerce Research and Applications, 11*(3), 205–217.
65. Salehan, M., & Kim, D. J. (2016). Predicting the performance of online consumer reviews: A sentiment mining approach to big data analytics. *Decision Support Systems, 81*, 30–40.

Flow in Knowledge Work: An Initial Evaluation of Flow Psychophysiology Across Three Cognitive Tasks

Karen Bartholomeyczik, Michael Thomas Knierim, Petra Nieken, Julia Seitz, Fabio Stano, and Christof Weinhardt

Abstract To accelerate the development of flow-adaptive IT in NeuroIS research, the present work aims to improve the automatic detection of flow in knowledge work-related situations by observing flow emergence across three controlled tasks. In a pretest, we manipulated the type and difficulty of task and recorded subjective (self-reports) as well as objective (EEG features) measures of flow and mental effort. Results indicate that a novel text typing task resembles the expertise of knowledge workers best which is reflected in elevated flow levels across tasks. Difficulty manipulations based on autonomously chosen task difficulty elicited contrasts in flow and mental effort, which was also reflected in the EEG data by Theta band power modulations. This further highlights the utility of autonomy for stimulating flow. We discuss limitations and improvements for the experiment and how this contributes to further research on flow-adaptive IT.

Keywords Flow · Psychophysiology · Knowledge work · Adaptive systems

K. Bartholomeyczik (✉) · M. T. Knierim · J. Seitz · F. Stano · C. Weinhardt
Karlsruhe Institute of Technology, Institute of Information Systems and Marketing, Karlsruhe, Germany
e-mail: karen.bartholomeyczik@kit.edu

M. T. Knierim
e-mail: michael.knierim@kit.edu

J. Seitz
e-mail: julia.seitz@kit.edu

F. Stano
e-mail: fabio.stano@student.kit.edu

C. Weinhardt
e-mail: weinhardt@kit.edu

P. Nieken
Karlsruhe Institute of Technology, Institute of Management, Karlsruhe, Germany
e-mail: petra.nieken@kit.edu

© The Author(s), under exclusive license to Springer Nature Switzerland AG 2022
F. D. Davis et al. (eds.), *Information Systems and Neuroscience*,
Lecture Notes in Information Systems and Organisation 58,
https://doi.org/10.1007/978-3-031-13064-9_3

1 Introduction

During flow, an intrinsically motivating state, people act with high involvement and a sense of control in an optimally challenging task [1]. As flow is associated with individual and organizational benefits (e.g. work performance, wellbeing), it is desirable for management and knowledge workers [2–4]. Convergingly, Neuro-Information-Systems (NeuroIS) scholars are increasingly investigating how flow experiences can be experimentally induced and observed using wearable sensors, for example to allow real-time flow classifications, to provide flow-fostering user experiences or to prevent flow-impeding interruptions [5–9]. Thus, they contribute to the design of information technology (IT) artifacts [10] by utilizing findings on the neurophysiological correlates of flow.

To acquire the necessary foundational knowledge on flow psychophysiology, experimental flow research has commonly manipulated task difficulty to establish the flow precondition of a skill-demand balance [11]. Cognitive tasks (e.g. mental arithmetic) have been employed in the knowledge work context due to their suggested conceptual closeness to the cognitive tasks commonly performed by knowledge workers [6, 12]. However, the artificial nature and focus on specific cognitive skills of these experimental tasks does not adequately and comprehensively capture tasks and skills common in knowledge work. Hence, our study aims to evaluate if newly designed experimental tasks can reproduce knowledge work more appropriately, and effectively elicit flow in the laboratory. By producing a closer correspondence between flow tasks in the laboratory and in the field, we strive to increase the external validity of experimental flow research. As autonomy is an essential determinant of intrinsic motivation [13–15], we further aim to evaluate if autonomous choice of the task difficulty and the task in general increases flow.

To accomplish these goals, we develop two novel computer-based tasks and conduct a pretest in which we examine the effectiveness of the novel tasks for subjectively and objectively eliciting flow based on self-reports and neurophysiological features in comparison with an already validated task. Our results indicate that especially the novel text typing task allows subjects to capitalize on high levels of (knowledge work-related) expertise. This is reflected in elevated flow levels in this task. We also find similar difficulty-related contrasts in self-reports of mental effort and EEG features.

Our study contributes to the NeuroIS field in three ways. First, we show if different tasks are equally effective for inducing flow in the laboratory. This helps other researchers to decide which task to apply in experimental designs. The provision of a novel text typing task with close resemblance to knowledge work also diminishes the concerns regarding artificiality and lack of engagement that were raised in response to earlier laboratory research [11, 16]. Second, we evaluate the potential of granting autonomy in task choice for increasing flow. The elicitation of increased flow levels in the laboratory is necessary to reach the flow-characteristic optimal state [1]. Third, we take on earlier attempts to unobtrusively measure flow and to identify (neuro-)physiological flow correlates (e.g. [6–9, 17–20]) by examining the applicability of mobile ear-EEG sensors. This validation of novel mobile sensors for analyzing flow psychophysiology enables the conduction of field studies with real-time flow classifications while minimizing interruptions criticized with regard to use of self-reports [21].

2 Theoretical Background and Related Work

Flow is an autotelic state characterized by high concentration, merging of action and awareness, loss of self-consciousness, sense of self-control, and distorted temporal experience [1]. In the work context, flow has been associated with positive individual, social, and organizational outcomes [22]. NeuroIS scholars have therefore proposed to develop IT artifacts that are flow-adaptive (e.g. preventing disruptions if a person is currently in flow, [5, 7]) or flow-enhancing (e.g. providing suitable material in e-learning [8]). Thus, IT artifacts can build on flow as an input or as a desired outcome. To enable IT artifacts to provide flow-adaptive interventions, real-time classifications of flow are necessary. Even though the comprehensive conceptualization and operationalization of the flow concept are still debated [23], the empirical field commonly agrees that a skill-demand balance is required for flow to emerge [24]. Hence, earlier research has extensively evaluated manipulations of this balance [11]. Although these difficulty manipulation (DM) paradigms have successfully induced flow in the laboratory [11], most of the applied tasks diverge from conventional knowledge work (e.g. requiring subjects to sum three or more numbers [6, 12]). The following lack of task-specific expertise requires additional attentional processes during task performance [25]. These processes might be obstructive for the experience of effortless attention, that is flow [25]. Also, not only the intrinsic enjoyability of these tasks remains questionable, but the given task difficulties diminish the autonomy of the subjects. According to the prominent Self-Determination Theory [14], autonomy is one of three basic psychological needs that determine motivation. Low autonomy due to the external difficulty determinations or lack of task choice could thereby limit flow elicitation in the laboratory. Hence, even though DM paradigms evidentially elicit contrasts [11], they suffer from limitations in external validity and might even restrict the maximum amount of elicited flow.

Based on these findings, we expect the highest perceived autonomy and flow for self-chosen optimal difficulty compared to easier and harder difficulties. As earlier studies have already differentiated flow from overload and boredom based on neurophysiological features (e.g. [6–9, 17–20]), we expect analogue contrasts in neurophysiological data.

3 Method

Experimental Procedure, Material, and Sample. We applied a 3×4 within-subject design with two independent variables: task (mental arithmetic, puzzle, text typing) and difficulty (moderate, optimal, easier, harder). Each subject performed each of the three tasks in a randomized order. All tasks were presented with each of the four difficulty levels in a fixed order. Lastly, subjects performed an additional round with a self-chosen task in the optimal difficulty level. This task choice increases autonomy and allows the performance of an activity that subjects perceive as enjoyable which is one of the core characteristics of flow [1]. The complete experimental procedure is outlined in Fig. 1. The experiment was implemented in oTree, a python-based framework for social science experiments, that allows to present surveys and experimental tasks on a common platform [26].

Our pretest sample consisted of $N = 6$ Ph.D. students (3 females, $M_{Age} = 28.5$) who belong to the category of knowledge workers.

Tasks. We developed two novel tasks (text typing and puzzle) to decrease artificiality, low engagement and lack of enjoyment criticized with regard to earlier experimental tasks [11, 16]. We designed the text typing task based on [27, 28] to resemble writing common in knowledge work. As explained above, the performance of a task closely related to the expertise of the subjects avoids the recruitment of additional attentional processes and enables the flow characteristic optimal and effortless state to emerge [1, 25]. The text typing task consisted of a set of pre-selected text segments of varied length that were presented to each subject. Subjects had to copy these segments sequentially in a given amount of time.

Our design of the puzzle task especially corresponded to the autotelic nature of flow experiences [1], in that game-like tasks are commonly administered to account for the enjoyable character of flow (e.g. [9, 18]). Also, solving puzzles provides direct feedback which is one of the preconditions of flow [29]. In contrast to [30], we used digital pictures to allow computer-based performance.

To instantiate all preconditions of flow [29], both novel tasks were optimally difficult (calibration see next paragraph) and provided clear goals (solving puzzles,

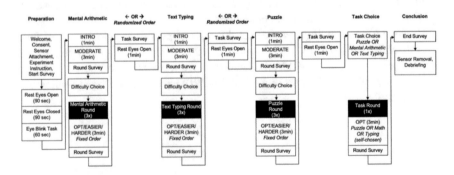

Fig. 1 Visualization of the experimental procedure

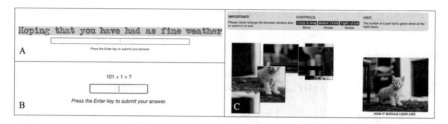

Fig. 2 Visualization of the tasks (A: Text Typing, B: Mental Arithmetic, C: Puzzle)

copying task in given time). As a reference for these novel tasks, we adapted an already validated mental arithmetic task from [6, 12]. In this task, subjects mentally sum two or more numbers in a given amount of time. Figure 2 depicts exemplary trials for all tasks.

DM Paradigm. To make task difficulties comparable, we drew on earlier findings that autonomous choice of optimal difficulty (i.e. with an optimal skill-demand balance) induces similar flow levels as external difficulty determinations [31]. Thus, based on a moderately difficult task round (moderate condition), subjects chose a difficulty level matching their skills (optimal condition). This optimal level was used for the difficulty calibration of the other conditions (easier, harder). Depending on the task, difficulty was adjusted by varying number of puzzle pieces, presented words, or required additions. Due to these predetermined ranges of possible difficulty levels, we could individualize the difficulty while allowing standardization across subjects.

Measures. After each round, subjects filled a survey with self-reports of flow (Flow Short Scale, FKS [3]), autonomy (three items [32]), and mental effort (rating scale, RSME, [33]). After each block, they answered a task survey including an item on perceived competence. In the end survey, enjoyment of each task was reported. All survey instruments (except for mental effort) used 7-point Likert scales. EEG data were collected with cEEGrids (printed Ag/AgCl electrodes arranged in a c-shaped array around the ear) [34]. These sensors allow to collect EEG data unobtrusively and have already been used in multiple studies (e.g. assessing workload or facial activities [35, 36]). The cEEGrids were connected to an OpenBCI biosignal acquisition board as in [36].

Analyses. We computed descriptive statistics and repeated-measures correlations for FKS, RSME, and autonomy. EEG data preprocessing involved mean-centering, band-pass filtering (FIR 3–30 Hz), and spherical interpolation of bad channels (flatness, amplitude, and high-frequency noise criteria). Channels were re-referenced to linked mastoids (L5 + R5). Theta (4–7 Hz), alpha (7–13 Hz), and beta (13–30 Hz) power were extracted for each channel after Welch's windowed PSD decomposition (1 s windows with 50% overlap) and finally median-aggregated for each subject.

4   Results

Psychometric variables. Descriptive statistics are shown in Tables 1 and 2. Internal consistency was good for FKS (standardized Cronbach's Alpha > 0.80) but questionable for autonomy (standardized Cronbach's Alpha $= 0.61$; item removal not possible).

As expected, mean values of autonomy were lower in the easier and harder compared to the optimal condition in the text typing and the puzzle task. Unexpectedly, mean values of autonomy were lower in the optimal compared to the moderate condition in the text typing and mental arithmetic task. Mean values of FKS scores were highest in the text typing and lowest in the puzzle task for each difficulty level. The expected inverted u-shape (highest mean FKS scores in the optimal condition) appeared in the mental arithmetic and the puzzle task but not in text typing. Mean values of RSME scores were highest in the mental arithmetic task (except for the easier condition). RSME scores increased from easier to harder as expected (with an exception in the moderately difficul mental arithmetic task). Competence and enjoyment were higher in the text typing task than in the other ones.

In the last task round, subjects chose the mental arithmetic or the text typing task with similar ratios. In this round, mean FKS scores were higher ($M = 5.3, SD = 0.9$) than in the first optimal mental arithmetic round ($M = 4.8, SD = 1.1$), but slightly lower than in the first optimal text typing round ($M = 5.5, SD = 1$).

Autonomy correlated positively and significantly with FKS scores, $r(57) = 0.52, p < 0.001$, and RSME scores, $r(57) = 0.43, p < 0.001$. There was no linear relationship between FKS and RSME scores, $r(57) = 0.02, p = 0.894$.

Neurophysiological variables. As the tasks elicited heterogenous contrast perceptions with substantial variances, we decided to forego a comparison of EEG powers across tasks and difficulty conditions and focused on a correlation analysis between self-reports and EEG powers instead. To account for the likely confounding influence of time, self-reports and frequency band powers were z-standardized within-participant and within-task. Repeated-measures correlations (Table 3) showed no significant relationships between frequency band powers and FKS scores or autonomy reports but a positive and significant relationship between the Theta band and RSME scores.

5   Discussion

5.1   Contributions

Our pretest indicates the effectiveness of all tasks for eliciting flow. Hence, the first major contribution of our study is the provision of two novel computer-based tasks that appear at least equally effective for inducing flow in the laboratory as the already validated mental arithmetic task [6, 12]. Primarily the text typing task

Table 1 Means and standard deviations (in parentheses) of FKS, autonomy, and RSME for each task and difficulty (E = easier, M = moderate, O = optimal, H = harder)

	Text typing				Mental arithmetic				Puzzle			
	E	M	O	H	E	M	O	H	E	M	O	H
FKS	4.4 (1.2)	5.4 (1.0)	5.5 (1.0)	5.6 (0.9)	4.2 (0.6)	4.6 (1.3)	4.8 (1.1)	4.4 (1.1)	3.6 (0.3)	5.1 (1.0)	4.5 (0.5)	4.3 (0.8)
Autonomy	3.3 (1.4)	4.0 (0.9)	3.7 (1.2)	3.6 (1.5)	3.0 (1.5)	4.2 (1.6)	3.7 (1.4)	3.7 (1.3)	2.4 (1.1)	4.1 (1.4)	4.3 (1.2)	3.2 (1.4)
RSME	28 (33)	37 (26)	48 (36)	56 (39)	14 (10)	82 (31)	75 (14)	80 (34)	11 (6)	43 (13)	49 (13)	72 (39)

Table 2 Means and standard deviations (in parentheses) of enjoyment and competence

	Text typing	Mental arithmetic	Puzzle
Enjoyment	4.5 (1.3)	3.8 (1.5)	3.8 (2.1)
Competence	4.8 (1.0)	3.0 (2.1)	4.4 (1.7)

Table 3 Repeated measures correlations between self-reports and frequency band powers

	Theta (4–7 Hz)	Alpha (7–13 Hz)	Beta (13–30 Hz)
FKS	$0.04, p = 0.787$	$0.20, p = 0.147$	$0.12, p = 0.376$
Autonomy	$0.07, p = 0.611$	$0.10, p = 0.481$	$0.09, p = 0.517$
RSME	$0.36, p = 0.008$	$0.22, p = 0.106$	$0.04, p = 0.765$

appears to elicit higher levels of flow. As perceived competence was also highest in this task, our findings are in line with the expertise effect (i.e. higher probability of flow experiences for higher levels of competence [37]). Hence, more externally valid evaluations of flow in knowledge work could be achieved by applying our novel text typing task because it might reproduce common tasks of knowledge workers that require their particular expertise.

Second, our study supports earlier findings that the autonomous choice of a challenging difficulty is at least as effective for eliciting a skill-demand balance as external determination of this balance [31]. Given that flow appeared as elevated in the self-chosen task, our findings reinforce that flow in the laboratory could generally be fostered by granting autonomy in experimental tasks [31, 38], thereby approximating the flow-characteristic optimal state [1]. The indicated importance of autonomous choices also contributes to the design of flow-enhancing IT artifacts because researchers might incorporate this reasoning in the design to leverage flow experiences in the usage of IT.

As DM elicited contrasts in subjective and objective measures of mental effort, the third contribution of our study is the indicated potential of using ear-EEG sensors for classifying mental effort levels even across multiple tasks. Based on a proposed inverted-u-shaped relationship between flow and mental effort [39] (see also the present non-significant linear association between these variables), this could be used as a starting point for applying these sensors for unobtrusive flow classifications as well. This is especially promising for the development of flow-adaptive IT artifacts that measure flow in real-time and adaptively employ interventions for fostering flow.

5.2   Limitations and Future Research

Due to the pretest format, the small sample size constitutes the major limitation of our study. Hence, we aim to replicate our study with a larger sample to allow more complex statistical analyses and to support our preliminary results. Since our findings did not indicate differences in perceived difficulty between the optimal and the harder condition for the text copying task, the contrast-based identification of (neuro-)physiological flow features is also impeded. To ensure that subjects will perceive the harder condition as overload, we will increase this task's difficulty.

Additionally, the small sample sizes prevent an investigation of individual differences. Based on our finding that flow was highest in the round with the self-chosen task, it would be interesting though to evaluate whether a person also experienced elevated levels of flow in the first block of this task that he or she later chose for the additional round. This could be masked by our comparison of flow levels between the different task blocks with all subjects included. A finding of this effect would imply that individual task preferences determine flow apart from the autonomy effect.

Generally, experimental flow research suffers from the limitation that the time periods during which the task is performed only last a few minutes. Hence, the recorded flow experiences might significantly deviate from flow experiences in the field in which task duration usually varies and is often terminated due to unforeseen interruptions [7]. On the same note, laboratory flow research based on self-reports requires an interruption of the subjects. This artificial cut-off might have an influence on perceived flow. Our application of the cEEGrids as an unobtrusive measure of flow psychophysiology is a first step towards addressing this limitation. Unfortunately, we cannot completely abstain from the collection of self-reports until the detection of stable (neuro-)physiological flow features. However, NeuroIS scholars are increasingly working on identifying these features (e.g. [6, 7, 9]) that hopefully will enable less invasive flow measurements in the future.

5.3   Conclusion

In sum, our study provides two novel tasks for investigating flow in knowledge work and supports the efficacy of granting autonomy for increasing flow. Especially the text typing task seems to resemble the expertise of knowledge workers, hence is particularly well suited for experimental flow research. Our results are especially promising for the NeuroIS field because they could serve as a basis for developing flow-adaptive IT artifacts, thereby targeting flow-associated positive outcomes like individual well-being and work performance.

Acknowledgements Funded as part of the Excellence Strategy of the German Federal and State Governments.

References

1. Csikszentmihalyi, M. (1975). *Beyond boredom and anxiety*. Jossey-Bass Publishers.
2. Bryce, J., & Haworth, J. (2002). Wellbeing and flow in sample of male and female office workers. *Leisure Studies, 21*, 249–263.
3. Engeser, S., & Rheinberg, F. (2008). Flow, performance and moderators of challenge-skill balance. *Motivation and Emotion, 32*, 158–172.
4. Zubair, A., & Kamal, A. (2015). Work related flow, psychological capital, and creativity among employees of software houses. *Psychological Studies, 60*, 321–331.
5. Rissler, R., Nadj, M., & Adam, M. T. P. (2017). Flow in information systems research: Review, integrative theoretical framework, and future directions. In J. M. Leimeister & W. Brenner (Eds.), *Proceedings der 13. Internationalen Tagung Wirtschaftsinformatik* (pp. 1051–1065). St. Gallen.
6. Knierim, M. T., Rissler, R., Hariharan, A., Nadj, M., & Weinhardt, C. (2019). Exploring flow psychophysiology in knowledge work. Lecture Notes in Information Systems and OrganisationIn. In F. Davis, R. Riedl, J. vom Brocke, P. M. Léger, & A. Randolph (Eds.), *Information systems and neuroscience* (Vol. 29, pp. 239–249). Springer.
7. Rissler, R., Nadj., M., Li, M. X., Knierim, M. T., & Maedche, A. (2018) Got flow? Using machine learning on physiological data to classify flow. In *Extended Abstracts of the 2018 CHI Conference on Human Factors in Computing Systems (CHI EA '18)* (pp. 1–6). Association for Computing Machinery, New York.
8. Wang, C. C., & Hsu, M. C. (2014). An exploratory study using in-expensive electroencephalography (EEG) to understand flow experience in computer-based instruction. *Information and Management, 51*, 912–923.
9. Tozman, T., Magdas, E. S., MacDougall, H. G., & Vollmeyer, R. (2015). Understanding the psychophysiology of flow: A driving simulator experiment to investigate the relationship between flow and heart rate variability. *Computers in Human Behavior, 52*, 408–418.
10. Riedl, R., Banker, R. D., Benbasat, I., Davis, F. D., Dennis, A. R., Dimoka, A., Gefen, D., Gupta, A., Ischebeck, A., Kenning, P. H., Müller-Putz, G. R., Pavlou, P. A., Straub, D. W., vom Brocke, J., & Weber, B. (2010). On the foundations of NeuroIS: Reflections on the Gmunden Retreat 2009. *Communications of the Association for Information Systems, 27*, 243–264.
11. Moller, A. C., Meier, B. P., & Wall, R. D. (2013). Developing an experimental induction of flow: Effortless action in the lab. In B. Bruya (Ed.), *Effortless attention: A new perspective in the cognitive sciene of attention and action* (pp. 191–204). The MIT Press.
12. Ulrich, M., Keller, J., Hoenig, K., Waller, C., & Grön, G. (2014). Neural correlates of experimentally induced flow experiences. *NeuroImage, 86*, 194–202.
13. Kukita, A., Nakamura, J., & Csikszentmihalyi, M. (2022). How experiencing autonomy contributes to a good life. *The Journal of Positive Psychology, 17*, 34–45.
14. Ryan, R. M., & Deci, E. L. (2017). *Self-determination theory: Basic psychological needs in motivation, development, and wellness*. Guilford Publishing.
15. Hackman, J. R., & Oldham, G. R. (1980). *Work redesign*. Addison-Wesley.
16. Delle Fave, A., Massimini, F., & Bassi, M. (2011). Instruments and methods in flow research. In A. Delle Fave, F. Massimini, & M. Bassi (Eds.), *Psychological selection and optimal experience across cultures* (pp. 59–87). Springer.
17. Peifer, C., & Tan, J. (2021). The psychophysiology of flow experience. In C. Peifer & S. Engeser (Eds.), *Advances in flow research* (pp. 191–230). Springer.
18. Huskey, R., Craighead, B., Miller, M. B., & Weber, R. (2018). Does intrinsic reward motivate cognitive control? A naturalistic-fMRI study based on the synchronization theory of flow. *Cognitive, Affective, & Behavioral Neuroscience, 18*, 902–924.
19. Peifer, C., Schulz, A., Schächinger, H., Baumann, N., & Antoni, C. H. (2014). The relation of flow-experience and physiological arousal under stress—Can u shape it? *Journal of Experimental Social Psychology, 53*, 62–69.

20. Yoshida, K., Sawamura, D., Inagaki, Y., Ogawa, K., Ikoma, K., & Sakai, S. (2014). Brain activity during the flow experience: A functional near-infrared spectroscopy study. *Neuroscience Letters, 573,* 30–34.
21. Csikszentmihalyi, M., & Hunter, J. (2003). Happiness in everyday life: The uses of experience sampling. *Journal of Happiness Studies, 4,* 185–199.
22. Peifer, C., & Wolters, G. (2021). Flow in the Context of Work. In C. Peifer & S. Engeser (Eds.), *Advances in flow research* (pp. 287–321). Springer.
23. Abuhamdeh, S. (2020). Investigating the "flow" experience: Key conceptual and operational issues. *Frontiers in Psychology, 11,* 158.
24. Barthelmäs, M., & Keller, J. Antecedents, boundary conditions and consequences of flow. In C. Peifer & S. Engeser (Eds.), *Advances in flow research* (pp. 71–107). Cham: Springer.
25. Hommel, B. (2010). Grounding attention in action control: The intentional control of selection. In B. Bruya (Ed.), *Effortless attention: A new perspective in the cognitive science of attention and action* (pp. 121–140). The MIT Press.
26. Chen, D. L., Schonger, M., & Wickens, C. (2016). oTree—An open-source platform for laboratory, online, and field experiments. *Journal of Behavioral Experimental Finance, 9,* 88–97.
27. Fest, S., Kvaloy, O., Nieken, P., & Schoettner, A. (2021). How (not) to motivate online workers: Two controlled field experiments on leadership in the gig economy. *The Leadership Quarterly, 32,* 101514.
28. Nieken, P. (2022). *Charisma in the gig economy: The impact of digital leadership and communication channels on performance.* Karlsruhe: Mimeo Karlsruhe Institute of Technology.
29. Nakamura, J., & Csikszentmihalyi, M. (2002). The concept of flow. In C. R. Snyder & S. J. Lopez (Eds.), *Handbook of positive psychology* (pp. 89–105). Oxford University Press.
30. Tse, D. C. K., Fung, H. H., Nakamura, J., & Csikszentmihalyi, M. (2016). Teamwork and flow proneness mitigate the negative effect of excess challenge on flow state. *The Journal of Positive Psychology, 13,* 284–289.
31. de Sampaio Barros, M. F., Araújo-Moreira, F. M., Trevelin, L. C., & Radel, R. (2018). Flow experience and the mobilization of attentional resources. *Cognitive, Affective, & Behavioral Neuroscience, 18,* 810–823.
32. Sheldon, K. M., & Hilpert, J. C. (2012). The balanced measure of psychological needs (BMPN) scale: An alternative domain general measure of need satisfaction. *Motivation and Emotion, 36,* 439–451.
33. Zijlstra, F. R. H. (1993). *Efficiency in work behaviour: A design approach for modern tools.* University Press.
34. Debener, S., Emkes, R., De Vos, M., & Bleichner, M. (2015). Unobtrusive ambulatory EEG using a smartphone and flexible printed electrodes around the ear. *Science and Reports, 5,* 1–11.
35. Wascher, E., Arnau, S., Reiser, J. E., Rudinger, G., Karthaus, M., Rinkenauer, G., Dreger, F., & Getzmann, S. (2019). Evaluating mental load during realistic driving simulations by means of round the ear electrodes. *Frontiers in Neuroscience, 13,* 1–11.
36. Knierim, M. T., Schemmer, M., & Perusquía-Hernández, M. (2021). Exploring the recognition of facial activities through around-the-ear electrode arrays (cEEGrids). Lecture Notes in Information Systems and OrganisationIn. In F. D. Davis, R. Riedl, J. vom Brocke, P. M. Léger, A. B. Randolph, & G. Müller-Putz (Eds.), *Information systems and neuroscience* (Vol. 52, pp. 47–55). Springer.
37. Rheinberg, F., & Engeser, S. (2018). Intrinsic motivation and flow. In J. Heckhausen & H. Heckhausen (Eds.), *Motivation and action* (pp. 579–622). Springer.
38. Knierim, M., Nadj, M., Li, M., & Weinhardt, C. (2019). Flow in knowledge work groups—Autonomy as a driver or digitally mediated communication as a limiting factor? *ICIS 2019 Proceedings, 7.*
39. Ewing, K. C., Fairclough, S. H., & Gilleade, K. (2016). Evaluation of an adaptive game that uses EEG measures validated during the design process as inputs to a biocybernetic loop. *Frontiers in Human Neuroscience, 10,* 223.

Biosignal-Based Recognition of Cognitive Load: A Systematic Review of Public Datasets and Classifiers

Julia Seitz and Alexander Maedche

Abstract Cognitive load is a user state intensively researched in the NeuroIS community. Recently, the interest in designing neuro-adaptive information systems (IS) which react to the user's current state of cognitive load has increased. However, its measurement through surveys is cumbersome and impractical. Alternatively, it was shown that by collecting biosignals and analysing them with supervised machine learning, it is possible to recognize cognitive load less obtrusively. However, data collection and classifier training are challenging. Specifically, large amounts of data are required to train a high-quality classifier. To serve this need and increase transparency in research, more and more datasets are publicly available. In this paper, we present our results of a systematic review of public datasets and corresponding classifiers recognizing cognitive load using biosignals. Thereby, we want to stimulate a discussion in the NeuroIS community on the role and potential of public datasets and classifiers for designing neuro-adaptive IS.

Keywords Cognitive load · NeuroIS · Biosignals · Datasets · Machine learning

1 Introduction

Cognitive load (CL), the amount of cognitive resources needed to conduct a task, is a state intensively researched in NeuroIS community [1–4]. To measure user's cognitive load, different measurement instruments have been proposed and used, e.g. scales such as NASA TLX or RMSE [2, 5, 6]. However, their key limitation is that they can only be applied after the cognitive load was experienced. Thus, they are poorly suited for real-time observation of cognitive load [7–10]. An alternative approach is to measure cognitive load via biosignals from the central and peripheral

J. Seitz (✉) · A. Maedche
Karlsruhe Institute of Technology (KIT), Institute of Information Systems and Marketing (IISM), Research Group Information Systems I, Kaiserstr. 89, 76133 Karlsruhe, Germany
e-mail: julia.seitz@kit.edu

A. Maedche
e-mail: alexander.maedche@kit.edu

© The Author(s), under exclusive license to Springer Nature Switzerland AG 2022
F. D. Davis et al. (eds.), *Information Systems and Neuroscience*,
Lecture Notes in Information Systems and Organisation 58,
https://doi.org/10.1007/978-3-031-13064-9_4

35

nervous system with electrocardiography (ECG), Electroencephalography (EEG), or eye tracking [3, 11–14]. As reviews and meta analyses indicate, cognitive load can be identified in certain biosignal characteristics [11, 12, 15]. The collected biosignals in combination with subjectively reported scales can be used as input for supervised machine learning (ML) algorithms that are able to derive a model that can automatically classify user's cognitive load states. Therefore, the biosignal characteristics identified in reviews and meta analyses can be used as a starting point for selected features as input to the ML algorithms. With the rise of wearable sensor technology and its embedding in everyday life, the collection of biosignals has increased and has become easier and more applicable in real-time [16]. The availability of sensor technology combined with powerful supervised ML techniques offers a unique opportunity for the field of NeuroIS to accelerate research in neuro-adaptive IS as well as passive brain-computer interfaces [17–19]. Specifically, cognitive load-adaptive IS could recognize cognitive load states in real-time and enable adaptation of tasks and information technology to the user accordingly. Further, passive brain-computer interfaces continuously monitor the user's cognitive state, exemplarily the user's cognitive load, and thereby enable feedback and insights on the user's reaction to the designed system in real-time [19–21]. However, building a classifier from scratch is challenging. Large amounts of labeled biosignal data are required as a foundation to successfully train ML classifiers. To mitigate this problem, reusing existing, labeled datasets, and existing classifiers is an interesting and promising approach. Reusing datasets seems to be especially promising since experiments often do have only small sample sizes which are not enough to reach the required amount of data needed to train a high-quality classifier. However, to be reusable, these datasets need to be publicly available and known to researchers. So far little work on public datasets and the application of supervised ML on these datasets has been published in the NeuroIS retreat field. With this paper we want to stimulate a discussion on the role and potential of public datasets and ML classifiers in NeuroIS. Specifically, we want to review existing datasets and corresponding classifiers for cognitive load, as an important and heavily researched user state. Thus, we articulate the following research question: *Which publicly available datasets and corresponding classifiers leveraging biosignals for recognizing cognitive load exist?* The remainder of the paper is as following: We first introduce the research method. Subsequently, we present and compare the identified datasets and classifiers. Finally, we end with a discussion of contributions, limitations and research avenues.

2 Method

To ensure that we identify all relevant datasets and classifiers, we conducted a systematic search. We chose the methodology of systematic literature reviews as suggested by [22] and followed a multi-step approach to discover existing datasets and corresponding classifiers. In the following, each step is shortly presented. As a first step, we defined the search strategy. This includes a search string, consisting of synonyms for cognitive load, relevant databases and corresponding exclusion criteria. We selected the databases mentioned in [23] in addition to other databases for datasets and classifiers, such as Kaggle, Google Data and Physionet to also cover non-academic and physiological datasets. In total, we used the databases IEEE dataport [24], Mendely data [25], FigShare [26], Kaggle [27], Google Data [28], Zenodo [29], datahub.io [30], Physionet [31], eu.dat [32], Fairdom [33], Dataverse [34], Dryad [35] and Dans Narcis [36]. As search string, we used the following words: *"cognitive load" OR workload OR "mental load" OR taskload OR "mental effort" OR "task load"*. The databases mostly did not provide the possibility to use an advanced or combined search string, so we applied each element on its own. However, this led to a vast amount of false positive results which were sorted out while reviewing the results of each database. After executing the search in February 2022, we screened the initial sample of all 12.741 datasets with a main focus on the exclusion criteria findability and accessibility as two core criteria of the FAIR principles for datasets [23, 37, 38]. FAIR thereby is an acronym standing for "Findable", "Accessible", "Interoperable" and "Reusable" [23, 37]. Findable and Accessible cover the dataset's ability to be found by humans and computers, exemplarily due to machine-readable metadata. We assumed that when datasets are found in our search, this requirement is fulfilled. Accessible refers to the user's knowledge on how to access the dataset after finding it. Datasets therefore should be available openly or after authentication and authorization and retrievable via a standardized communications protocol [37]. We also include datasets accessible after request or login which we found in our search. This resulted in 79 datasets. After duplicate removal, a final set of 20 datasets evolved which we checked for scientific publications describing the dataset and classifiers trained on these datasets and, if applicable, other datasets mentioned in the paper. We used Google Scholar to ensure that publications from all related research fields are covered. This resembles the backward-forward search in a classic systematic literature review and resulted in 11 additional datasets and 34 papers covering classifiers. The final next steps cover the descriptive and conceptual analysis of the selected datasets, which we outline in Chap. 3.

3 Results

To ensure a literature-grounded comparison, we compared the 31 identified public datasets based on selected requirements for reference datasets published by [39]. We chose the general requirement categories mentioned by the authors (population, stimuli, modality, self-reported information, sensors) and additionally checked for publications describing the dataset. The detailed evaluation per dataset is visible in Table 1 (✓; when criteria is fulfilled, x when not). Subsequently, we highlight the summarized results:

Population. When comparing the researched populations, we see a focus in most datasets on a specific age range (i.e., young adults between 20 and 26 years).

Table 1 Overview on datasets and related classifiers

	Description of dataset and classifier						Requirements for reference datasets										
No	Ground Truth	Biosignal	Task	# Participants	Environment	Classifier No	Population			Effective stimuli	Multimodal	Self-report			Sensor		
							Diff. age	Diff. gender	Gender balance			Scale used	Add. internal	Add. external	Device specified	Noise ddescribed	Calibration
1	TLX	EEG, ECG, ACC, RESP, GSR, TEMP, BVP	MATB-II, physical activity	48	L	32, 33	x	✓	x	✓	✓	✓	✓	✓	✓	x	x
2[a]	TLX	ECG, EDA, EMG, RESP	Math task	18	L	34	x	–	–	✓	✓	✓	✓	x	✓	✓	✓
3	Task	EEG	Operating model build in solid works	8	L	35	x	x	x	x	x	x	✓	x	✓	✓	x
4	Rating scale	EYE, GSR, PPG, FNIRS	Lego puzzle	22	L	36	✓	✓	✓	x	✓	✓	x	x	✓	x	x
5	Task	ECG, EDA, PPG, ACC	Math problem, logic problem, stroop test	62	L	37–39	✓	✓	x	✓	✓	x	x	x	✓	x	✓
6[a]	Task	EDA, ECG	Reading, math task	40	L	40–41	x	✓	x	✓	✓	x	x	x	✓	x	✓
7[a]	Task	ECG, RESP, MOVEMENT	Marksmannship	8	L	42	x	✓	x	x	✓	x	x	✓	✓	x	x
8[a]	TLX	ECG, EYE	Word matching task	41	L	43	x	✓	✓	x	✓	✓	-	x	✓	✓	x
9	TLX	GSR, HR, TEMP	Driving	10	F		✓	✓	x	x	✓	✓	x	x	✓	x	✓

(continued)

Table 1 (continued)

No	Ground Truth	Biosignal	Task	# Participants	Environment	Classifier No	Population			Effective stimuli	Multimodal	Self-report			Sensor	Noise described	Calibration
							Diff. age	Diff. gender	Gender balance			Scale used	Add. internal	Add. external	Device specified		
10	TLX, RSME	EDA, ECG, BODY, FACE	Knowledge work	25	L		x	✓	x	✓	✓	✓	✓	✓	✓	x	x
11	TLX,	PPG, ECG, GSR	N-back task	22	L	36, 44	x	✓	x	✓	✓	✓	✓	x	✓	✓	✓
12	TLX	TEMP, ACC, EDA, RR	Games	23	L	45–51	✓	x	x	✓	✓	✓	✓	x	✓	✓	✓
13	TLX	TEMP, ACC, EDA, RR	Psychological task, n-back task	23	L	45–51	x	✓	x	✓	✓	✓	✓	x	✓	✓	✓
14	No labels	HRV, RR	Driving, work, physical activity	1	F		x	x	x	x	x	x	x	x	✓	x	x
15	TLX	ECG, RESP	MATB-II	26	L	52	x	✓	x	✓	✓	✓	✓	x	✓	x	✓
16[a]	Task, TLX	EYE, PUPIL, HR, HRV, PPG	Multitasking, vigilance task, arithmetic task	100	L	53	✓	✓	x	✓	✓	✓	✓	x	✓	✓	✓
17	Task	FNIRS	N-back task	68	L	54	x	✓	x	✓	x	x	✓	x	✓	✓	✓

(continued)

Table 1 (continued)

| | Description of dataset and classifier | | | | | | Requirements for reference datasets | | | | | | | | | | |
| | | | | | | | Population | | | Effective sstimuli | Multimodal | Self-report | | | Sensor | | |
No	Ground Truth	Biosignal	Task	# Participants	Environment	Classifier No	Diff. age	Diff. gender	Gender balance			Scale used	Add. internal	Add. external	Device specified	Noise ddescribed	Calibration
18	Task	EEG, FNIRS	N-back task, response, word generation	26	L	55–57	×	✓	✓	✓	✓	×	×	×	✓	×	✓
19	Task, performance	FNIRS; TCD	N-back task	14	L		–	✓	×	✓	✓	×	×	×	✓	✓	✓
20	Task, TLX	EEG, PUPIL, ECG, PPG	Digit span task and resting	86	L		×	✓	×	✓	✓	✓	✓	×	✓	×	✓
21	Rating scale	EEG	Multitasking test	48	L	58–63	×	×	×	–	×	✓	×	×	✓	✓	✓
22[a]	Task	EEG	MATB-II	15	L	64	×	✓	×	✓	✓	×	×	×	✓	×	✓
23	ISA, FISA	EEG	Simulated flight, n-back task	35	L		×	✓	×	✓	×	✓	✓	×	×	×	×
24	TLX	EYE	Math task, span task	13	L		×	✓	✓	×	×	✓	×	×	✓	×	✓
25	Annotated	EYE	Learning	1	F		×	×	×	×	×	×	×	×	✓	×	×
26	Annotated	EYE	Learning	1	F		×	×	×	✓	×	×	×	×	✓	×	×

(continued)

Table 1 (continued)

No	Description of dataset and classifier						Requirements for reference datasets											
	Ground Truth	Biosignal	Task	# Participants	Environment	Classifier No	Population			Effective sstimuli	Multimodal	Self-report			Sensor		Calibration
							Diff. age	Diff. gender	Gender balance			Scale used	Add. internal	Add external	Device specified	Noise ddescribed	
27	TLX, task	FNIRS	Office work: reading, writing	20	F		x	✓	x	x	x	✓	x	x	✓	✓	✓
28	TLX	EYE	Surgery task	8	L	65	x	✓	✓	✓	x	✓	✓	x	✓	✓	x
29	TLX	ECG, PPG, ACC, PULSE, RESP,	N-back task, walking	13	L		–	✓	✓	✓	✓	x	x	✓	✓	✓	✓
30	TLX, task	ECG, EYE, RESP	Driving	9	F		✓	✓	x	x	✓	✓	✓	x	✓	x	✓
31	Task	EDA, RESP, HR, TEMP	Bo test	88	L		x	✓	x	✓	✓	x	✓	x	✓	✓	✓

ISA: instantaneous self assessment, RSME: rating scale mental effort, TLX: Nasa-TLX, EEG: electroencephalography, ECG: electrocardiography, fNIRS:functional near-infrared spectroscopy, RESP: respiration rate, HR: heart rate data, HRV: heart rate variability, TEMP: temperature, ACC: accelerometer, EOG: electrooculography, TCD: transcranial doppler, EDA: electrodermal activity, GSR: galvanic skin response, BVP: blood volume pressure, x: no, ✓: yes, –:no information, ᵃavailable on request or login, L = laboratory, F = field

Researchers mostly tried to account for different genders, however an equal distribution was only achieved in 7 cases. The sample size has not been reported to be statistically estimated and varies between 1 and 100, in average 10 to 40 participants.

Stimuli. We see a strong focus on established tasks for achieving cognitive load, such as MATB-II, math tasks, n-back task, stroop task, or digit span task which we considered as effective stimuli in our study. Even though this helps to induce cognitive load, these tasks mostly do not resemble tasks conducted in a real-world setting (i.e., relevant tasks according to [34]), where neuro-adaptive IS would expected to be used. Some datasets address this gap by gathering data during simulated flights, surgery, driving or knowledge work and thereby combine the relevant task with an effective stimulus. Apart from 6 datasets, the tasks were all performed in a laboratory environment.

Self-reported information. 17 datasets used an established rating scale (NASA TLX, RSME, ISA) as ground truth. Otherwise, cognitive load was labeled based on task level (8 times), an own rating scale (2 times), or otherwise (3 times). 14 times internal factors related to cognitive load were reported, covering personal information (i.e., age, gender, handedness, personality) or related user states (i.e., fatigue or sleepiness). 4 datasets reported external factors (i.e., physical activity).

Modality. In 21 out of 31 cases, multimodal biosignals were recorded. A high focus was set on biosignals from the peripheral nervous system (e.g., via ECG/PPG (18 times), respiration (10 times), EDA/GSR (11 times), eye data (5 times)), from brain activity (via EEG or fNIRS, in total 5 times), or both. Monomodal datasets mostly recorded brain activity (6 times), or eye data (4 times).

Sensors. The sensors were mentioned in the datasets except once and are widely used sensors, especially wearables, such as Empatica e4, Shimmer 3, or Bioharness 3 [40–42]. Baseline data were collected in most experiments and details on calibration or sensor attachment are reported in most datasets, in contrast to the findings of [39]. However, collected baselines were partially not included in the datasets. In case calibration information was provided in the dataset or corresponding publication, it described laboratory conditions or detailed sensor attachment. We assume that for creating the datasets, preprocessing steps have been conducted when the use of signal processing techniques, such as the application of bandwidth filters, has been reported in either the description of the dataset or the related publication. In case signal or sensor noise was indicated, datasets included raw and already preprocessed data (e.g. [43, 44]) or mentioned preprocessing steps in the corresponding publication to the dataset.

Publication. We were also interested in publications describing the experimental task conducted to gain the dataset. For 23 datasets, we identified papers which provide further information on the experiment, statistically validate the data or evaluate first ML models trained on the datasets. Publications are linked in the table references.

Classifiers: Finally, we summarized the published classifiers for cognitive load trained on the datasets and linked them in the table. Due to space constraints, classification details are not attached, but can be requested from the authors. For 13 datasets, we did not find any published classifiers for cognitive load. We found two datasets that were part of a classification challenge [45, 46] and thus showed a high number and range of different classifiers. Each paper compared various classifiers and mostly used established supervised ML algorithms. Most popular algorithms were support vector machine (21/34 papers), random forest (15/34) and k-nearest neighbor (12/34). Selected features predominantly covered established biosignal characteristics (e.g., SDNN for ECG data; blink activity for eye data) (28/34) and less focus was set on generic time series features or raw data as input. When datasets consist of multimodal data streams, classifiers were mostly trained on multimodal data and compared to monomodal ones.

4 Discussion and Conclusion

We perform a systematic review and provide an overview of publicly available datasets and corresponding classifiers measuring cognitive load via biosignals. We compared the datasets based on established criteria for reference datasets. Our work comes with three major limitations: First, further datasets could exist which were not found with our search strategy. This especially includes datasets which do cover cognitive load measurements but do not mention our key words from the search string explicitly in their searchable meta-data and thus have not been found with our applied search string or have been published after we conducted our search. Further, we perceive the construct of cognitive load itself as a distinct user state in this work. This view may be limited, as it can be argued that a wider view on the construct incorporating cognitive load-related constructs such as working memory, mind-wandering, or task vigilance may be beneficial. By including these terms and taking a broader view on the construct, more datasets can again be found. Besides, datasets that have been published outside the databases we have selected may exist as well and are not reported in this work. Second, more classifiers may exist not linked to the datasets and thus not presented in our overview. Third, with regards to our results, the criteria we used to compare our dataset are not exhaustive. There might be additional interesting criteria which allow a more precise evaluation and comparison of the datasets, especially with regards to sensors used, such as the specific devices used per dataset, or a more in-depth evaluation of data cleaning and pre-processing steps conducted. Further work might expand these criteria. Also, we assumed that no pre-preprocessing has been conducted when it was not indicated. When in doubt and reusing the datasets, we strongly encourage researchers to contact the authors or run an in-depth analysis on the data to identify if pre-processing has been conducted. With our work, we aim to stimulate a discussion on publishing datasets and ways for using them. As a starting point, we therefore suggest that future work could focus on three avenues, which we derived based on our main findings: (1) **Publish more diverse datasets** with adequate licensing rights in popular dataset databases. As the comparison shows, diversity is needed in future datasets regarding the sample population in terms of diverse age groups, ensured gender balance, and diverse participant backgrounds and health conditions. Thereby we were able to translate the findings of [39] to cognitive load datasets and support their call for establishing reference datasets fulfilling each criteria. Also, increased reporting of internal factors (esp. personality, further user states), external factors and baseline/calibration information is favorable to ensure similar conditions when reusing the datasets. Gaining such detailed insights on internal factors is relevant to train personalized ML models. Further, more diverse tasks should be conducted when collecting data. Exemplarily, so far, few datasets on collaboration or office work tasks except from math tasks exist. By using less obtrusive devices and considering the related challenges, collecting more datasets in the field, and thus on more relevant tasks may be possible and help to increase applicability of the dataset and classifier to real-world scenarios of neuro-adaptive IS, as also addressed in the second avenue. (2) **Continue the evaluation of existing datasets and their trained classifiers based on their applicability** to other tasks and domains (see [47–49]). This again sheds light on tasks and domains in which datasets should especially be gathered and opens the way to design neuro-adaptive IS applicable in a real-world context. In this context, researchers may also face the challenges of generalization when reusing datasets and classifiers trained on other individuals than the participants of a possible experiment or users of a real-world application [50]. Therefore, further research exploring the reusability of datasets and classifiers on new individuals and new tasks is, from our point of view, needed to accelerate the creation and evaluation of adaptive systems. (3) **Use existing datasets to train publicly available classifiers**, and, based on them, advance the design of neuro-adaptive IS and passive brain-computer interfaces. As visible in our overview, for some datasets, we did not find published classifiers at all. This may have diverse reasons. However, it should be ensured that this is not due to missing awareness about the existence of datasets since as it can be seen for the challenge datasets, such awareness can lead to the application of a diverse range of advanced ML approaches. Overall, we believe that public datasets increase overall transparency and knowledge transfer and enable better classifiers for neuro-adaptive IS design. Thus, we encourage NeuroIS researchers to publish their datasets and classifiers in established databases and make use of existing datasets. As a community, NeuroIS Society and its members should discuss the opportunity to create a community-focused proprietary dataset repository as an alternative to actively support an existing dataset database, already visible to a broader audience and targeting on bio-signals.

Acknowledgements Funded by the Deutsche Forschungsgemeinschaft (DFG, German Research Foundation)—GRK2739/1—Project Nr. 447089431—Research Training Group: KD2 School—Designing Adaptive Systems for Economic Decisions.

References

1. Fischer, T., Davis, F. D., & Riedl, R., et al. (2019). NeuroIS: A survey on the status of the field. In R. Riedl & J. vom Brocke (Eds.), *Davis FD* (pp. 1–10). Springer International Publishing.
2. Vanneste, P., Raes, A., Morton, J., et al. (2021). Towards measuring cognitive load through multimodal physiological data. *Cognition, Technology & Work, 23*, 567–585. https://doi.org/10.1007/S10111-020-00641-0/TABLES/5

3. Gwizdka, J. (2021). "Overloading" cognitive (work)load: What are we really measuring? *Lecture Notes in Information Systems and Organization LNISO, 52*, 77–89. https://doi.org/10.1007/978-3-030-88900-5_9

4. Riedl, R., Fischer, T., Léger, P. M., & Davis, F. D. (2020). A decade of neurois research. *Data Base for Advances in Information Systems, 51*, 13–54. https://doi.org/10.1145/3410977.3410980

5. Paas, F., Tuovinen, J. E., Tabbers, H., & Van Gerven, P. W. M. (2003). Cognitive load measurement as a means to advance cognitive load theory. *Educational Psychology, 38*, 63–71. https://doi.org/10.1207/S15326985EP3801_8

6. Hart, S. G. (2006). Nasa-Task Load Index (NASA-TLX); 20 years later. In *Proceedings of the Human Factors and Ergonomics Society Annual Meeting, 50*, 904–908. https://doi.org/10.1177/154193120605000909

7. Zhou, T., Cha, J. S., Gonzalez, G., et al. (2020). Multimodal physiological signals for workload prediction in robot-assisted surgery. *ACM Transactions on Human-Robot Interaction, 9*, 1–26. https://doi.org/10.1145/3368589

8. Matthews, G., De Winter, J., & Hancock, P. A. (2020). What do subjective workload scales really measure? Operational and representational solutions to divergence of workload measures. *Theoretical Issues in Ergonomics Science, 21*, 369–396. https://doi.org/10.1080/1463922X.2018.1547459

9. Young, M. S., Brookhuis, K. A., Wickens, C. D., & Hancock, P. A. (2015). State of science: Mental workload in ergonomics. *Ergonomics, 58*, 1–17. https://doi.org/10.1080/00140139.2014.956151

10. Mital, A., & Govindaraju, M. (1999). Is it possible to have a single measure for all work? *International Journal of Industrial Engineering, Applications and Practice, 6*, 190–195.

11. Tao, D., Tan, H., Wang, H., et al. (2019). A systematic review of physiological measures of mental workload. *International Journal of Environmental Research and Public Health*. https://doi.org/10.3390/IJERPH16152716

12. Charles, R. L., & Nixon, J. (2019). Measuring mental workload using physiological measures. *Applied Ergonomics, 74*, 221–232. https://doi.org/10.1016/j.apergo.2018.08.028

13. Perkhofer, L., & Lehner, O. (2019). Using gaze behavior to measure cognitive load. *Lecture Notes in Information Systems and Organization, 29*, 73–83. https://doi.org/10.1007/978-3-030-01087-4_9

14. Riedl, R., Fischer, T., Léger, P.-M., & Davis, F. D. (2020). A decade of neurosis research. *Data Base for Advances in Informations and Systems, 51*, 13–54. https://doi.org/10.1145/3410977.3410980

15. Chikhi, S., Matton, N., & Blanchet, S. (2022). EEG power spectral measures of cognitive workload: A meta-analysis. *Psychophysiology* e14009. https://doi.org/10.1111/PSYP.14009

16. Zheng, Y.-L., Ding, X.-R., Poon, C. C. Y., et al. (2014). Unobtrusive sensing and wearable devices for health informatics. *IEEE Transactions on Biomedical Engineering, 61*, 1538–1554. https://doi.org/10.1109/TBME.2014.2309951

17. Fairclough, S. H. (2009). Fundamentals of physiological computing. *Interacting with Computers, 21*, 133–145. https://doi.org/10.1016/J.INTCOM.2008.10.011

18. Hettinger, L. J., Branco, P., Encarnacao, L. M., & Bonato, P. (2003). Neuroadaptive technologies: Applying neuroergonomics to the design of advanced interfaces. *Theoretical Issues in Ergonomics Science, 4*, 220–237.

19. Zander, T. O., & Kothe, C. (2011). Towards passive brain-computer interfaces: Applying brain-computer interface technology to human-machine systems in general. *Journal of Neural Engineering*. https://doi.org/10.1088/1741-2560/8/2/025005

20. Randolph, A. B., Labonté-Lemoyne, É., Léger, P. M., et al. (2015). Proposal for the use of a passive BCI to develop a neurophysiological inference model of IS constructs. *Lecture Notes in Informations and System Organization, 10*, 175–180. https://doi.org/10.1007/978-3-319-18702-0_23/TABLES/1

21. vom Brocke, J., Riedl, R., & Léger, P.-M. (2013). Application strategies for neuroscience in information systems design science research. *The Journal of Computer Information Systems, 53*, 1–13.

22. vom Brocke, J., Simons, A., & Nievhaves, B., et al. (2009). Reconstructing the giant: On the importance of Rigour in documenting the literature search process. In *ECIS 2009 Proceedings* (pp 2206–2217).

23. Wilkinson, M. D., Dumontier, M., & Aalbersberg, I. J., et al (2016). The FAIR guiding principles for scientific data management and stewardship. *Scientific Data, 31*(3), 1–9. https://doi.org/10.1038/sdata.2016.18

24. IEEE (2022*). Dataset storage and dataset search platform.* IEEE DataPort. https://ieee-dataport.org/. Accessed 3 Feb 2022.

25. Mendeley (2022). Mendeley data. https://data.mendeley.com/. Accessed 3 Feb 2022.

26. Figshare (2022). Figshare—credit for all your research. https://figshare.com/. Accessed 3 Feb 2022.

27. Kaggle. Your machine learning and data science community. https://www.kaggle.com/. Accessed 3 Feb 2022.

28. Google (2022.) Google data studio overview. https://datastudio.google.com/overview. Accessed 3 Feb 2022

29. Zenodo (2022). Zenodo—Research. Shared. https://zenodo.org/. Accessed 2 Feb 2022.

30. DataHub (2022). DataHub—frictionless data. https://datahub.io/. Accessed 3 Feb 2022.

31. Goldberger, A., Amaral, L., Glass, L., Hausdorff, J., Ivanov, P. C., Mark, R., & Stanley, H. E. (2000). PhysioBank, PhysioToolkit, and PhysioNet: Components of a new research resource for complex physiologic signals. *Circulation, 101* (23). [Online]

32. EUDAT (2022). EUDAT—research data services, expertise & technology solutions. https://www.eudat.eu/. Accessed 8 Mar 2022.

33. Wolstencroft, K., Krebs, O., Snoep, J. L., et al. (2017). FAIRDOMHub: A repository and collaboration environment for sharing systems biology research. *Nucleic Acids Research, 45*, D404–D407. https://doi.org/10.1093/NAR/GKW1032

34. Harvard (2022). Harvard dataverse. https://dataverse.harvard.edu/. Accessed 8 Mar 2022.

35. Dryad (2022). Dryad home—publish and preserve your data. https://datadryad.org/stash/. Accessed 8 Mar 2022.

36. Narcis D (2022). NARCIS. https://www.narcis.nl/?Language=en. Accessed 2 Feb 2022.

37. Initative, G.F. (2022). FAIR principles—GO FAIR. https://www.go-fair.org/fair-principles/. Accessed 2 Feb 2022.

38. Cuno, A., Condori-Fernandez, N., Mendoza, A., & Lovon, W. R. (2020). A FAIR evaluation of public datasets for stress detection systems. In *Proceedings of the international conference on Chilean computer science society (SCCC)*. https://doi.org/10.1109/SCCC51225.2020.9281274

39. Mahesh, B., Prassler, E., Hassan, T., & Garbas, J. U. (2019). Requirements for a reference dataset for multimodal human stress detection. In *2019 IEEE international conference on pervasive computing and communications workshops* (PerCom Workshops) (pp. 492–498). https://doi.org/10.1109/PERCOMW.2019.8730884

40. Zephyr Technology (2012). BioHarness 3.0 User Manual.

41. Empatica (2022). E4 wristband | Real-time physiological signals | Wearable PPG, EDA, Temperature, Motion sensors. https://www.empatica.com/en-eu/research/e4/. Accessed 8 Mar 2022.

42. Shimmer (2022). Shimmer3 GSR+ Unit - Shimmer Wearable Sensor Technology. https://shimmersensing.com/product/shimmer3-gsr-unit/. Accessed 8 Mar 2022.

43. Koldijk, S., Sappelli, M., & Verberne, S., et al. (2014). The SWELL knowledge work dataset for stress and user modeling research. In *Proceedings of the 16th international conference on multimodal interaction* (pp. 291–298). ACM.

44. Huang, Z., Wang, L., & Blaney, G., et al. (2021). The Tufts fNIRS mental workload dataset & benchmark for brain-computer interfaces that generalize. In *Proceedings of the Neural Information Processing Systems (NeurIPS) track on datasets and benchmarks.*

45. Hinss, M. F., Darmet, L., & Somon, B., et al. (2021). An EEG dataset for cross-session mental workload estimation. In *Passive BCI competition of the neuroergonomics conference 2021*. https://doi.org/10.5281/ZENODO.4917218

46. Gjoreski, M., Kolenik, T., & Knez, T., et al. (2020). Datasets for cognitive load inference using wearable sensors and psychological traits. *Applied Science, 10*, 3843. https://doi.org/10.3390/APP10113843
47. Shu, L., Xie, J., Yang, M., et al. (2018). A review of emotion recognition using physiological signals. *Sensors, 18*, 2074. https://doi.org/10.3390/s18072074
48. Larradet, F., Niewiadomski, R., Barresi, G., et al. (2020). Toward emotion recognition from physiological signals in the wild: Approaching the methodological issues in real-life data collection. *Frontiers in Psychology.* https://doi.org/10.3389/fpsyg.2020.01111
49. Hasnul, M. A., Aziz, N. A. A., Alelyani, S., et al. (2021). Electrocardiogram-based emotion recognition systems and their applications in healthcare—A review. *Sensors, 21*, 5015. https://doi.org/10.3390/s21155015
50. Brouwer, A. M., Zander, T. O., van Erp, J. B. F., et al. (2015). Using neurophysiological signals that reflect cognitive or affective state: Six recommendations to avoid common pitfalls. *Frontiers in Neuroscience.* https://doi.org/10.3389/FNINS.2015.00136/PDF

Datasets Identified in Systematic Search and Mentioned in the Table

1. Albuquerque, I., Tiwari, A., Parent, M., Cassani, R., Gagnon, J. F., Lafond, D., Tremblay, S., & Falk, T. H. (2020). WAUC: A multi-modal database for mental workload assessment under physical activity. *Frontiers in Neuroscience, 14*, 1037. https://doi.org/10.3389/FNINS.2020.549524/BIBTEX
1a. Bernstein, S. (2022). MuSAE Lab: Multimodal/Multisensory Signal Analysis & Enhancement. https://musaelab.ca/. Accessed 8 Mar 2022.
2. Ding, Y., Cao, Y., Duffy, V. G., Wang, Y., & Zhang, X. (2020). Measurement and identification of mental workload during simulated computer tasks with multimodal methods and machine learning. *Ergonomics, 63*(7), 896–908. https://doi.org/10.1080/00140139.2020.1759699
3. Anand, V., Ahmed, Z., & Sreeja, S. (2019). Cognitive-mental-workload-estimation-using-ML: An automated approach for task evaluation using EEG signals. https://github.com/ivishalanand/Cognitive-Mental-Workload-Estimation-Using-ML. Accessed 8 Mar 2022.
4. Dolmans, T. C., Poel, M., van't Klooster, J. W., & Veldkamp, B. P. (2021). Perceived mental workload classification using intermediate fusion multimodal deep learning. *Frontiers in Human Neuroscience, 14*, 581. https://doi.org/10.3389/FNHUM.2020.609096/BIBTEX
4a. Dolmans, T. C., Poel, M., van 't Klooster, J. W., & Veldkamp, B. P. (2020). Perceived mental workload detection using multimodal physiological... (https://www.narcis.nl). Dataset. https://www.narcis.nl/dataset/RecordID/doi%3A10.4121%2F12932801. Accessed 8 Mar 2022.
5. Markova, V., Ganchev, T., & Kalinkov, K. (2019). CLAS: A database for cognitive load, affect and stress recognition. In *Proceedings of the International Conference on Biomedical Innovations and Applications, BIA.* https://doi.org/10.1109/BIA48344.2019.8967457
5a. Markova, V. (2022). Database for cognitive load affect and stress recognition. IEEE Dataport. https://ieee-dataport.org/open-access/database-cognitive-load-affect-and-stress-recognition. Accessed 8 Mar 2022.
6. Mijic, I., Sarlija, M., & Petrinovic, D. (2019). MMOD-COG: A database for multimodal cognitive load classification. In *International symposium on image and signal processing and analysis, ISPA* (pp. 15–20). https://doi.org/10.1109/ISPA.2019.8868678
7. Rao, H. M., Smalt, C. J., Rodriguez, A., Wright, H. M., Mehta, D. D., Brattain, L. J., Edwards, H. M., Lammert, A., Heaton, K. J., & Quatieri, T. F. (2020). Predicting cognitive load and operational performance in a simulated marksmanship task. *Frontiers in Human Neuroscience, 14*, 222. https://doi.org/10.3389/FNHUM.2020.00222/BIBTEX
8. Ahmad, M. I., Keller, I., Robb, D. A., & Lohan, K. S. (2020). A framework to estimate cognitive load using physiological data. *Personal and Ubiquitous Computing.* https://doi.org/10.1007/S00779-020-01455-7/FIGURES/4

8a. Brutus TT. (2021). Repositories. GitHub. https://github.com/BrutusTT?tab=repositories. Accessed 8 Mar 2022.

9. Schneegass, S., Pfleging, B., Broy, N., Schmidt, A., & Heinrich, F. (2013). A data set of real world driving to assess driver workload. In *Proceedings of the 5th international conference on automotive user interfaces and interactive vehicular applications – automotive UI'13*. https://doi.org/10.1145/2516540

9a. HCILab. (2013). Automotive user interfaces driving data set. https://www.hcilab.org/automotive/. Accessed 8 Mar 2022.

10. Koldijk, S., Sappelli, M., Verberne, S., Neerincx, M. A., & Kraaij, W. (2014). The SWELL knowledge work dataset for stress and user modeling research. In *Proceedings of the 16th international conference on multimodal interaction* (pp. 291–298). https://doi.org/10.1145/2663204

10a. Kraaij, W., Koldijk, S., & Sappelli, M. (2014). The SWELL knowledge work dataset for stress and user modeling research. *DANS*. https://doi.org/10.17026/dans-x55-69zp

11. Beh, W. K., Wu, Y. H., & Wu, A. Y. (2021). MAUS: A dataset for mental workload assessment on N-back task using wearable sensor. https://github.com/rickwu11/MAUS. Accessed 8 Mar 2022.

11a. Beh, W. K., Wu, Y. H., & Wu, A. Y. (2021). MAUS: A dataset for mental workload assessment on N-back task using wearable sensor. ArXiv. https://doi.org/10.48550/arxiv.2111.02561

12. Gjoreski, M., Kolenik, T., Knez, T., Luštrek, M., Gams, M., Gjoreski, H., & Pejović, V. (2020). Datasets for cognitive load inference using wearable sensors and psychological traits. *Applied Sciences, 10*(11), 3843. https://doi.org/10.3390/APP10113843

12a. Gjoreski, M. (2020). *CogDatasets*. GitHub. https://github.com/MartinGjoreski/martingjoreski.github.io/blob/master/files/CogDatasets.rar. Accessed 8 Mar 2022.

13. Gjoreski, M. (2020). *CogDatasets*. GitHub. https://github.com/MartinGjoreski/martingjoreski.github.io/blob/master/files/CogDatasets.rar. Accessed 8 Mar 2022.

14. Jota, G. (2021). HRV Field Environment | Kaggle. Kaggle. https://www.kaggle.com/gustavojota/hrv-dataset/code. Accessed 8 Mar 2022.

15. Kalatzis, A., Teotia, A., Prabhu, V. G., & Stanley, L. (2021). A Database for cognitive workload classification using electrocardiogram and respiration signal. *Lecture Notes in Networks and Systems, 259*, 509–516. https://doi.org/10.1007/978-3-030-80285-1_58

15a. HilabMSU (2021). CW-Database. GitLab. https://gitlab.com/hilabmsu/cw-database. Accessed 8 Mar 2022.

16. Siegel, E. H., Wei, J., Gomes, A., Oliviera, M., Sundaramoorthy, P., Smathers, K., Vankipuram, M., Ghosh, S., Horii, H., Bailenson, J., & Ballagas, R. (2021). HP Omnicept Cognitive Load Database (HPO-CLD)—developing a multimodal inference engine for detecting real-time mental workload in VR. https://developers.hp.com/omnicept/omnicept-open-data-set-abstract. Accessed 8 Mar 2022.

16a. Siegel, E. H., Wei, J., Gomes, A., Oliviera, M., Sundaramoorthy, P., Smathers, K., Vankipuram, M., Ghosh, S., Horii, H., Bailenson, J., & Ballagas, R. HP Omnicept Cognitive Load Database (HPO-CLD)—developing a multimodal inference engine for detecting real-time mental workload in VR. Technical Report, HP Labs, Palo Alto, CA. Available at: https://developers.hp.com/omnicept/omnicept-open-data-set-abstract

17. Tufts HCI Lab (2021). The Tufts fNIRS to mental workload dataset | Tufts HCI Lab. https://tufts-hci-lab.github.io/code_and_datasets/fNIRS2MW.html. Accessed 8 Mar 2022.

17a. Huang, Z., Wang, L., Blaney, G., Slaughter, C., McKeon, D., Zhou, Z., Jacob, R. J. K., & Hughes, M. C. (2021). The Tufts fNIRS mental workload dataset & benchmark for brain-computer interfaces that generalize. In *Proceedings of the neural information processing systems (NeurIPS) track on datasets and benchmarks*. https://openreview.net/pdf?id=QzNHE7QHhut

18. Shin, J., & Al, E. (2017). Open access dataset for EEG+ NIRS single-trial classification. *IEEE Transactions on Neural Systems and Rehabilitation Engineering, 25*(10).

18a. Shin, J., (2022). Simultaneous acquisition of EEG and NIRS during cognitive tasks for an open access dataset. TU Berlin Homepage. http://doc.ml.tu-berlin.de/simultaneous_EEG_NIRS/. Accessed 8 Mar 2022.

19. Mukli, P., Yabluchanskiy, A., & Csipo, T. (2021). Mental workload during n-back task captured by TransCranial Doppler (TCD) sonography and functional Near-Infrared Spectroscopy (fNIRS) monitoring (p. v1). PhysioNet. https://doi.org/10.13026/zfb2-1g43. Accessed 8 Mar 2022.

19a. Csipo, T., Lipecz, A., Mukli, P., Bahadli, D., Abdulhussein, O., Owens, C. D., Tarantini, S., Hand, R. A., Yabluchanska, V., Mikhail Kellawan, J., Sorond, F., James, J. A., Csiszar, A., Ungvari, Z. I., & Yabluchanskiy, A. (2021). Increased cognitive workload evokes greater neurovascular coupling responses in healthy young adults. *PLOS ONE, 16*(5), e0250043. https://doi.org/10.1371/JOURNAL.PONE.0250043

20. Pavlov, Y. G., Kasanov, D., Kosachenko, A. I., & Kotyusov, A. (2021). EEG, pupillometry, ECG and photoplethysmography, and behavioral data in the digit span task and rest. OpenNeuro. https://doi.org/10.12751/g-node.1ecsf4. Accessed 8 Mar 2022.

21. Lim, W. L., Sourina, O., & Wang, L. (2018). STEW: Simultaneous task EEG workload dataset. IEEE Dataport. https://doi.org/10.21227/44r8-ya50. Accessed 8 Mar 2022.

21a. Lim, W. L., Sourina, O., & Wang, L. P. (2018). STEW: Simultaneous task EEG workload data set. *IEEE Transactions on Neural Systems and Rehabilitation Engineering, 26*(11), 2106–2114. https://doi.org/10.1109/TNSRE.2018.2872924

22. Hinss, M. F., Darmet, L., Somon, B., Jahanpour, E., Lotte, F., Ladouce, S., & Roy, R. N. (2021). An EEG dataset for cross-session mental workload estimation. In *Passive BCI competition of the neuroergonomics conference 2021*. https://doi.org/10.5281/ZENODO.4917218. Accessed 8 Mar 2022.

23. Hamann, A. (2021). Dataset for: Under pressure: Mental workload-induced changes in cortical oxygenation and frontal theta activity during simulated flights. *PsychArchives Dataset*. https://doi.org/10.23668/psycharchives.5291. Accessed 8 Mar 2022.

24. Krejtz, K., Duchowski, A. T., Niedzielska, A., Biele, C., & Krejtz, I. (2018). Eye tracking cognitive load using pupil diameter and microsaccades with fixed gaze. *PLOS ONE, 13*(9), e0203629. https://doi.org/10.1371/journal.pone.0203629

24a. Krejtz, K., Duchowski, A. T., Niedzielska, A., Biele, C., & Krejtz, I. (2018). Eye tracking cognitive load using pupil diameter and microsaccades with fixed gaze. FigShare. https://figshare.com/articles/dataset/Eye_tracking_cognitive_load_using_pupil_d iameter_and_microsaccades_with_fixed_gaze/7089800/. Accessed 8 Mar 2022.

25. Prieto, L. P., Sharma, K. (2015). JDC2014—An eyetracking dataset from facilitating a semi-authentic multi-tabletop lesson. https://doi.org/10.5281/ZENODO.16515. Accessed 8 Mar 2022.

25a. Prieto, L., Sharma, K., Wen, Y., & Dillenbourg, P. (2015). The burden of facilitating collaboration: Towards estimation of teacher orchestration load using eye-tracking measures. CSCL. https://www.semanticscholar.org/paper/The-Burden-of-Facilitating-Collab oration%3A-Towards-Prieto-Sharma/6c5a38d8dbab3f95a638d907e0966a168953bd70

26. Prieto, L. P., & Sharma, K. (2016). JDC2015—A multimodal dataset from facilitating multi-tabletop lessons in an open-doors day. https://doi.org/10.5281/ZENODO.198709. Accessed 8 Mar 2022.

26a. Prieto, L. P., Sharma, K., Kidzinski, Ł., Dillenbourg, P. (2018). Orchestration load indicators and patterns: In-the-wild studies using mobile eye-tracking. *IEEE Transactions on Learning Technologies, 11*(2), 216–229. https://doi.org/10.1109/TLT.2017.2690687

27. Midha, S., Maior, H. A., Wilson, M. L., & Sharples, S. (2021). Measuring mental workload variations in office work tasks using fNIRS. *International Journal of Human Computer Studies*. https://doi.org/10.1016/J.IJHCS.2020.102580

27a. Midha, S. (2021). Measuring mental workload variations in office work tasks using fNIRS. Dataset. https://rdmc.nottingham.ac.uk/handle/internal/9123. Accessed 8 Mar 2022.

28. Wu, C., Cha, J., Sulek, J., Zhou, T., Sundaram, C. P., Wachs, J., & Yu, D. (2020). Eye-tracking metrics predict perceived workload in robotic surgical skills training. *Human Factors: The Journal of the Human Factors and Ergonomics Society, 62*(8), 1365–1386. https://doi.org/10.1177/0018720819874544

28a. Wu, C., Cha, J., Sulek, J., Zhou, T., Sundaram, C., Wachs, J., & Yu, D. (2019). Eye-tracking metrics predicts perceived workload in robotic surgical skills training. Purdue University Research Repository. https://doi.org/10.4231/0EVK-9P12. Accessed 8 Mar 2022.

29. Vollmer, M., Bläsing, D., Reiser, J. E., Nisser, M., & Buder, A. (2020). Simultaneous physiological measurements with five devices at different cognitive and physical loads (version 1.0.0). *PhysioNet*. https://doi.org/10.13026/chd5-t946. Accessed 8 Mar 2022.

30. Heikoop, D., de Winter, J., & van Arem, B., Stanton, N. A. (Neville). (2019). Supplementary data for the article: Acclimatizing to automation: Driver workload and stress during partially automated car following in real traffic. 4TU.ResearchData. Dataset. https://doi.org/10.4121/uuid:cbe0f5c6-0439-4533-86c2-faeebedf5662. Accessed 8 Mar 2022.

30a. Heikoop, D. D., de Winter, J. C. F., van Arem, B., & Stanton, N. A. (2019). Acclimatizing to automation: Driver workload and stress during partially automated car following in real traffic. *Transportation Research Part F: Traffic Psychology and Behaviour*, *65*, 503–517. https://doi.org/10.1016/J.TRF.2019.07.024

31. Filo, P., & Janoušek, O. (2022). The relation between physical and mental load, and the course of physiological functions and cognitive performance. *Theoretical Issues in Ergonomics Science*, *23*(1), 38–59. https://doi.org/10.1080/1463922X.2021.1913535

31a. Filo, P. (2021). The relation between physical and mental load, and the course of physiological functions and cognitive performance. *Mendeley Data*, *1*. https://doi.org/10.17632/7DJSBZ6PRV.1. Accessed 8 Mar 2022.

Classifiers Identified in Systematic Search and Mentioned in the Table

32. Albuquerque, I., Tiwari, A., Parent, M., Cassani, R., Gagnon, J. F., Lafond, D., Tremblay, S., & Falk, T. H. (2020). WAUC: A multi-modal database for mental workload assessment under physical activity. *Frontiers in Neuroscience, 14*, 1037. https://doi.org/10.3389/FNINS.2020.549524/BIBTEX

33. Rosanne, O., Albuquerque, I., Cassani, R., Gagnon, J. F., Tremblay, S., & Falk, T. H. (2021). Adaptive filtering for improved EEG-based mental workload assessment of ambulant users. *Frontiers in Neuroscience, 15*, 341. https://doi.org/10.3389/FNINS.2021.611962/BIBTEX

34. Ding, Y., Cao, Y., Duffy, V. G., Wang, Y., & Zhang, X. (2020). Measurement and identification of mental workload during simulated computer tasks with multimodal methods and machine learning. *Ergonomics, 63*(7), 896–908. https://doi.org/10.1080/00140139.2020.1759699

35. Anand, V., Ahmed, Z., & Sreeja, S. (2019). Cognitive-mental-workload-estimation-using-ML: An automated approach for task evaluation using EEG signals. https://github.com/ivishalanand/Cognitive-Mental-Workload-Estimation-Using-ML. Accessed 8 Mar 2022.

36. Dolmans, T. C. (2020). Code for: perceived mental workload classification using intermediate fusion multimodal deep learning. https://doi.org/10.5281/ZENODO.4043058

37. Beh, W. K., Wu, Y. H., & Wu, A. Y. (2021). Robust PPG-based mental workload assessment system using wearable devices. *IEEE Journal of Biomedical and Health Informatics*. https://doi.org/10.1109/JBHI.2021.3138639

38. Feradov, F., Ganchev, T., & Markova, V. (2020). Automated detection of cognitive load from peripheral physiological signals based on Hjorth's parameters. In *Proceedings of the International Conference on Biomedical Innovations and Applications, BIA, 2020*, 85–88. https://doi.org/10.1109/BIA50171.2020.9244287

39. Lisowska, A., Wilk, S., & Peleg, M. (2021). Is it a good time to survey you? Cognitive load classification from blood volume pulse. In *Proceedings—IEEE symposium on computer-based medical systems*, 2021-June (pp. 137–141). https://doi.org/10.1109/CBMS52027.2021.00061

40. Mijic, I., Sarlija, M., & Petrinovic, D. (2019). MMOD-COG: A database for multimodal cognitive load classification. In *International symposium on image and signal processing and analysis, ISPA*, 2019-September, 15–20. https://doi.org/10.1109/ISPA.2019.8868678

41. Kesedzic, I., Sarlija, M., Bozek, J., Popovic, S., & Cosic, K. (2021). Classification of cognitive load based on neurophysiological features from functional near-infrared spectroscopy and electrocardiography signals on n-back task. *IEEE Sensors Journal, 21*(13), 14131–14140. https://doi.org/10.1109/JSEN.2020.3038032

42. Rao, H. M., Smalt, C. J., Rodriguez, A., Wright, H. M., Mehta, D. D., Brattain, L. J., Edwards, H. M., Lammert, A., Heaton, K. J., & Quatieri, T. F. (2020). Predicting cognitive load and operational performance in a simulated marksmanship task. *Frontiers in Human Neuroscience, 14*, 222. https://doi.org/10.3389/FNHUM.2020.00222/BIBTEX

43. Ahmad, M. I., Keller, I., Robb, D. A., & Lohan, K. S. (2020). A framework to estimate cognitive load using physiological data. *Personal and Ubiquitous Computing*, 1–15. https://doi.org/10.1007/S00779-020-01455-7/FIGURES/4

44. Beh, W. K., Wu, Y. H., & Wu, A. Y. (2021). MAUS: A dataset for mental workload assessment on N-back task using wearable sensor. Retrieved March 3, 2022, from https://github.com/ric kwu11/MAUS

45. Gjoreski, M., Kolenik, T., Knez, T., Luštrek, M., Gams, M., Gjoreski, H., & Pejović, V. (2020). Datasets for cognitive load inference using wearable sensors and psychological traits. *Applied Sciences, 10*(11), 3843. https://doi.org/10.3390/APP10113843

46. Jaiswal, D., Chatterjee, D., Gavas, R., Ramakrishnan, R. K., & Pal, A. (2021). Effective assessment of cognitive load in real-world scenarios using wrist-worn sensor data. In *BodySys 2021—proceedings of the 2021 ACM workshop on body centric computing systems* (pp. 7–12). https://doi.org/10.1145/3469260.3469666

47. Salfinger, A. (2020). Deep learning for cognitive load monitoring: A comparative evaluation. In *UbiComp/ISWC 2020 Adjunct - Proceedings of the 2020 ACM international joint conference on pervasive and ubiquitous computing and proceedings of the 2020 ACM international symposium on wearable computers* (pp. 462–467). https://doi.org/10.1145/3410530.3414433

48. Li, X., & De Cock, M. (2020). Cognitive load detection from wrist-band sensors. In *UbiComp/ISWC 2020 adjunct—proceedings of the 2020 ACM international joint conference on pervasive and ubiquitous computing and proceedings of the 2020 acm international symposium on wearable computers* (pp. 456–461). https://doi.org/10.1145/3410530.3414428

49. Borisov, V., Kasneci, E., & Kasneci, G. (2021). Robust cognitive load detection from wrist-band sensors. *Computers in Human Behavior Reports, 4*, 100116. https://doi.org/10.1016/J.CHBR.2021.100116

50. Tervonen, J., Pettersson, K., & Mäntyjärvi, J. (2021). Ultra-short window length and feature importance analysis for cognitive load detection from wearable sensors. *Electronics, 10*(5), 613. https://doi.org/10.3390/ELECTRONICS10050613

51. Gjoreski, M., Mahesh, B., Kolenik, T., Uwe-Garbas, J., Seuss, D., Gjoreski, H., Luštrek, M., Gams, M., & Pejović, V. (2020). *Cognitive load monitoring with wearables-lessons learned from a machine learning challenge.* https://doi.org/10.1109/ACCESS.2021.3093216

52. Kalatzis, A., Teotia, A., Prabhu, V. G., & Stanley, L. (2021). A database for cognitive workload classification using electrocardiogram and respiration signal. In *Lecture Notes in Networks and Systems* (Vol. 259, pp. 509–516). https://doi.org/10.1007/978-3-030-80285-1_58

53. Siegel, E. H., Wei, J., Gomes, A., Oliviera, M., Sundaramoorthy, P., Smathers, K., Vankipuram, M., Ghosh, S., Horii, H., Bailenson, J., & Ballagas, R. (2021). HP Omnicept Cognitive Load Database (HPO-CLD)—developing a multimodal inference engine for detecting real-time mental workload in VR. https://developers.hp.com/omnicept/omnicept-open-data-set-abstract

54. Huang, Z., Wang, L., Blaney, G., Slaughter, C., Mckeon, D., Zhou, Z., Jacob, R. J. K., & Hughes, M. C. (2021). The Tufts fNIRS mental workload dataset & benchmark for brain-computer interfaces that generalize. Retrieved March 18, 2022, from https://tufts-hci-lab.git hub.io/code_and_datasets/fNIRS2MW.html

55. Van Eyndhoven, S., Bousse, M., Hunyadi, B., De Lathauwer, L., & Van Huffel, S. (2018). Single-channel EEG classification by multi-channel tensor subspace learning and regression. In *IEEE international workshop on Machine Learning for Signal Processing (MLSP)*. https://doi.org/10.1109/MLSP.2018.8516927

56. Ergun, E., & Aydemir, O. (2018). Decoding of binary mental arithmetic based near-infrared spectroscopy signals. In *UBMK 2018—3rd international conference on computer science and engineering* (pp. 201–204). https://doi.org/10.1109/UBMK.2018.8566462

57. Shao, S., Wang, T., Song, C., Wang, Y., & Yao, C. (2021). Fine-grained and multi-scale motif features for cross-subject mental workload assessment using bi-LSTM. *Journal of Mechanics in Medicine and Biology, 21*(5). https://doi.org/10.1142/S0219519421400200

58. Lim, W. L., Sourina, O., & Wang, L. P. (2018). STEW: Simultaneous task EEG workload data set. *IEEE Transactions on Neural Systems and Rehabilitation Engineering, 26*(11), 2106–2114. https://doi.org/10.1109/TNSRE.2018.2872924

59. Mohdiwale, S., Sahu, M., Sinha, G. R., & Bajaj, V. (2020). Automated cognitive work-load assessment using logical teaching learning-based optimization and PROMETHEE multi-criteria decision making approach. *IEEE Sensors Journal, 20*(22), 13629–13637. https://doi.org/10.1109/JSEN.2020.3006486

60. Pandey, V., Choudhary, D. K., Verma, V., Sharma, G., Singh, R., & Chandra, S. (2020). Mental workload estimation using EEG. In *Proceedings—2020 5th international conference on research in computational intelligence and communication networks, ICRCICN* (pp. 83–86). https://doi.org/10.1109/ICRCICN50933.2020.9296150

61. Zhu, G., Zong, F., Zhang, H., Wei, B., & Lui, F. (2021). Cognitive load during multitasking can be accurately assessed based on single channel electroencephalography using graph methods. IEEE Access. https://ieeexplore.ieee.org/stamp/stamp.jsp?arnumber=9350659

62. Belsare, S., Kale, M., Ghayal, P., Gogate, A., & Itkar, S. (2021). Performance comparison of different EEG analysis techniques based on deep learning approaches. In *2021 international conference on emerging smart computing and informatics* (ESCI) (pp. 490–493). https://doi.org/10.1109/ESCI50559.2021.9396856

63. Das Chakladar, D., Dey, S., Roy, P. P., & Dogra, D. P. (2020). EEG-based mental workload estimation using deep BLSTM-LSTM network and evolutionary algorithm. *Biomedical Signal Processing and Control, 60*, 101989. https://doi.org/10.1016/J.BSPC.2020.101989

64. Sedlar, S., Benerradi, J., Le Breton, C., Deriche, R., Papdopoulo, T., & Wilson, M. (2021). Rank-1 CNN for mental workload classification from EEG. HAL Science Ouverte. https://hal.archives-ouvertes.fr/hal-03357020, linked in Neuroergonomics Grand Challenge: Hackathon Passive BCI – Neuroergonomics Conference 2021. Retrieved March 18, 2022, from https://www.neuroergonomicsconference.um.ifi.lmu.de/pbci/

65. Skaramagkas, V., Ktistakis, E., Manousos, D., Tachos, N. S., Kazantzaki, E., Tripoliti, E. E., Fotiadis, D. I., & Tsiknakis, M. (2021). Cognitive workload level estimation based on eye tracking: A machine learning approach. In *BIBE 2021—21st IEEE international conference on bioinformatics and bioengineering, proceedings*. https://doi.org/10.1109/BIBE52308.2021.9635166

Enhancing Wireless Non-invasive Brain-Computer Interfaces with an Encoder/Decoder Machine Learning Model Pair

Ernst R. Fanfan, Joe Blankenship, Sumit Chakravarty, and Adriane B. Randolph⬤

Abstract This project follows a design science research approach to demonstrate a proof-of-concept for developing a means to remove the wires from non-invasive, electroencephalographic brain-computer interface systems while maintaining data integrity and increasing the speed of transmission. This paper uses machine learning techniques to develop an encoder/decoder pair. The encoder learns the important information from the analog signal, reducing the amount of data encoded and transmitted. The decoder ignores the noise and expands the transmitted data for further processing. This paper uses one channel from a non-invasive BCI and organizes the analog signal in 500 datapoint frames. The encoder reduces the frames to seventy-five datapoints and after noise injection, the decoder successfully expands them back to virtually-indistinguishable frames from the originals. The hopes are for improved overall efficiency of non-invasive, wireless brain-computer interface systems and improved data collection for neuro-information systems.

Keywords Electroencephalograph · Brain-computer interface · Machine learning · Data compression · NeuroIS

1 Introduction

As technology is evolving, so does the way we interface with it. Interface developers have tried to remove discomfort throughout the years and connect devices to users

E. R. Fanfan · J. Blankenship · S. Chakravarty · A. B. Randolph (✉)
Kennesaw State University, Kennesaw, GA 30144, USA
e-mail: arandol3@kennesaw.edu

E. R. Fanfan
e-mail: efanfan@students.kennesaw.edu

J. Blankenship
e-mail: jblank31@students.kennesaw.edu

S. Chakravarty
e-mail: schakra2@kennesaw.edu

© The Author(s), under exclusive license to Springer Nature Switzerland AG 2022
F. D. Davis et al. (eds.), *Information Systems and Neuroscience*,
Lecture Notes in Information Systems and Organisation 58,
https://doi.org/10.1007/978-3-031-13064-9_5

intuitively. A system connected to the brain would remove many barriers for an intuitive system, hence the wide range of brain-computer interface (BCI) research.

Although BCI research has been around for nearly fifty years, its extensions have more recently been included in neuro-information systems (neuroIS) and often follow a design science research approach [1]. It has facilitated paralyzed people to move on their own accord, use a computer, and send email [2]. Unfortunately, because of the limitations and invasiveness of the technology, BCIs were often more viable in highly-regulated medical studies or as a novelty [3], but they can do more when considering their non-invasive forms. This work focuses on non-invasive (NI) BCIs, which refers to technology that does not require surgery for its use. Through ongoing research, NI-BCIs will continue to elevate the quality of life of patients with extremely limited mobility [3].

Based on previous experience and the literature review, some of the most significant limitations of BCI technology are related to wires and noise [4]. The encephalogram (EEG) electrodes pick up electrical signals, which means they can detect electrical signals in the environment, too. This noise affects our ability to classify intent, especially with motor imagery (MI) applications [4]; it makes classification slow and inaccurate and can increase a user's frustration and reduce concentration [3]. Besides using machine learning (ML) to aid in the classification, the next best thing is to use invasive BCIs comparable to the one Neuralink is developing [5].

To eliminate the wires, the encoder/decoder (E/D) algorithms must minimize the number of bytes to compromise the transmission speed. The previous must also be robust enough to account for data degradation. Here, we consider use of ML to develop such an E/D pair that may transmit over Bluetooth without compromising speed and classification. This research will consider using borrowed techniques from Long Short-Term Memory (LSTM) and other Deep Neural Networks. This research will especially focus on the backpropagation of these ML techniques and must modify these techniques to fit our goals. These ML techniques are not one-size-fits-all and will be at the core of the success of this project.

We propose training an E/D pair as a means to improve the speed of NI-BCIs. It is possible that such a pair can help to enhance the precision and reliability of neuroIS methods as encouraged by Riedl et al. [6]. The result may be more authentic data collection, especially in contexts where wires are not ideal such as recording EEG while people are mobile or walking [7, 8]. Machine learning techniques will self-adjust and isolate the crucial bites for classification. This triage will only codify and transmit what is needed.

In the following sections, we present the literature review that led to the inception of this study. We outline in the research design section how we plan to test the E/D pair. Then we present the techniques used and an overview of the dataset. Afterward, we discuss the potential implication of the research by speculating on the future of BCI technology.

2 Brain-Computer Interfaces and Machine Learning Techniques

Brain-computer interfaces have been around for over five decades and focused on real-world applications in more recent times. In McFarland and Wolpaw [4] and Wolpaw et al. [9], the role of BCIs in control and communication was discussed. The features of BCI and its crucial parts were presented. Furthermore, the different sorts of BCI based on utilization of electrophysical signals were described, and the critical problems in BCI-based control and communication systems were highlighted including noise and artifacts. Although there have been advances in BCIs where some systems use active over passive electrodes and others transmit their data wirelessly, many systems still must overcome challenges that often arise from movement of the user or environment.

McFarland and Wolpaw [4] focused on feature extraction using ML techniques. One paradigm described was the use of MI-EEG, a self-controlled brain signal that does not involve any external stimulus. In the MI-oriented BCI mechanism, the subject is urged to imagine moving distinct parts of the body for triggering neuronal activities in particular brain regions that are linked with the movements.

Chaudhary et al. [2] explained the role of BCIs in communication and motor rehabilitation. This study discussed BCIs for communication in individuals suffering from locked-in disorder or paralysis. They also described BCI use in motor rehabilitation after spinal cord impairment and severe stroke. This study reported the promising advantages of BCIs in clinical applications.

In Asieh et al. [10], the authors discussed the different presentation methods for EEG-based communication. They compared them to determine a means to increase the communication speed. They compared word-based, letter-based, and icon-based augmentative and alternative communication (AAC), event-related potential (ERP), and rapid serial visual presentation (RSVP). They also experimented with combinations of the previously listed techniques.

Fanfan et al. [3] proposed using an NI-BCI in an information system as a communication aid. This study focused on a specific medical application. They researched how to improve the quality of life of locked-in patients using advances in the Internet of Things (IoT). Also, their proposed system would aid in the decision-making process of caregivers. Such system may be part of a design science artifact as envisioned by Randolph et al. [11].

Rasheed [12] presented a review of research involving the application of ML in the BCI arena. The author covered topics ranging from ERP, RSVP, AAC, mental state, MI, and EEG, to selection classification. This paper compared the results obtained using Support Vector Machines (SVM), Artificial Neural Networks (ANN), K-Nearest Neighbor (KNN), linear regression, and many more. In Müller et al. [13], the ML approach was proposed for EEG signal analysis in real-time. It even discussed the significance of ML schemes for mental condition monitoring and EEG-oriented BCI applications. The previous has the potential to assist as a diagnostic tool.

In Lotte et al. [14], the researchers investigated several classification schemes for EEG-BCI systems. Additionally, they identified numerous challenges for further strengthening the EEG categorization performance in BCI. These are considered further for this work. On another team, Lotte et al. [15] reviewed different classification approaches for EEG-oriented BCIs. This study reviewed five classification approaches, namely, nearest neighbor schemes, non-linear Bayesian schemes, neural networks, linear classifiers, and fusions of classifiers. This study revealed that among five categories, fusions of classifiers seemed very practical for contemporaneous BCI experiments.

Unfortunately, even as more wireless BCIs now exist, there is limited literature available on EEG radio transmission. Hence, this project offers another pathway for exploring improved algorithms and hardware for non-invasive wireless EEG signal transmission. The above led us to develop the technique expressed in the introduction section. Training the encoder/decoder pair using ML techniques will increase transmission speed and maintain data integrity. The model will isolate crucial information and encode what is needed for classification. The previous will minimize the number of bytes transmitted. Also, when the algorithm introduces noise, the decoder will recover the data, and the classifier should maintain its performance.

3   Research Design

We use ML techniques to develop an encoder/decoder (E/D) pair. The encoder compresses the multichannel EEG signals to be transmitted wirelessly, and the decoder decompresses the data. Eventually, the classifier will label the signal, and errors will quantify the E/D pair's performance.

We will use the dataset to train the classifier. Then the classifier will label the testing portion of the dataset to get control. The same test portion will pass through the encoder once without noise and once with noise. The first pass will serve to determine if the whole algorithm works. The second pass will determine if our E/D pair works under simulated wireless conditions. The noise will be present at the encoder, and we will adjust its level to test the limits of the E/D pair. After the classifier labels each batch, we will compare the results. The above is an experiment group. We must use the same testing dataset for each experiment group to better understand the performance of the E/D pair. Then, the experiment operator will make the necessary adjustments and repeat the above steps to maximize the accuracy of the classifier with new testing datasets.

The code will split the dataset according to the standard 70% training and 30% testing ratio. Furthermore, the algorithm will organize the data into 500 data points frames per channel. Each frame will go through each transformative step and require an input size of 500×16. We chose five hundred (500) because it worked best during the single-channel test performed.

4 Proof-of-Concept

The original autoencoder experiment was to find a way to compress EEG signal code and introduce noise to the hidden layer. The goal was to have a noise-free way of compressing EEG signals. Once compressed, the important data would be prioritized and make the signal easier to process. The goal of the next experiment is to create a generative adversarial network with a variational autoencoder. The hope is that this method will provide a stronger network than the autoencoder alone. The variational autoencoder generative adversarial network would have the same goal as the autoencoder, but the adversarial network would also ensure that the compressed data can be decompressed more accurately.

Autoencoders are neural networks that take input vectors, compress them down in a hidden layer, and expand them back to their original size as accurately as possible [16]. The idea is to take the input vectors and process them into a smaller hidden layer to accomplish the compression process. Then a decoder will reverse the process. The backpropagation is one of the essential pieces of this puzzle. The previous is responsible for the learning process, and without it, our experiment and this proof would look completely different. We decided to use Root Mean Square Error (RSME) for this experiment.

The encoder contains a feature input layer followed by two fully connected layers with ReLU as activation functions and, finally, a regression layer. Since we are only using one EEG channel for this proof, the encoder has an input size of 500 vectors and reduces it to 75. The decoder does the reverse. It takes the 75 vectors from the encoder and expands it to 500. The decoder uses a feature input layer, a single fully-connected layer with a sigmoid for activation, and a regression layer.

The proof-of-concept used a fully connected layer to convert the noise into vectors, then injected it into the encoder. Since we do not want that code to learn, we set the learning rate factor and bias to zero [16]. This proof firstly passed the EEG data clean and then passed it with noise injected into the hidden layers. This subproject compared the input to the output. The closer the output graph resembles the input graph, the more robust the E/D pair. This proof-of-concept demonstrated that training an E/D pair is viable to develop optimized compression for EEG Bluetooth transmission. The following steps will expand from one channel to 16, then 32 channel NI-BCI signals, and add the classification step.

5 Future Dataset

This research needs a labeled multichannel EEG signal dataset acquired using an NI-BCI. Preferably the NI-BCI will contain 16 channels or more. The best dataset should come from a previous experiment. The previous is essential to have a baseline performance for the classifier.

In future iterations, we plan to select a dataset from the Patient Repository for EEG Data and Computational Tools [15]. These datasets are well-curated and contain various EEG data of various neurological conditions. The over-the-air deep learning-based radio signal classification data set from DeepSig [17] contains a repository of various radio signals that could interfere with Bluetooth signals. The algorithm will use the previous dataset to inject noise and simulate transmission. This will help us test how robust the modeled encoder/decoder pair are. After injecting noise into the process, if the classifier can maintain the performance, that will prove the robustness of the model.

The best set of data are EEG signals transmitted via radio. Unfortunately, datasets fitting the precious description do not seem to exist. This project is considering using a Generative Adversarial Network (GAN) to construct a repository dataset for our use and the use of the scientific community. As more interest in the subject grows, this repository will be essential for future experiments.

We anticipate the data we will gain access to will be already processed. It was a part of a similar classifying experiment. To be sure, we will review the data and adjust if necessary. Removing excess channels and organizing the data into training, testing, and evaluation groups are the data manipulations we anticipate. We must also code how the algorithm will build the frames to pass to the E/D pair. We intend to use the lessons learned from the proof-of-concept and minimize issues during the experiment.

6 Conclusion

If this project can develop a faster and more robust E/D pair, we will increase the processing speed of the NI-BCI signal and help improve system design and data collection for neuroIS studies. Increasing the processing speed of the signal will increase the possible applications of NI-BCIs and lead to further improvement of the quality of life of locked-in patients. The speed at which current systems process NI-BCI signals has limited the application of this technology. Continuing to remove the wires without losing speed is the goal. Then, we are looking at lighter and less cumbersome designs for NI-BCIs. The more comfortable the patient or participant feels wearing the devices, the more they want to use them, and the longer they may wear them. This work represents advancements that may take place to enhance the overall NI-BCI system.

References

1. Vom Broke, J., Riedl, R., & Léger, P. M. (2013). Application strategies for neuroscience in information systems design research. *Journal of Computer Information Systems, 53*(3), 1–13.
2. Chaudhary, S., Taran, S., Bajaj, V., & Sengur, A. (2019). Convolutional neural network based approach towards motor imagery tasks EEG signal classification. *IEEE Sensor, 19*(12), 4494–4500.
3. Fanfan, E. R., Randolph, A., & Suo, K. (2020). Design of a healthcare monitoring and communication system for locked-in patients using machine learning, IOTs, and brain-computer interface technologies. In *SAIS 2020 Proceedings.*
4. McFarland, D. J., & Wolpaw, J. R. (2011, May 01). Brain-computer interfaces for communication and control. *Communications of the ACM, 54*(5), 60–66.
5. Neuralink. Neuralink, 2021 [Online]. Retrieved 2021 from https://neuralink.com/
6. Riedl, R., Davis, F. D., & Hevner, A. R. (2014). Towards a NeuroIS research methodology: Intensifying the discussion on methods, tools, and measurement. *Journal of the Association for Information Systems, 15*(10).
7. Léger, P.-M., Labonté-Lemoyne, E., Fredette, M., Cameron, A.-F., Bellavance, F., Lepore, F., Faubert, J., Boissonneault, E., Murray, A., Chen, S., & Sénécal, S. (2020). Task switching and visual discrimination in pedestrian mobile multitasking: Influence of IT mobile task type. In *Information systems and neuroscience* (pp. 245–251). Springer.
8. Courtemanche, F., Labonté-LeMoyne, E., Léger, P.-M., Fredette, M., Senecal, S., Cameron, A.-F., Faubert, J., & Bellavance, F. (2019). Texting while walking: An expensive switch cost. *Accident Analysis & Prevention, 127*, 1–8.
9. Wolpaw, J., Birbaumer, N., McFarland, D., Pfurtscheller, G., & Vaughan, T. (2002). Brain-computer interfaces for communication and control. *Clinical Neurophysiology, 113*(6), 767–791.
10. Asieh, A., Mohammad, M., & Deniz, E. (2018). Language-model assisted and icon-based communication through a brain-computer interface with different presentation paradigms. *IEEE Transactions on Neural Systems and Rehabilitation Engineering, 26*(9), 1835–1844.
11. Randolph, A. B., Petter, S. C., Storey, V. C., & Jackson, M. M. (2022). Context-aware user profiles to improve media synchronicity for individuals with severe motor disabilities. *Information Systems Journal, 32*(1), 130–163.
12. Rasheed, S. (2021). A review of the role of machine learning techniques towards brain–computer interface applications. *Machine Learning & Knowledge Extraction, 3*, 835–862.
13. Müller, K., Tangermann, M., Dornhege, G., Krauledat, M., Curio, G., & Blankertz, B. (2008). Machine learning for real-time single-trial EEG-analysis: From brain–computer interfacing to mental state monitoring. *Journal of Neuroscience Methods, 167*(1), 82–90.
14. Lotte, F., Bougrain, L., Cichocki, A., Clerc, M., Congedo, M., Rakotomamonjy, A., & Yger, F. (2018). A review of classification algorithms for EEG-based brain–computer interfaces: A 10 year update. *Journal of Neural Engineering, 15*(3), 031005.
15. Lotte, F., Congedo, M., Lécuyer, A., Lamarche, F., & Arnaldi, B. (2007). A review of classification algorithms for EEG-based brain-computer interfaces. *Journal of Neuro Engineering, 4*(2), 24.
16. Blankenship, J. (2021). *Wireless autoencoders with hidden channel as communication channel*, 2.021.
17. DeepSig. RF datasets for machine learning, 2018 [Online]. Retrieved 2021 from https://www.deepsig.ai/datasets

How Does the Content of Crowdfunding Campaign Pictures Impact Donations for Cancer Treatment

Andreas Blicher, Rob Gleasure, Ioanna Constantiou, and Jesper Clement

Abstract This study investigates how visual stimuli influence cancer-related charitable online giving. Particularly, the study investigates how different types of crowdfunding campaign pictures affect decision making. We gathered crowdfunding campaigns from GoFundMe and divided them according to the main picture used in each campaign, i.e., cancer-related pictures vs. non-cancer-related pictures and pictures of individuals versus pictures of groups. We then conducted an experiment using physiological measures. The results from the experiment show that cancer-related pictures receive more money and more immediate attention and arousal than non-cancer-related pictures. Furthermore, group pictures receive more money and more total attention than individual pictures. The physiological measures from the experiment provide valuable knowledge about the underlying emotional mechanisms involved in the donation process.

Keywords Crowdfunding · Decision Making · Cancer · NeuroIS · GoFundMe

1 Introduction

Most people will suffer from serious illness at some point in their lives, at which point they have to rely on others in society for help. Health systems are intended to cover most of these scenarios, yet there are times when individuals cannot access the care they need, for example because they cannot afford the treatments. These types of systematic gaps in public funding are often addressed, at last partly, by acts of charity [1]. Notably, charitable crowdfunding has been shown to reduce rates of medical bankruptcy in vulnerable populations [2]. However, it is less clear whether charitable crowdfunding is actually providing support to those who most need it.

A. Blicher (✉) · R. Gleasure · I. Constantiou
Department of Digitalization, Copenhagen Business School, Copenhagen, Denmark
e-mail: abs.digi@cbs.dk

J. Clement
Department of Marketing, Copenhagen Business School, Copenhagen, Denmark

© The Author(s), under exclusive license to Springer Nature Switzerland AG 2022
F. D. Davis et al. (eds.), *Information Systems and Neuroscience*,
Lecture Notes in Information Systems and Organisation 58,
https://doi.org/10.1007/978-3-031-13064-9_6

It is not always obvious why people give money to others—others whom they may never meet or know. It has been suggested that charity is essentially irrational, and better explained with emotions [3]. This appears especially likely for charitable crowdfunding, where donors may have limited access to the information needed for rational judgements. Hence, donors seem to rely on emotional judgements, based on a range of heuristics and cognitive biases [4, 5]. These emotional judgements and biases in charitable behaviors often lead to the systematic misallocations of funds [6].

Thus, it appears that, while charitable crowdfunding may have the potential to address inequalities in contemporary health systems, it also has the potential to create new inequalities if we do not understand, and account for, donors' emotional decision-making processes and cognitive biases. These types of emotional processes and biases are difficult to study via self-report, as individuals may be unaware of them, or prone to social desirability biases in their responses. They are also difficult to study with behavioral studies alone, where cognitive elements must be inferred from limited cognitive measurement.

This paper therefore investigates health-related charitable giving using NeuroIS tools and theories. We focus on the role of pictures in fundraising campaigns. This is because previous research suggests that semantic content alone does not account for charitable giving, as neither negative nor positive text features are related to resource sharing. Only photographs reveal a significant relationship between arousal and giving [7]. Thus, we ask:

Q1: Which types of pictures are more likely to help charitable crowdfunding campaigns attract donations?

Q2: Which types of cognitive biases lead these pictures to attract more donations?

The next section provides a brief background on charitable crowdfunding and rationalistic theories of altruism, culminating in two research hypotheses. Next, we present a laboratory experiment that combines behavioral measures with measures of eye movements and skin conductance. The results suggest campaigns attract more donors when they use pictures that depict illness and when they use pictures with multiple individuals.

2 Literature and Theory Development

2.1 Donation-Based Crowdfunding

Crowdfunding platforms allow fundraisers to make open appeals for financial contributions from the public. Crowdfunding platforms operate on a spectrum between altruistically and strategically motivated [8]. On the one end of the spectrum, fundraisers provide donors with repayment, equity, or valuable material rewards in exchange for their contribution. On the other end of the spectrum, fundraisers ask donors to contribute without promising any material or financial returns.

Altruism or pro-social motivations appears to play a role on most types of crowdfunding platforms. For example, Dai and Zhang [9] showed that Kickstarter projects raise more money as they approach their fundraising target, particularly if those projects appeal to prosocial motivations. Similarly, Du et al. [10] showed that recipients of peer-to-peer loans are more likely to repay their loans when reminded of lenders' positive expectations, while reminders of negative consequences for non-repayment have limited impact. Even in equity crowdfunding, certain investors may be more likely to invest in ventures with social benefits, such as those focused on sustainability [11].

Perhaps unsurprisingly, it is charitable crowdfunding where research has observed most evidence of altruism [12–16]. These studies demonstrate the existence of altruism and social motivations in donors. However, the specific types of altruism, and the potential biases they enable, have received limited attention.

Scientific research has commonly explained altruism according with three rationalistic perspectives: *egoistic*, *egocentric*, and *altercentric* [17]. The *egoistic* perspective explains altruism by assuming actors foresee some expected social benefit, such as reciprocity or reputational gains. Whether or not this behavior actually constitutes altruism is debatable, given the strategic nature of *egoistic* helping decisions. However, while there are signs of reciprocity among donors to crowdfunding campaigns [18], the tendency for many contributors to hide their identity on donation crowdfunding platforms suggests other motivations are also in effect. We therefore explore the *egocentric* and *altercentric* perspectives in more detail in the next sections.

2.2   Egocentric Altruism in Charitable Crowdfunding

The *egocentric* view of altruism suggests that individuals will help others because they participate in the resulting joy or alleviation of suffering or distress. This perspective relies on emotional contagion to explain helping or comforting behaviors [19]. This lays the foundation for empathy, which allows people to relate to the emotional states of other people. Although cognition is also an important part of empathy, it is a secondary component. The primary component of empathy is that one person's emotional or arousal state affects another person [19]. This primary underlying emotional mechanism of empathy provides one person (the subject) with access to the subjective state of another person (the object) through the subject's own neural and bodily representations [20].

Emotional contagion has been shown to occur via at least three mechanisms [21]. The first is mimicry, in which an emotional expression activates synchronous behavior on the part of the perceiver, which in turn activates affective processes [22, 23]. The second mechanism is category activation, in which exposure to emotional expressions primes an emotion category, which in turn leads to activation of specific emotional processes [24, 25]. Finally, the third mechanism is social appraisal, in which individuals use the emotions of others as a guide for their own emotion appraisals, leading to similar emotional experiences [26, 27].

For these reasons, emotional contagion is often involuntary, meaning an individual will adopt the emotional state of another person without necessarily intending to do so [28]. These processes are nonetheless complex and interwoven with conscious evaluations and decision making. Notably, the psychological literature distinguishes sympathy from personal distress [19]. Personal distress makes the affected party selfishly seek to alleviate its own distress, a distress which mimics that of the object [29], while sympathy is defined as an affective response that consists of feelings of sorrow or concern for a distressed or needy other, with less emphasis on sharing the emotion of the other [30].

These nuances become important when trying to compare and contrast the influence of positive and negative affect on charitable giving [3]. On the one hand, a number of research findings reveal that negative affect evoked by empathic pain can increase charitable giving [31–33]. This suggests fundraising would be more effective if it focuses on the suffering of the person in need. On the other hand, a number of research findings reveal that positive affect (i.e., "warm glow") evoked by anticipation of giving can increase charitable behavior [34–36]. This suggests fundraisers should avoid focusing on the suffering of the person in need; instead highlighting the quality of life they hope to re-establish. While both are feasible, there is considerable evidence that negative emotion is a more powerful motivator of online behavior [37]. Thus, we predict images that emphasize the negative state of the person in need will be more effective for fundraising.

H1: Charitable crowdfunding campaigns with a picture that depicts illness are more likely to attract donors than campaigns with a picture that does not.

2.3 Altercentric Altruism in Charitable Crowdfunding

The *altercentric* view of altruism suggests that individuals help others in need because they have evolved a pro-social trait or "moral gene" [17]. The benefit of this trait is not linked to any one interaction; rather it emerges because altruistic individuals tend to group together, and these groups are more successful than less altruistic groups.

This type of altruism replaces the focus on emotional contagion in *egocentric* altruism with a focus on 'kinship', not in the genetic sense but in the preference and recognition of some share social and moral norms [38]. This makes sense for charitable crowdfunding platforms, which are relatively easy to avoid for reactive and guilt-driven potential donors who would prefer to 'avoid the ask' [cf. 39]. Hence, these charitable crowdfunding platforms are often characterized by a shared moral imperative [14, 40].

This results in some confusion as to whether individuals or groups are more likely to elicit charitable giving. There is evidence that donations to large numbers of victims are typically muted relative to donations to a single identified victim, as an individual is generally more personally relatable to each individual donor [41, 42]. This suggests fundraising should depict solely the person in need. However, other research shows that people donate more to large numbers of victims, provided these victims are perceived as entitative—comprising a single, coherent unit [43]. This suggests fundraising should show the person in need along with others who also display shared social and moral norms, such as friends and/or family who are sharing the experience. These alternative predictions are, once again, both feasible. The key differentiating criterion appears to be the ability for donors to cognitively relate multiple individuals as entitative. However, given the strong community element of charitable crowdfunding platforms, we assume potential donors should find it comparatively easy to establish a sense of kinship. Thus, we predict images with multiple individuals will be more effective in stimulating *altercentric* altruism.

> *H2: Charitable crowdfunding campaigns with a picture that depicts multiple individuals are more likely to attract donors than campaigns with a picture that depicts a single individual.*

3 Methods

3.1 Participants

Participants for the experiment were recruited at the authors' University. Data from 22 participants ($M_{age} = 26.77$, $SD = 8.22$; 14 female, 64%) has been collected and the results are reported in the present paper. The participants were all given a movie ticket ($15) for their participation.

3.2 Measures

The crowdfunding task measures crowdfunding choices and contains 48 crowdfunding campaigns consisting of a picture and a text. The task includes four conditions: (i) individual picture, (ii) group picture, (iii) cancer-related picture, and (iv) non-cancer-related picture. On each trial, two crowdfunding campaigns are presented and the participant is asked to choose which campaign to support. Thus, the four possible combinations are (i) individual cancer-related vs. group cancer-related, (ii) individual non-cancer-related vs. group non-cancer-related, (iii) individual cancer-related vs. individual non-cancer-related (iv) group cancer-related vs. group non-cancer-related. The experiment is created using Qualtrics software (www.qualtrics.com), with iMotions software (www.imotions.com) used to coordinate and integrate physiological measures. The crowdfunding campaigns are taken from the GoFundMe website (www.gofundme.com). Each crowdfunding campaign contains a picture and a text (100–200 words). During the task, a Tobii Pro Nano records eye movements and a Shimmer3 GSR + records skin conductance.

3.3 Procedure

Upon arrival at the laboratory the participant is given verbal and written information about the experiment. First, the shimmer is attached to the wrist of the non-dominant hand of the participant. Then, a two-minute baseline is conducted. No stimuli are presented during the baseline. Before the experiment, the participant is told "You are given $24 to donate to crowdfunding". However, "You can only support one of the two crowdfunding campaigns" and "Click on the donate button under the crowdfunding campaign you would like to support". The participant is told that "$1 is donated for every choice you make". The participant is then presented with 24 crowdfunding choices each containing two new options. Upon completion of the experiment, the participant is debriefed and thanked for their participation.

The eye-tracking data analysis was conducted on the first fixation duration data, the last fixation duration data, and the total fixation duration data, respectively. Areas of interest (AOIs) were created for each stimulus set. For each stimulus set two 450 × 255 px rectangles were created. The rectangles covered the pictures of the two crowdfunding campaigns presented on each trial. An I-VT fixation filter was created. The fixation filter parameters were: 20 ms window length, 30°/second velocity threshold, 75 ms max gap length, 60 ms minimum fixation duration, 75 ms max time between fixations, and 0.5° max angle between fixations. The first fixation duration data, the last fixation duration data, and the total fixation duration data were analyzed for each AOI.

The galvanic skin response data analysis was conducted on the number of peaks data. Peak detection of the first five seconds after stimulus onset were calculated for each stimulus set. The peak detection parameters were: 8000 ms phasic filter length, 5 Hz lowpass filter cutoff frequency, 0.01 microSiemens peak onset threshold, 0 microSiemens peak offset threshold, 0.005 microSiemens peak amplitude threshold, 500 ms minimum peak duration, and 4000 ms gap interpolation length threshold.

4 Results

4.1 Decision-Making Data

A 2 × 2 analysis of variance (ANOVA) with cancer (cancer-related vs. non-cancer-related) and group (group vs. individual) as within-subject factors was conducted on the decision-making data. The analysis revealed a significant main effect of cancer, $F(1, 21) = 134.54, p < 0.001, \eta^2 = 0.87$, and a significant main effect of group, $F(1, 21) = 6.13, p = 0.022, \eta^2 = 0.23$. The cancer × group interaction, $F(1, 21) = 0.40, p = 0.535, \eta^2 = 0.02$, did not reach significance. Post hoc dependent samples t-tests revealed a significant difference between individual cancer-related vs. individual non-cancer-related pictures, $t(21) = 10.80, p < 0.001, d = 0.92$, a significant difference between group cancer-related vs. group non-cancer-related pictures, $t(21) = 6.06$,

$p < 0.001$, $d = 0.79$, and a significant difference between group non-cancer-related vs. individual non-cancer-related pictures, $t(21) = 2.27$, $p = 0.034$, $d = 0.44$. The difference between group cancer-related vs. individual cancer-related pictures, $t(21) = 2.03$, $p = 0.056$, $d = 0.40$, did not reach significance. The results suggest that more money were donated to the individual cancer-related ($M = 4.5$, $SD = 0.67$) compared to the individual non-cancer-related pictures ($M = 1.5$, $SD = 0.67$), more money were donated to the group cancer-related ($M = 4.2$, $SD = 0.98$) compared to the group non-cancer-related pictures ($M = 1.7$, $SD = 0.98$), and more money were donated to the group non-cancer-related ($M = 3.6$, $SD = 1.22$) compared to the individual non-cancer-related pictures ($M = 2.4$, $SD = 1.22$). More money were not donated to the group cancer-related ($M = 3.6$, $SD = 1.36$) compared to the individual cancer-related pictures ($M = 2.4$, $SD = 1.36$).

4.2 Eye-Tracking Data

Three 2×2 analysis of variance (ANOVA) with cancer (cancer-related vs. non-cancer-related) and group (group vs. individual) as within-subject factors were conducted on the first fixation duration data, the last fixation duration data, and the total fixation duration data, respectively.

The first fixation duration analysis revealed a significant main effect of cancer, $F(1, 21) = 20.23$, $p < 0.001$, $\eta^2 = 0.49$, and a significant main effect of group, $F(1, 21) = 7.95$, $p = 0.010$, $\eta^2 = 0.28$. The cancer $\times$ group interaction, $F(1, 21) = 1.10$, $p = 0.306$, $\eta^2 = 0.05$, did not reach significance. The results suggest that the first fixation duration was longer for the cancer-related ($M = 255.31$, $SD = 45.81$) compared to the non-cancer-related pictures ($M = 220.87$, $SD = 43.13$), the first fixation duration was longer for the individual ($M = 261.18$, $SD = 67.49$) compared to the group pictures ($M = 215.01$, $SD = 41.34$).

The last fixation duration analysis revealed a significant main effect of group, $F(1, 21) = 37.02$, $p < 0.001$, $\eta^2 = 0.64$. The main effect of cancer, $F(1, 21) = 1.41$, $p = 0.248$, $\eta^2 = 0.06$, and the cancer $\times$ group interaction, $F(1, 21) = 0.05$, $p = 0.824$, $\eta^2 = 0.01$, did not reach significance. The results suggest that the last fixation duration was longer for the individual ($M = 342.52$, $SD = 47.13$) compared to the group pictures ($M = 284.22$, $SD = 46.90$). The last fixation duration was not longer for the cancer-related ($M = 321.73$, $SD = 58.49$) compared to the non-cancer-related pictures ($M = 305.01$, $SD = 46.54$).

The total fixation duration analysis revealed a significant main effect of group, $F(1, 21) = 82.06$, $p < 0.001$, $\eta^2 = 0.80$. The main effect of cancer, $F(1, 21) = 0.02$, $p = 0.898$, $\eta^2 = 0.01$, and the cancer $\times$ group interaction, $F(1, 21) = 0.53$, $p = 0.474$, $\eta^2 = 0.03$, did not reach significance. The results suggest that the total fixation duration was longer for the group ($M = 2302.57$, $SD = 1059.22$) compared to the individual pictures ($M = 1551.02$, $SD = 810.58$). The total fixation duration was not longer for the cancer-related ($M = 1933.81$, $SD = 927.75$) compared to the non-cancer-related pictures ($M = 1919.79$, $SD = 984.99$).

4.3   Galvanic Skin Response Data

A $2 \times 2 \times 2$ analysis of variance (ANOVA) with cancer (cancer-related vs. non-cancer-related) and group (group vs. individual) and side (left vs. right) as within-subject factors were conducted on the number of peaks data. The analysis revealed a significant main effect of cancer, $F(1, 21) = 4.67, p < 0.042, \eta^2 = 0.18$. The main effect of group, $F(1, 21) = 1.09, p = 0.308, \eta^2 = 0.05$, the cancer $\times$ side interaction, $F(1, 21) = 0.04, p = 0.853, \eta^2 = 0.01$, the group $\times$ side interaction, $F(1, 21) = 0.10$, $p = 0.747, \eta^2 = 0.01$, and the cancer $\times$ group $\times$ side interaction, $F(1, 21) = 0.50, p = 0.485, \eta^2 = 0.02$, did not reach significance. The results suggest that the number of peaks was higher for the cancer-related ($M = 2.74, SD = 2.18$) compared to the non-cancer-related pictures ($M = 2.56, SD = 2.07$). The number of peaks was not higher for the group ($M = 2.70, SD = 2.20$) compared to the individual pictures ($M = 2.59, SD = 2.06$).

5   Discussion

This work aims to present a deep analysis of health-related online fundraising that is conspicuously absent in existing literature. The physiological measures from the experiment provide valuable knowledge about the underlying emotional mechanisms involved in the donation process.

The results from the experiment suggest that the participants donated more money to crowdfunding campaigns with cancer-related pictures than campaigns with non-cancer-related pictures. Interestingly, the eye-tracking results suggest that the participants' first fixation duration was longer for cancer-related pictures than for non-cancer-related pictures. Furthermore, the galvanic skin response results suggest that the participants' number of peaks during the first five seconds after stimulus onset was higher for crowdfunding campaigns with cancer-related pictures than campaigns with non-cancer-related pictures. The results from the experiment also suggest that the participants donated more money to crowdfunding campaigns with group pictures than campaigns with individual pictures and that the participants' total fixation duration was longer for group pictures than for individual pictures.

These results provide a valuable starting point to better understand donation behaviors on charitable crowdfunding platforms. However, they also raise some concerns about the generalizability of a controlled laboratory experiment to actual crowdfunding platforms; platforms where many donors already have established relationships with those in need and much communication happens via other channels. Thus, this study is part of a series of experiments that shift the balance from controlled intervention to behavioral naturalism. More precisely, we have tested the same hypotheses with two additional studies.

(i) An online behavioral experiment (n = 100) that mirrors the protocol of this study. This provided robustness and greater statistical confidence in the behavioral results.

(ii) A quasi-experiment [44] of observational data from GoFundMe, in which we gather a large dataset of campaigns and compare treatments using coarsened exact matching [45]. This allowed us to test whether the observed effects maintained a significant impact when other competing influences are present.

Collectively, these results help to explain some of the peculiarities of charitable crowdfunding. They also provide a means to 'level the playing field' for those in need. If charitable crowdfunding is to fill in the gaps in contemporary health systems, then campaign designers need to be aware of donor biases. Otherwise, oversights and gaps will persist, based on disproportionately influential decisions like which picture to use for a fundraiser, and donations will be less likely to make it those who need it most.

Acknowledgements The authors wish to thank the Carlsberg Foundation, grant number CF20-0254, and everyone who participated in the study.

Declaration of Conflicting Interests The authors declared no potential conflicts of interest with respect to the research, authorship, and/or publication of this article.

Ethical Approval The study was approved by the institutional review board and was carried out in accordance with the provisions of the World Medical Association Declaration of Helsinki.

References

1. List, J. A. (2011). The market for charitable giving. *Journal of Economic Perspectives, 25*(2), 157–180.
2. Burtch, G., & Chan, J. (2018). Investigating the relationship between medical crowdfunding and personal bankruptcy in the United States: Evidence of a digital divide. *Management Information Systems Quarterly, 43*(1), 237–262.
3. Genevsky, A., Västfjäll, D., Slovic, P., & Knutson, B. (2013). Neural underpinnings of the identifiable victim effect: Affect shifts preferences for giving. *Journal of Neuroscience, 33*(43), 17188–17196.
4. Chen, S., Thomas, S., & Kohli, C. (2016). What really makes a promotional campaign succeed on a crowdfunding platform?: Guilt, utilitarian products, emotional messaging, and fewer but meaningful rewards drive donations. *Journal of Advertising Research, 56*(1), 81–94.
5. Sasaki, S. (2019). Majority size and conformity behavior in charitable giving: Field evidence from a donation-based crowdfunding platform in Japan. *Journal of Economic Psychology, 70*, 36–51.
6. Baron, J., & Szymanska, E. (2011). Heuristics and biases in charity. In D. M. Oppenheimer & C. Y. Olivola (Eds.), The science of giving: Experimental approaches to the study of charity (pp. 215–236). Taylor & Francis Group.
7. Genevsky, A., & Knutson, B. (2015). Neural affective mechanisms predict market-level microlending. *Psychological Science, 26*(9), 1411–1422.

8. Berns, J. P., Figueroa-Armijos, M., da Motta Veiga, S. P., & Dunne, T. C. (2020). Dynamics of lending-based prosocial crowdfunding: Using a social responsibility lens. *Journal of Business Ethics, 161*(1), 169–185.

9. Dai, H., & Zhang, D. J. (2019). Prosocial goal pursuit in crowdfunding: Evidence from kickstarter. *Journal of Marketing Research, 56*(3), 498–517.

10. Du, N., Li, L., Lu, T., & Lu, X. (2020). Prosocial compliance in P2P lending: A natural field experiment. *Management Science, 66*(1), 315–333.

11. Vismara, S. (2019). Sustainability in equity crowdfunding. *Technological Forecasting and Social Change, 141*, 98–106.

12. Bagheri, A., Chitsazan, H., & Ebrahimi, A. (2019). Crowdfunding motivations: A focus on donors' perspectives. *Technological Forecasting and Social Change, 146*, 218–232.

13. Gleasure, R., & Feller, J. (2016). Does heart or head rule donor behaviors in charitable crowdfunding markets? *International Journal of Electronic Commerce, 20*(4), 499–524.

14. Gleasure, R., & Feller, J. (2018). What kind of cause unites a crowd? Understanding crowdfunding as collective action. *Journal of Electronic Commerce Research, 19*(3), 223–236.

15. Liu, L., Suh, A., & Wagner, C. (2018). Empathy or perceived credibility? An empirical study on individual donation behavior in charitable crowdfunding. *Internet Research*.

16. Snyder, J., Crooks, V. A., Mathers, A., & Chow-White, P. (2017). Appealing to the crowd: Ethical justifications in Canadian medical crowdfunding campaigns. *Journal of medical ethics, 43*(6), 364–367.

17. Khalil, E. L. (2004). What is altruism? *Journal of economic psychology, 25*(1), 97–123.

18. André, K., Bureau, S., Gautier, A., & Rubel, O. (2017). Beyond the opposition between altruism and self-interest: Reciprocal giving in reward-based crowdfunding. *Journal of Business Ethics, 146*(2), 313–332.

19. de Waal, F. B. (2008). Putting the altruism back into altruism: The evolution of empathy. *Annual Review of Psychology, 59*, 279–300.

20. Preston, S. D., & de Waal, F. B. (2002). Empathy: Its ultimate and proximate bases. *Behavioral and Brain Sciences, 25*(1), 1–20.

21. Goldenberg, A., & Gross, J. J. (2020). Digital emotion contagion. *Trends in cognitive sciences, 24*(4), 316–328.

22. Barsade, S. G., Coutifaris, C. G., & Pillemer, J. (2018). Emotional contagion in organizational life. *Research in Organizational Behavior, 38*, 137–151.

23. Hess, U., & Fischer, A. (2014). Emotional mimicry: Why and when we mimic emotions. *Social and personality psychology compass, 8*(2), 45–57.

24. Peters, K., & Kashima, Y. (2015). A multimodal theory of affect diffusion. *Psychological Bulletin, 141*(5), 966.

25. Niedenthal, P. M., Winkielman, P., Mondillon, L., & Vermeulen, N. (2009). Embodiment of emotion concepts. *Journal of personality and social psychology, 96*(6), 1120.

26. Manstead, A. S., & Fischer, A. H. (2001) Social appraisal. *Appraisal Processes in Emotion: Theory, Methods, Research,* 221–232.

27. Clément, F., & Dukes, D. (2017). Social appraisal and social referencing: Two components of affective social learning. *Emotion Review, 9*(3), 253–261.

28. Kramer, A. D., Guillory, J. E., & Hancock, J. T. (2014). Experimental evidence of massive-scale emotional contagion through social networks. *Proceedings of the National Academy of Sciences, 111*(24), 8788–8790.

29. Batson, C. D. (1991). *The Altruism question: Toward a social-psychological answer*. Erlbaum.

30. Eisenberg, N. (2000). Empathy and sympathy. In M. Lewis, & J. M. Haviland-Jones (Eds.), *Handbook of emotion* (2nd edn, pp. 677–691). Guilford.

31. Small, D. A., & Verrochi, N. M. (2009). The face of need: Facial emotion expression on charity advertisements. *Journal of Marketing Research, 46*(6), 777–787.

32. Hein, G., Silani, G., Preuschoff, K., Batson, C. D., & Singer, T. (2010). Neural responses to ingroup and outgroup members' suffering predict individual differences in costly helping. *Neuron, 68*(1), 149–160.

33. Masten, C. L., Morelli, S. A., & Eisenberger, N. I. (2011). An fMRI investigation of empathy for 'social pain' and subsequent prosocial behavior. *NeuroImage, 55*(1), 381–388.
34. Andreoni, J. (1990). Impure altruism and donations to public goods: A theory of warm-glow giving. *The Economic Journal, 100*(401), 464–477.
35. Andreoni, J. (1995). Warm-glow versus cold-prickle: The effects of positive and negative framing on cooperation in experiments. *The Quarterly Journal of Economics, 110*(1), 1–21.
36. Harbaugh, W. T., Mayr, U., & Burghart, D. R. (2007). Neural responses to taxation and voluntary giving reveal motives for charitable donations. *Science, 316*(5831), 1622–1625.
37. Stieglitz, S., & Dang-Xuan, L. (2013). Emotions and information diffusion in social media—sentiment of microblogs and sharing behavior. *Journal of management information systems, 29*(4), 217–248.
38. Richerson, P. J., & Boyd, R. (2001). The evolution of subjective commitment to groups: A tribal instincts hypothesis. In R. M. Nesse (Ed.), *Russell sage foundation series on trust. Evolution and the capacity for commitment* (Vol. III, pp. 186–220). Russell Sage Foundation.
39. Andreoni, J., Rao, J. M., & Trachtman, H. (2017). Avoiding the ask: A field experiment on altruism, empathy, and charitable giving. *Journal of political Economy, 125*(3), 625–653.
40. Choy, K., & Schlagwein, D. (2016). Crowdsourcing for a better world: On the relation between IT affordances and donor motivations in charitable crowdfunding. *Information Technology & People, 29*(1), 21–247.
41. Slovic, P. (2007). If I look at the mass I will never act: Psychic numbing and genocide. *Judgment and Decision Making, 2*(2), 1–17.
42. Small, D. A., & Loewenstein, G. (2003). Helping a victim or helping the victim: Altruism and identifiability. *Journal of Risk and Uncertainty, 26*(1), 5–16.
43. Smith, R. W., Faro, D., & Burson, K. A. (2013). More for the many: The influence of entitativity on charitable giving. *Journal of Consumer Research, 39*(5), 961–976.
44. Lazer, D., Pentland, A., & Adamic, L. (2008). Social science. Computational social science. *Science, 323*(5915), 721–723.
45. Iacus, S. M., King, G., & Porro, G. (2012). Causal inference without balance checking: Coarsened exact matching. *Political analysis, 20*(1), 1–24.

All Eyes on Misinformation and Social Media Consumption: A Pupil Dilation Study

Mahdi Mirhoseini, Spencer Early, and Khaled Hassanein

Abstract The research on misinformation has shown that those users who spend cognitive resources while reading news on social media are more likely to identify fake headlines. Although various behavioral and neurophysiological measures have been used in the literature to examine this hypothesis, the association between pupil dilation, which has been established as a measure of cognitive load in the NeuroIS field, and users' performance in judging the accuracy of headlines has yet to be studied. A within subject experiment using different types of news headlines is designed in which users rate the accuracy of 80 Facebook posts. Consistent with the heuristic-systematic model of information processing (HSM), our results suggest that pupil dilation is positively linked with users' accuracy rate.

Keywords Fake news · Heuristic-systematic model · Eye-tracking · Pupil dilation · Pupillometry · Misinformation · Dual-process models

1 Introduction

Since social media has become a major source for citizens to read and share news, the spread of false information, regardless of intentionality, has manifested its adverse effects on users and society. Users lose their trust in scientific findings [1, 2] and the free flow of information is disrupted, which weakens the quality of democracies [3]. The fake news phenomenon, defined as "fabricated information that mimics news media content in form but not in organizational process or intent" [4], has gained the

M. Mirhoseini (✉)
John Molson School of Business, Concordia University, Montréal, Canada
e-mail: mahdi.mirhoseini@concordia.ca

S. Early · K. Hassanein
DeGroote School of Business, McMaster University, Hamilton, Canada
e-mail: earlys@mcmaster.ca

K. Hassanein
e-mail: hassank@mcmaster.ca

© The Author(s), under exclusive license to Springer Nature Switzerland AG 2022
F. D. Davis et al. (eds.), *Information Systems and Neuroscience*,
Lecture Notes in Information Systems and Organisation 58,
https://doi.org/10.1007/978-3-031-13064-9_7

interest of researchers across multiple disciplines to study how and why individuals interact with fake news, and what can be done to counter it.

In the Information Systems field, researchers have looked mainly at technological (e.g., intervention methods [5, 6] and presentation formats [7]), and behavioral factors [8]. The NeuroIS field has the potential to complement these efforts by shedding light on the neurophysiological indicators that help us understand how users believe misinformation.

Based on the dual process model of cognition [9], cognitive load—defined as the set of working memory resources used to perform a task [10]—is an indicator of utilizing System 2 resources, which enables users to deliberate and ultimately identify misinformation. Although pupillometry has been established as a measure of cognitive load in the NeuroIS field [11], to the best of our knowledge, its association with belief in misinformation has not been investigated yet. Therefore, our goal in the current paper is to explore whether there is any association between pupil dilation and users' performance in identifying misinformation.

We designed a laboratory experiment in which users rate 80 news headlines while an eye-tracker records their pupil metrics. This study can potentially contribute to our understanding of whether pupillometry can be an indicator of users' analytical thinking in the face of fake headlines. It also adds to the existing literature on the association between cognitive load and performance in detecting fake news. The remainder of this paper continues with a brief literature review on eye-tracking studies in the misinformation research, dual process model of cognition, and pupil response and cognitive load. Then we outline the research hypotheses, details on methodology, and present the results. We conclude by discussing the results, contributions of our work, and limitations.

2 Literature Review and Hypotheses

2.1 Eye-Tracking Studies

To identify all relevant articles that used any eye-tracking measures in studying the phenomenon of fake news, we conducted a systematic literature review using the following search string: ("eye-track*" OR "eye track*") AND ("fake news" OR "misinform*" OR "disinform*" OR "false news") on the SCOPUS and Web of Science databases. We included both journal and conference articles and only limited our search to the publications in English. The search on both databases was done within the title, abstract, and the keywords. Furthermore, the search was not limited to any time period. In total, 23 papers were found. Numerous gaze related measures such as dwell time on different message components [12], eye fixations as a measure of cognitive load [13, 14], fixation duration as a measure of user engagement with cognitive reasoning [15], as a control method for assuring that users attend to designated flags [16], and different types of fixation durations to predict false/true headlines [17]. However, no study investigated the relationship between pupil dilation and user behavior or performance in the fake news domain.

2.2 Cognitive Load and Misinformation

Based on dual-process models of cognition [9], there are two distinct cognitive systems that contribute to human decision making: 1—a fast and intuitive process (System1) that is effortless to use, and 2—an analytical and effortful process (System2). Falling under dual process models, the heuristic-systematic model of information processing (HSM) suggests that System1 simplifies the decision making process by using shortcuts and heuristics, and is therefore more susceptible to different types of decision making biases [18]. In contrast, System2 relies on analyzing and considering the message content [19] making it less susceptible to such biases.

In the fake news domain, the dual process models predict that using System2 cognitive processes increases users' performance in identifying fake news articles. A number of misinformation studies have found support for this theory by investigating whether using cognitive load results in identifying fake headlines. For instance, consistent with this framework, Pennycook and Rand [20] found that more analytical thinking is associated with higher performance in identifying misinformation. They concluded that the main reason that people fall for the misinformation trap is their laziness on social media. An important outcome of finding support for dual process theories in the misinformation context is that creating a condition for users or encouraging them to utilize System2 cognitive processes reduces their error. An interesting study by Bago et al. [21] showed that giving individuals an opportunity to re-think and deliberate, increases their accuracy.

2.3 Pupil Response and Cognitive Load

Pupil dilation has been established as a reliable measure of cognitive load [22]. Human's pupil is larger in size as more cognitive resources including perception, memory, attention, and reasoning are being used by the brain [23]. Within the NeuroIS domain, it has been used as a measure of cognitive load [24–26].

A study by Ladeira and Dalmoro [27] investigated the impact of real versus fake news on various eye tracking measures. They found that due to low heuristic availability, cognitive load (measured by gaze duration, fixation frequency, and pupil diameter) is higher when users read fake news compared to real news. Similarly, Sümer and Bozkir [28] found that the number of fixations and the number of saccades are more on the fake content compared to real news. In the current study, we expect that reading news headlines that refer to political and controversial subjects, consume more attention, memory, and reasoning resources, which leads to the enlargement of users' pupils.

H1: reading political content leads to more pupil dilation compared to neutral content.

Consistent with the heuristic-systematic model of information processing, we expect that those who use more cognitive resources (i.e. pupil dilation) are more likely to identify false news headlines. Therefore:

H2: Pupil dilation is positively associated with performance on identifying misinformation.

3   Methods

The experiment was designed utilizing behavioral measures and pupillometry metrics to quantify the diameter of a participant's pupil and build an underlying behavioral profile. News headlines were constructed and categorized into three unique categories. Control headlines, in which the content of the statement was obviously true or false (e.g., In 2018, the average adult human's height was 10 feet). Neutral headlines, in which the content was derived from pop culture and other non-divisive topics (e.g., New McDonald's restaurant opened every 14.5 h globally). Finally, political headlines, in which the content was deemed politically or socially divisive (e.g., Donald Trump declares himself "Second coming of God"). Each participant was presented with a total of 80 headlines, 20 control, 20 neutral, and 40 political. Half of each headline category were true headlines and half were false headlines.

An initial survey was utilized to categorize a list of topics and phrases into political/neutral groupings. Faculty and staff were asked to rate how divisive a topic (e.g., Donald Trump, Abortion, Netflix, etc.) was on a 5-point Likert scale. The topics were then converted into length adjusted headlines (~10–12 words). This was done by leveraging factchecking websites, such as Snopes.com, to extract objectively true news headlines based on the specified topic or phrase. Half of the factchecked true headlines were then manually adjusted to shift the premise of the news headline to objectively false. This process was done for both the neutral and political headline groupings. The control group headlines were constructed based on wild assertions regarding obvious facts that most people should be able to recognize.

Presentation format for each headline was constructed to resemble the appearance of a news article within the Facebook timeline (See Fig. 1). Each post was presented in the center of the screen, dimensionality and positioning was consistent across all 80 headlines. A blank reference screen was presented for two seconds prior to each post to assess a baseline of pupil dilation activity.

Thirty subjects participated in the study (n = 30), 60% of participants were male and 40% were female. 34% of the participants self-identified as conservative and 66% self-identified as liberal. 16% has a high school or equivalent degree, 53% had a Bachelor's degree, 26% had a Master's degree, and 3% had a Doctorate degree. 7% were between the ages of 18 and 24, 46% between 25 and 34, 20% between 35 and

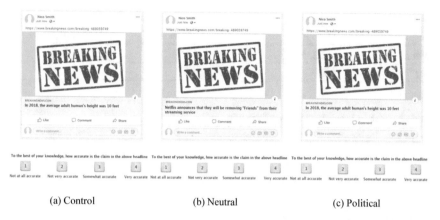

(a) Control   (b) Neutral   (c) Political

Fig. 1 News headlines format

44, 15% between 45 and 55, and 12% over 55. Participants could have corrected-to-normal vision but could not wear eyeglasses to maintain eye tracking integrity. Upon completion of the task, or formally withdrawn, participants were financially compensated. Participants were asked to rate the accuracy of each presented headline on a 4-point scale, ranging from "not at all accurate" to "very accurate", enabling a simple classification threshold for gauging correct/incorrect responses. Participants were asked to only use the number keys 1–4 to respond in order to reduce head movement and alignment for pupil diameter calculations. Demographic information was collected in a pre-assessment survey while behavioral constructs were collected in a post-assessment survey. Seven subjects were removed from the analysis because of synchronization issues. Therefore, a final sample of 23 was used to run the analysis.

The observation computer was fitted with a non-invasive Tobii Pro X2-60 screen-based eye tracker. The device was recording at a 60 Hz sampling rate, providing a diameter calculation for both the left and right eyes. Tobii Pro Lab software (Version) was utilized for all eye metric processing, including a preprocessing calibration method to ensure participants were sitting correctly in front of the screen and that the software was able to locate both eyes. The presentation screen was a 24-inch LED monitor with a 1920 by 1080 resolution. Observation room lighting and seating distance from the eye tracker was consistent across all participants. Additionally, the headline presentation was formatted to be centered on the screen to reduce head movement. There was no programmed input delay between headline presentation and the participants ability to rate the headline. In postprocessing, mean pupil dilation calculations were executed for each reference screen, recording the average diameter of each eye over the 2 s duration. Mean pupil dilation calculations for headline response were recorded within the timeline of presentation onset to participant input response. Pupil diameter samplings that exceeded three standard deviations of the participants total session average were removed from calculations to minimize outlier observations impacting response averages. Additionally, samplings in which the eye

Table 1 Performance means

Headline	Control	Neutral	Political
Mean	91.15%	61.92%	58.46%
SD	0.07	0.11	0.08

Table 2 Descriptive statistics of pupil dilation

Headline	Control	Neutral	Political
Mean % (pupil dilation)	7.94%	8.63%	9.23%
SD	3.18	3.76	3.82

tracker was unable to detect one or both eyes, including blinks, were removed from the mean calculations for both the reference screen and headline response. As suggested by [11], we calculated the relative change of pupil diameter compared to the last reference screen.

4 Results

Our results show that the participant had a much higher performance on the control headlines compared to the Neutral and Political headlines (See Table 1).

To test H1, we conducted Analysis of variance for repeated measures in SPSS to test whether pupil dilation is different among the headline types (Table 2).

The results show there is a difference among the three types of headlines ($F = 9.82$, $p < 0.01$), supporting H1. Furthermore, pairwise comparisons show that pupil dilation is difference between control and neutral ($F = 3.84$, $p < 0.05$), control and political ($F = 9.82$, $p < 0.001$), and neutral and political ($F = 3.02$, $p < 0.05$).

To analyze H2, we used panel regression in STATA to see whether pupil dilation predicts users' performance in identifying fake headlines. The results demonstrate a positive association between pupil dilation and users' performance ($b = 1.01$, $P < 0.001$), providing support for H2.

5 Discussion and Concluding Remarks

We designed an experiment to investigate whether pupil dilation, as a proxy for using cognitive resources by users, can be linked to their performance in identifying fake news headlines on social media. Our results, in line with the heuristic-systematic model of information processing, suggest that when users put cognitive effort in analyzing news headlines, they are more likely to discern true from fake news. Other research in this area have found similar results but using measures other than pupil dilation. For instance, Bago, Rand [21] found that deliberation, regardless of whether the news headline confirms or rejects users' ideology.

Implications regarding future work and research direction can be divided into two primary categories, model validation with respect to the underlying sample and collection procedure. As well as refinement and extension of the media presentation and the modes through which participants interact with news headlines/articles. Pupillometry is one dimension within the classification of cognitive load, however other metrics including Electroencephalogram (EEG), Electrocardiogram (EKG), and skin conductance can also be assessed to provide a measure of cognitive load. Extending the data collection procedure to include these additional metrics would provide a more holistic view of cognitive load within the experiment. Additionally, the media presentation method could be modified. The format was designed to replicate the Facebook timeline; however, consumption of news media is much more dynamic, and the user has more agency in their decision making regarding what they consume. Integrating the rating system into an experiment that allows participants choice in the progression of news consumption may provide greater insight and a more tangible view of cognitive load within a more realistic consumption environment.

References

1. Vraga, E. K., & Bode, L. (2017). Leveraging institutions, educators, and networks to correct misinformation: A commentary on Lewandosky, Ecker, and Cook.
2. Scheufele, D. A., & Krause, N. M. (2019). Science audiences, misinformation, and fake news. *Proceedings of the National Academy of Sciences, 116*(16), 7662–7669.
3. Persily, N. (2017). The 2016 US election: Can democracy survive the internet? *Journal of Democracy, 28*(2), 63–76.
4. Lazer, D. M., et al. (2018). The science of fake news. *Science, 359*(6380), 1094–1096.
5. Kim, A., Moravec, P. L., & Dennis, A. R. (2019). Combating fake news on social media with source ratings: The effects of user and expert reputation ratings. *Journal of Management Information Systems, 36*(3), 931–968.
6. Moravec, P. L., Kim, A., & Dennis, A. R. (2020). Appealing to sense and sensibility: System 1 and system 2 interventions for fake news on social media. *Information Systems Research, 31*(3), 987–1006.
7. Kim, A., & Dennis, A. R. (2019). Says who? The effects of presentation format and source rating on fake news in social media. *MIS Quarterly, 43*(3), 1025–1039.
8. Moravec, P., Minas, R., & Dennis, A. R. (2018). Fake news on social media: People believe what they want to believe when it makes no sense at all. *Kelley School of Business research paper*, (18-87).
9. Kahneman, D. (2011). *Thinking, fast and slow*. Macmillan.
10. DeStefano, D., & LeFevre, J.-A. (2007). Cognitive load in hypertext reading: A review. *Computers in Human Behavior, 23*(3), 1616–1641.
11. Gwizdka, J. (2021). "Overloading" cognitive (work) load: What are we really measuring? In *NeuroIS retreat*. Springer.
12. Chou, W.-Y.S., et al. (2020). How do social media users process cancer prevention messages on Facebook? An eye-tracking study. *Patient Education and Counseling, 103*(6), 1161–1167.
13. Lutz, B., et al. (2020). Identifying linguistic cues of fake news associated with cognitive and affective processing: Evidence from NeuroIS. In *NeuroIS retreat*. Springer.

14. Lutz, B., et al. (2020). Affective information processing of fake news: Evidence from NeuroIS. *Information systems and neuroscience* (pp. 121–128). Springer.
15. Ebnali, M., & Kian, C. (2019). Nudge users to healthier decisions: A design approach to encounter misinformation in health forums. In *International Conference on Applied Human Factors and Ergonomics*. Springer.
16. Figl, K., et al. (2019). *Fake news flags, cognitive dissonance, and the believability of social media posts.*
17. Hansen, C., et al. (2020). Factuality checking in news headlines with eye tracking. In *Proceedings of the 43rd International ACM SIGIR Conference on Research and Development in Information Retrieval.*
18. Chen, S., Duckworth, K., & Chaiken, S. (1999). Motivated heuristic and systematic processing. *Psychological Inquiry, 10*(1), 44–49.
19. Tandoc, E. C., et al. (2021). Falling for fake news: The role of political bias and cognitive ability. *Asian Journal of Communication, 31*(4), 237–253.
20. Pennycook, G., & Rand, D. G. (2019). Lazy, not biased: Susceptibility to partisan fake news is better explained by lack of reasoning than by motivated reasoning. *Cognition, 188*, 39–50.
21. Bago, B., Rand, D. G., & Pennycook, G. (2020). Fake news, fast and slow: Deliberation reduces belief in false (but not true) news headlines. *Journal of experimental psychology: General, 149*(8), 1608.
22. Siegle, G. J., Ichikawa, N., & Steinhauer, S. (2008). Blink before and after you think: Blinks occur prior to and following cognitive load indexed by pupillary responses. *Psychophysiology, 45*(5), 679–687.
23. Klingner, J., Tversky, B., & Hanrahan, P. (2011). Effects of visual and verbal presentation on cognitive load in vigilance, memory, and arithmetic tasks. *Psychophysiology, 48*(3), 323–332.
24. Perkhofer, L., & Lehner, O. (2019). Using gaze behavior to measure cognitive load. *Information systems and neuroscience* (pp. 73–83). Springer.
25. Sénécal, S., et al. (2018). How product decision characteristics interact to influence cognitive load: An exploratory study. *Information systems and neuroscience* (pp. 55–63). Springer.
26. Giroux, F., et al. (2020). Hedonic multitasking: The effects of instrumental subtitles during video watching. In *NeuroIS Retreat*. Springer.
27. Ladeira, W. J., et al. (2021). Visual cognition of fake news: the effects of consumer brand engagement. *Journal of Marketing Communications*, 1–21.
28. Sümer, Ö., et al. (2021). FakeNewsPerception: An eye movement dataset on the perceived believability of news stories. *Data in Brief, 35*, 106909.

Increased Audiovisual Immersion Associated with Mirror Neuron System Enhancement Following High Fidelity Vibrokinetic Stimulation

Kajamathy Subramaniam, Jared Boasen, Félix Giroux, Sylvain Sénécal, Pierre-Majorique Léger, and Michel Paquette

Abstract Haptic technologies are widely used in multimedia entertainment to increase the immersiveness of the experience. Studies regarding the psychological effects of haptics during audiovisual (AV) entertainment support this notion. However, the neurophysiological mechanism by which haptics increase AV immersion remains unclarified. Using between groups exploratory comparisons of whole-brain source-localized electroencephalographic theta (5–7 Hz), alpha (8–12 Hz), and beta (15–29 Hz) band activity, the present study analyzed the effect of high fidelity vibrokinetic (HFVK) stimulation on cortical brain activity and self-perceived immersion during the viewing of cinematic AV stimuli. Our results revealed that HFVK increased immersiveness potentially via enhanced top-down processing within sensorimotor areas in the mirror neuron system.

Keywords Immersion · Haptic · Somatosensory · Audiovisual · Mirror neuro system · EEG

1 Introduction

Haptic technology applies the use of tactile, vibrations and motion on users to stimulate the somatosensory and vestibular system [1]. One type of haptics called, high-fidelity vibrokinetics (HFVK), has been increasingly used in audiovisual (AV) multimedia entertainment such as cinema and video games to enhance the immersiveness of the experience, a notion that is supported by numerous psychological studies [2–7]. However, while emerging neurophysiological evidence suggests that

K. Subramaniam · J. Boasen (✉) · F. Giroux · S. Sénécal · P.-M. Léger
Tech3Lab, Department of Information Technology, HEC Montréal, Montréal, Canada
e-mail: jared.boasen@hec.ca

J. Boasen
Faculty of Health Sciences, Hokkaido University, Sapporo, Japan

M. Paquette
D-BOX Technologies Inc., Longueuil, Canada

© The Author(s), under exclusive license to Springer Nature Switzerland AG 2022
F. D. Davis et al. (eds.), *Information Systems and Neuroscience*,
Lecture Notes in Information Systems and Organisation 58,
https://doi.org/10.1007/978-3-031-13064-9_8

HFVK can induce sustained changes in cortical activation and connectivity during AV multimedia experiences [8–10], the neurophysiological mechanism by which HFVK moderates immersion remains almost wholly unexplored. Clarifying this mechanism would permit more effective design of HFVK stimuli not only for entertainment, but also for generalized information systems (IS) use, thereby making this topic of interest to the NeuroIS community.

Immersion is a psychological construct strongly linked to attention [11, 12], and hence HFVK may moderate immersion via attentional mechanisms. Externally-directed attention, which would be required for bottom-up processing of an AV multimedia stimulus, is classically associated with occipitoparietal event-related alpha-band (8–12 Hz) and beta-band desynchronization [13]. Internally-directed attention, such as that engaged during top-down processing of imagery or mediation, is associated with synchronization of alpha and beta-band activity [14]. Decreased sustained attention towards task-related stimuli has been associated with increased occipital theta (4–7 Hz) and decreased occipital and temporal spectral beta (15–30 Hz) power preceding the response [15].

HFVK stimuli may also moderate immersion via enhanced sensory processing [1]. Theta (4–7 Hz) rhythms are thought to oscillate at speeds that are similar to the information rates of natural stimuli. Event-related synchronization (ERS) of theta oscillatory activity in auditory areas has frequently been observed in response to rhythmic elements of auditory/speech stimuli [16, 17]. ERS of occipital theta activity is also reported following presentation of visual stimuli and is thought to reflect encoding of visual information into memory [18]. Theta activity is also thought to be important for vestibular-related spatial processing such as during virtual navigation [19]. Meanwhile, somatosensory stimulation and sensorimotor integration processing, particularly during perception of actions, are widely observed to trigger what is known as rolandic alpha (~10 Hz) and beta (~20 Hz) responses in the motor and somatosensory (S1) cortices [20]. The phenomenon has been attributed to activation in what is known as the mirror neuron system (MNS) [21].

To clarify the neurophysiological mechanism underlying how HFVK stimulation affects the immersiveness of AV stimulus experiences, the present (ongoing) study recorded EEGs in participants during a cinematic viewing experience and compared cortical activity and self-perceived measures of immersion between participants who were stimulated with HFVK and those who were not. We used a whole-brain exploratory approach targeting brain activity in the theta (5–7 Hz), alpha (8–12 Hz) and beta (15–29 Hz) frequency bands. We hypothesized that HFVK stimulation would increase self-perceived immersion, and that the underlying neurophysiological mechanism associated with this increase would be enhanced sensory or sensory integration processing, and/or increased task-directed attentional processing.

2 Methods and Results

2.1 Participants

Thus far, 16 healthy right-handed participants have been recruited to participate in this ongoing study, and randomly assigned into either a group that received HFVK stimulation (HFVK group; F: 5, M: 2; mean age: 24.4 ± 6.0 years), and a group that did not (Control group; F:6, M:3; mean age: 27.6 ± 5.1 years). The subjects were recruited via social media and our institution's research participant recruitment panel. Exclusion criteria were prior HFVK experience, and more than two times prior viewing experience of any one of the AV stimuli. The study has been ethically approved by our institutional review board and written informed consent was obtained from all participants prior to the experiment.

2.2 Stimuli

Fifteen clips from eight different mainstream movies with commercially available HFVK stimuli (D-BOX Technologies Inc., Longueuil, Canada) served as experimental stimuli. The clips ranged from 103 to 288 s in length. Each clip conveyed a standalone narrative of a single or multiple protagonists. Approximately 20 s at the beginning and ending of each clip were HFVK-free and were respectively used as baseline and post-stimulus periods of brain activity in EEG analyses. The middle part of the clips was presented either with or without HFVK stimulation in accordance with the group assignment of the participant. The clips were played via a proprietary media player (D-BOX Technologies Inc., Longueuil, Canada) run on a Windows 7 laptop. Video was displayed on a 70 × 120 cm high-definition Samsung TV, and audio was amplified on a Pioneer VSX-324 AV receiver and played in 5.1 surround sound on Pioneer S-11A-P speakers. The HFVK stimuli were manifested on actuators embedded in a cushioned recliner seat [1, 6]. Aside from one clip, all HFVK stimuli in the present study were designed to coincide with physical forces acting upon the protagonist/s which were implied by the visual or ambient environmental elements of the viewed scene. For example, the swaying of waves, the vibration of a vehicle on the road, or the movements of dancing, etc. The presentation onset of all clips was precisely synchronized to the EEG recording using the method described in [3].

2.3 Procedure

A 32-electrode, gel EEG system (actiCAP, BrainProducts, GmbH, Munich, Germany) was installed according to the 32ch Standard Cap layout for actiCAP.

Electrode positions were digitized using CapTrak (BrainProducts, GmbH, Munich, Germany). Participants were then seated in the HFVK seat, which was installed in a blackened room, and instructed to sit as still as possible facing the TV during clip playback. Participants watched the clips in the same order, and answered a validated immersion questionnaire [11] after each of them. The experiment lasted around 2 h.

2.4   EEG Recordings and Pre-processing

EEGs were recorded raw at a 500 Hz sampling rate using BrainVision recorder (company info), and processed using Brainstorm on MATLAB 20a (MathWorks, Natick, MA, USA). Raw signals were bandpass filtered (1–40 Hz) and cleaned using independent component analysis. Baseline and post-stimulus periods of each clip were manually marked with events at three-second intervals, and then epoched at − 1000 to 4000 ms relative to event onset. EEG electrode positions and fiducials were co-registered to a template brain. Minimum-norm estimation was used to calculate cortical currents without dipole orientation constraints. Cortical activity was time–frequency decomposed at the voxel level into theta (5–7 Hz), alpha (8–12 Hz), and beta (15–29 Hz) bands and envelopes calculated via Hilbert transform. Cortical activity envelopes in each frequency band were then averaged across epochs, and then over the period of 0–3000 ms for the baseline and post-stimulus periods separately for each participant. Finally, mean post-stimulus cortical activity was normalized as a percent deviation of mean baseline cortical activity in each participant separately and used in our statistical analysis.

2.5   Statistical Analyses

The difference in self-perceived immersion between the HFVK and control groups was tested via an independent t-test using SPSS 26 (IBM, Armonk, NY, USA) based on the mean total immersion scores. Differences in normalized post-stimulus brain activity between the HFVK and Control groups were compared using permutation cluster analysis across all three frequency bands using FieldTrip in Brainstorm. The significance threshold for all statistical tests was set at $p \leq 0.05$.

3   Results

Mean immersion scores in the HFVK group (N = 7) were significantly higher than those of the control group (N = 9) according to the independent t test (mean ± SEM: 29.95 ± 2.08 vs. 38.18 ± 0.74, respectively; $p = 0.004$). As for the permutation cluster analysis of normalized post-stimulus brain activity, the control group (N = 8,

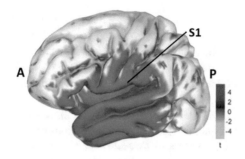

Fig. 1 Cortical t map of beta (15–29 Hz) activation illustrating areas of higher activity centered in the left hemisphere around S1 for the HFVK group (N = 7) vs. control group (N = 8). A: anterior, P: posterior, S1: somatosensory cortex, HFVK: high fidelity vibrokinetics

with one participant excluded due to excessive noise) was significantly different from the HFVK group (N = 7) ($p = 0.043$), with the largest differences broadly exhibited in the left hemisphere around S1 where beta activity was higher in the HFVK group compared to the Control group (see Fig. 1).

4 Discussion

The present study explored the neurophysiological mechanism underlying how HFVK stimulation in synchrony with AV multimedia stimulation increases self-perceived immersion. The study revealed significantly higher levels of self-perceived immersion in the HFVK group compared the control group, corroborating prior reports [15]. This result was furthermore associated with significant differences in cortical brain activity between groups according to our permutation cluster analysis.

The significance of the differences of a cluster analysis cannot be localized. However, that the largest differences were centered around S1 suggests that, rather than attentional processing, the HFVK may have moderated immersion via enhancement of the MNS. However, beta activity was higher in the HFVK than the control group, whereas MNS related rolandic beta responses are often observed as desynchronization [13, 22]. A potential explanation for this paradox is offered by [21, 22] who revealed that prior experience with somatosensory stimulation during an action translated to increased rolandic beta activity during later observation of an action that had an expected somatosensory component compared to an action which did not. Based on this evidence, the comparatively higher rolandic beta activity exhibited by the HFVK group over the control group may indicate that HFVK stimulation induced participants in the HFVK group to imagine internal representations of somatosensory and vestibular stimuli even during the post-HFVK stimulus periods of the clips. This would be congruent with observations that beta ERS is stronger than ERD during internal ideation or imagery [23]. In other words, the participants in the HFVK group may have comparatively been more strongly imagining themselves inside the cinematic worlds they were viewing, implying a mechanism HFVK-driven bottom-up enhancement via the MNS, leading to induced enhancement of top-down sensorimotor processing even in the absence of HFVK. However, this explanation alone does not explain why beta differences were so strongly left-hemispheric, whereas the general consensus is that the MNS is bilateral.

One possible explanation for the left laterality of our results is the involvement of language. Almost all of the post-HFVK stimulus periods of the movie clips featured dialog between the main protagonist/s and other characters. Left lateralization of language processing function and rolandic alpha/beta activity is frequently observed in right-handed people, which our subjects all were [24, 25]. In their review on speech perception, [26] postulated that left auditory and motor processing areas are recruited during temporally predictable auditory processing (i.e. auditory processing that involves a visual component) via cross-frequency coupling between delta/theta responses originating in sensory areas and rolandic beta. Thus, a logical next step in our study would be to compare delta/theta to beta cross-frequency coupling between the HFVK and the control groups to verify if this idea is supported. Presuming that it is, our results may well have been a sign that the sustained enhancement of MNS activation due to induced somatosensory and vestibular imagery was dominated by language pathway processing.

There are several limitations that should be acknowledged with the present study. First, this is a work in progress, and the sample size is low. Additionally, although permutation cluster analysis is a robust and appropriate neurophysiological statistical test, it does not statistically validate the beta activity differences we have discussed here.

In conclusion, the results of the present study support the ability of HFVK stimulation to increase the immersiveness of AV experiences. Moreover, the result of cortical brain activity analyses suggests a possible mechanism for this increased immersiveness involving sustained enhancement of top-down sensorimotor processing in areas of the MNS. The result may be driven by increased intensity of internal representations of somatosensory and vestibular stimuli in conjunction with the actions and dialog of the character featured in the experienced AV stimuli. Simply put, HFVK stimulation may have induced participants to more strongly imagine themselves within the scene, which is precisely the goal of HFVK designers. Overall, the results point to the success of an HFVK design strategy targeting elements of the scene that the protagonist might see or feel, thereby creating a sensory bridge between the viewer and the protagonist. Going forward, we hope that these results will drive further development of HFVK stimuli for generalized IS use.

Acknowledgements The authors would like to thank David Brieugne, Emma Rucco and all the technical staff at the Tech3Lab for their effort and assistance in executing this study. This study was financially supported by NSERC and Prompt (IRCPJ/514835-16, and 61_Léger-Deloitte 2016.12, respectively).

References

1. Petersen, C. G. (2019). The address of the ass: d-box motion code, personalized surround sound, and focalized immersive spectatorship. *Journal of Film and Video, 71,* 3–19.
2. Giroux, F., Boasen, J., Sénécal, S., Fredette, M., Tchanou, A. Q., Ménard, J. F., Léger, P. M., et al. (2019). Haptic stimulation with high fidelity vibro-kinetic technology psychophysiologically enhances seated active music listening experience. In *2019 IEEE World Haptics Conference* (pp. 151–156). IEEE.
3. Boasen, J., Giroux, F., Duchesneau, M. O., Senecal, S., Leger, P. M., & Menard, J. F. (2020). High-fidelity vibrokinetic stimulation induces sustained changes in intercortical coherence during a cinematic experience. *Journal of Neural Engineering, 17.*
4. Gardé, A., Léger, P. M., Sénécal, S., Fredette, M., Chen, S. L., Labonté-Lemoyne, É., & Ménard, J. F. (2018). Virtual reality: Impact of vibro-kinetic technology on immersion and psychophysiological state in passive seated vehicular movement. In *International Conference on Human Haptic Sensing and Touch Enabled Computer Applications* (pp. 264–275). Springer.
5. Gardé, A., Léger, P. M., Sénécal, S., Fredette, M., Labonté-Lemoyne, E., Courtemanche, F., & Ménard, J. F. (2018). The effects of a vibro-kinetic multi-sensory experience in passive seated vehicular movement in a virtual reality context. In *Extended Abstracts of the 2018 CHI Conference on Human Factors in Computing Systems* (pp. 1–6).
6. Pauna, H., Léger, P. M., Sénécal, S., Fredette, M., Courtemanche, F., Chen, S. L., Ménard, J. F., et al. (2017). The psychophysiological effect of a vibro-kinetic movie experience: The case of the D-BOX movie seat. *Information Systems and Neuroscience: Lecture Notes in Information Systems and Organisation, 25,* 1–7.
7. Waltl, M., Timmerer, C. & Hellwagner, H. (2010). Improving the quality of multimedia experience through sensory effects. In *2010 2nd International Workshop on Quality of Multimedia Experience QoMEX 2010—Proceedings* (pp. 124–129).
8. Jackman, A. H. (2015). 3-D cinema: immersive media technology. *GeoJournal, Neuroscience, Psychology, and Economics,* 18–33.
9. Kang, D., Kim, J., Jang, D.-P., Cho, Y. S., & Kim, S.-P. (2015). Investigation of engagement of viewers in movie trailers using electroencephalography. *Brain-Computer Interfaces,* 193–201.
10. Fornerino, M., Helme-Guizon, A., & Gotteland, D. (2008). Movie consumption experience and immersion: impact on satisfaction. *Recherche et Applications en Marketing,* 93–110.
11. Jennett, C., Cox, A. L., Cairns, P., Dhoparee, S., Epps, A., Tijs, T., & Walton, A. (2008). Measuring and defining the experience of immersion in games. *International Journal of Human-Computer Studies,* 641–661.
12. Burns, C. G., & Fairclough, S. H. (2015). Use of auditory event-related potentials to measure immersion during a computer game. *International Journal of Human-Computer Studies,* 107–114.
13. Friese, U., Daume, J., Göschl, F., König, P., Wang, P., & Engel, A. K. (2016). Oscillatory brain activity during multisensory attention reflects activation, disinhibition, and cognitive control. *Scientific Reports, 6,* 32775.
14. Fink, A., Graif, B., & Neubauer, A. C. (2009). Brain correlates underlying creative thinking: EEG alpha activity in professional vs. novice dancers. *NeuroImage,* 854–862.
15. Ko, L. W., Komarov, O., Hairston, W. D., Jung, T. P., & Lin, C. T. (2017). Sustained attention in real classroom settings: An EEG study. *Frontiers in Human Neuroscience,* 1–10.
16. Luo, H., & Poeppel, D. (2012). Cortical oscillations in auditory perception and speech: Evidence for two temporal windows in human auditory cortex. *Frontiers in Psychology, 3,* 170–170.
17. Hyafil, A., Fontolan, L., Kabdebon, C., Gutkin, B., & Giraud, A. L. (2015). Speech encoding by coupled cortical theta and gamma oscillations. *eLife, 4,* 1–45.
18. Klimesch, W., Fellinger, R., & Freunberger, R. (2011). Alpha oscillations and early stages of visual encoding. *Frontiers in Psychology, 2.*
19. Kober, S. E., & Neuper, C. (2011). Sex differences in human EEG theta oscillations during spatial navigation in virtual reality. *International Journal of Psychophysiology, 79,* 347–355.

20. Caetano, G., Jousmäki, V., & Hari, R. (2007). Actor's and observer's primary motor cortices stabilize similarly after seen or heard motor actions. *Proceedings of the National Academy of Sciences, 9058–9062.*

21. Rizzolatti, G., & Craighero, L. (2004). The mirror-neuron system. *Annual Review of Neuroscience, 169–192.*

22. Quandta, L. C., Marshall, P., Bouquet, C. A., & Shipley, T. F. (2013). Somatosensory experiences with action modulate alpha and beta power during subsequent action observation. *Brain Research,* 55–65.

23. Tariq, M., Trivailo, P. M., & Simic, M. (2020). Mu-Beta event-related (de)synchronization and EEG classification of left-right foot dorsiflexion kinaesthetic motor imagery for BCI. *PLoS ONE,* 1–20.

24. Caplan, R., & Dapretto, M. (2001). Making sense during conversation: An fMRI study. *NeuroReport, 12,* 3625–3632.

25. Hickok, G., & Poeppel, D. (2004). Dorsal and ventral streams: A framework for understanding aspects of the functional anatomy of language. *Cognition, 92,* 67–99.

26. Biau, E., & Kotz, S. A. (2018). Lower beta: A central coordinator of temporal prediction in multimodal speech. *Frontiers in Human Neuroscience, 12,* 1–12.

Recognizing Polychronic-Monochronic Tendency of Individuals Using Eye Tracking and Machine Learning

Simon Barth, Moritz Langner, Peyman Toreini, and Alexander Maedche

Abstract Eye tracking technology is a NeuroIS tool that provides non-invasive and rich information about cognitive processes. Recently, it has been demonstrated that eye movement analysis using machine learning algorithms also represents a promising approach to recognize user characteristics and states as a foundation for designing neuro-adaptive information systems. Polychronicity, an individual's attitude towards multitasking work, is a user characteristic tightly related to cognitive processes and therefore a potential candidate to be recognized with eye tracking technology. However, existing research to the best of our knowledge did not yet investigate automatic recognition of the user's polychronic-monochronic tendency. In this study, we leverage eye movement data analysis and machine learning to recognize the user's level of polychronicity. In a lab experiment, eye tracking data of 48 participants was collected and subsequently the users's polychronic-monochronic tendency was predicted.

Keywords Polychronicity · Machine Learning · Eye Tracking

S. Barth · M. Langner (✉) · P. Toreini · A. Maedche
Karlsruhe Institute of Technology (KIT), Institute of Information Systems and Marketing (IISM),
Karlsruhe, Germany
e-mail: moritz.langner@kit.edu

S. Barth
e-mail: simon.barth@student.kit.edu

P. Toreini
e-mail: peyman.toreini@kit.edu

A. Maedche
e-mail: alexander.maedche@kit.edu

© The Author(s), under exclusive license to Springer Nature Switzerland AG 2022
F. D. Davis et al. (eds.), *Information Systems and Neuroscience*,
Lecture Notes in Information Systems and Organisation 58,
https://doi.org/10.1007/978-3-031-13064-9_9

1 Introduction

Designing neuro-adaptive systems is one of the four grand research areas in NeuroIS [1]. Neuro-adaptive systems recognize user characteristics as well as states and, on this basis, adapt themselves to the user's current needs. The recognition of user characteristics and states leverages biological, psychophysiological, or behavioral signals collected by NeuroIS tools, like EEG, ECG and eye tracking [2]. Subsequently, machine learning (ML) algorithms are applied to train classifiers to recognize user's characteristics and states based on the recorded data. Recently, researchers leveraged various signals and ML algorithms to successfully recognize a diverse set of user characteristics such as personality, working memory capacity as well as user states such as emotions, mental workload, flow [3–7].

A user characteristic highly relevant in today's attention economy is the polychronic-monochronic tendency. Individuals' polychronic-monochronic tendency refers to their attitude towards multitasking work and being involved in two or more activities during the same block of time [8]. When a person engages in two or more activities during the same block of time, it is considered as polychronicity, while when a person engages in one activity at a time, it is considered as monochronicity [8]. Previous studies have shown that the fit between multitasking requirements in jobs and the polychronic or monochronic tendency of personnel could potentially lead to higher job satisfaction, reduced stress, better performance, etc. [9, 10]. Rather polychronic individuals were also reported to perceive multiple team memberships as less emotionally exhausting, experiencing higher role efficacy without increased role stress [11]. Moreover, users with a higher polychronicity tend to use dual screens [12]. Polychronic-monochronic tendency is tightly related to cognitive processes and therefore a potential candidate to be recognized on the basis of eye tracking technology [13]. Previous research has indicated that polychronicity tendency affects switching patterns or gaze duration while users were multitasking [14]. Multitasking and attention shifts are considered as the core of polychronicity while the attention allocation is related to users' eye-movements following the eye-mind hypothesis by [15]. However, to the best of our knowledge, no study has so far investigated the recognition of the user's polychronic-monochronic tendency via eye movement data and machine learning algorithms. An automatic recognition of polychronic-monochronic tendency would allow to design neuro-adaptive systems that adapt themselves to the user's characteristics. Recently, the NeuroIS community called for the usage of eye-tracking devices to design intelligent systems [16, 17] and integrate it with ML [18]. Therefore, in this study, we investigate the potential to recognize the user's polychronic-monochronic tendency by using eye tracking technology in combination with supervised ML algorithms. In this study, we aim to answer the following research question:

> *RQ: How to recognize the user's polychronic-monochronic tendency based on eye-movement data in combination with supervised machine learning algorithms?*

In this paper, we present the results of a comparison of four supervised ML algorithms and Linear Regression that use labeled eye-tracking data to determine the user's polychronic-monochronic tendency while conducting an information processing task. We collected eye-movements of 48 users that were exploring four simplified information dashboard (in total 48 * 4 samples). Moreover, we captured the user's polychronic-monochronic tendency by collecting survey-based data using the polychronicity-monochronicity tendency scale (PMTS) [8]. The goal of this research is to analyze if the polychronic-monochronic tendency can be recognized based on this training data. Overall, we extracted 58 features from the collected data and investigated the F1-Score of established algorithms (kNN, Random Forests, Support Vector Machines, Multi-Layer Perceptron and Linear Regression) in terms of their ability to predict the users' polychronic-monochronic tendency. Our findings show that eye tracking data can indeed be leveraged to predict polychronic-monochronic tendency. Random Forest classifiers provided on average the best results (F1-Score = 0,693) for predicting polychronic-monochronic tendency.

2 Methodology

To investigate the research question of this study, we used the data collected within another controlled laboratory experiment. We recorded data from 48 participants (22 females, 26 males) recruited from the university laboratory pool with an average age of 22.72 (SD = 2.24) years. Participants received four different information dashboards for 30 s each to perform an information processing task. The task duration of 30 s is mapped to the goal of this study. Between the information processing tasks, they were busy with other interrupting tasks that were related to the goal of the conducted experiment. Therefore, there was a time gap before receiving the next dashboard. The graphs on the four dashboards had different content (sales, marketing, customer service, and human resource) but the same design. During the dashboard exploration, the eye movement data of participants was collected via a Tobii Eye Tracker 4C. As each participant investigated four different information dashboards, in total, we could successfully collect 192 (4 * 48) eye-movement samples that were used for recognizing polychronic-monochronic tendency.

After the experiment, the participants level of polychronic-monochronic tendency was captured through the Polychronic-Monochronic Tendency Scale (PMTS) [8]. The PMTS consists of 5 single questions which are considered as the underlying indicators for polychronic-monochronic behavior. These questions are: (1) I prefer to do two or more activities at the same time. (2) I typically do two or more activities at the same time. (3) Doing two or more activities at the same time is the most efficient way to use my time. (4) I am comfortable doing more than one activity at the same time. (5) I like to juggle two or more activities at the same time. Answers were given by participants on a 7-point Likert scale with 1 meaning "I strongly disagree" and 7 corresponding to "I strongly agree". In the following, polychronic and monochronic individuals were classified based on their self-reported average score of the PMTS. This means the average of all answers on the PMTS is calculated for each participant and compared to neutral position on the PMTS which is 4 points on the 7-point

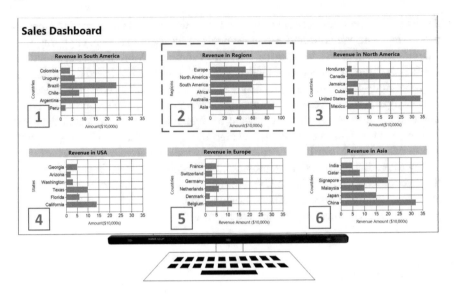

Fig. 1 Example of an information dashboard

Likert scale. With this approach we applied a binary classification as users with an average higher than 4 are considered as polychrons and lower than 4 as monochrons. In total 21 participants were classified as polychronic whereas 27 participants were considered as monochronic based on their self-reported data.

The explored dashboards used in this study were simple static dashboards that were designed to control elements that influence stimulus-driven attention, such as color-coding, size, chart type, and were used in previous IS studies [19, 20]. An example of such a dashboard can be seen in Fig. 1. The dashboard comprises six bar graphs, which are defined as an AOI (shown with blue square). Furthermore, each graph has six chunks of information, and all elements are in grey. Therefore, the six AOIs of the dashboard have the same complexity regarding the design. The participants received these dashboards in random order during the experiment.

3 Results

First, we calculated various eye tracking features out of the recorded data such as block duration, number of fixations, fixation {duration, rate}, number of saccades, saccade {amplitude, velocity, rate, absolute angle, relative angle}, saccade-fixation ratio, fixation-saccade ratio and pupil diameter. As done by [21], we extended the eye tracking features with statistical indicators for a better comparison. This includes the minima, the maxima, the means, the medians, the standard deviation, the skew, the kurtosis, and the range of each users' eye movement (not for counts, rates and bock

Table 1 Mean F1-scores of classifiers

	kNN	RF	MLP	SVM	LR
Mean F1-score	0.683	**0.693**	0.677	0.681	0.681

duration). This extended the total amount of features to 56 eye tracking features plus gender and age. These features were then reduced to a maximum amount of 15 based on relative importance to each classifier via the SelectFromModel method provided by scikit-learn. Participants were classified binary as either polychronic or monochronic.

In order to recognize the user's polychronic-monochronic tendency, we leveraged the python package scikit-learn including model selection and evaluation [22] and five supervised ML algorithms [23]. The supervised ML algorithms k-Nearest Neighbor (kNN), Random Forest (RF), Multilayer Perceptron (MLP), Support Vector Machine (SVM) and Linear Regression (LR) were trained to predict the polychronic-monochronic tendency of participants based on a set of max 15 features from the extracted 58 features. Hyperparameter tuning, where applicable, was conducted on a trial-and-error basis using tenfold grid search cross-validation. Table 1 shows the mean F1 scores of the best performing parameters of each classifier. Random Forest (F1-Score = 0,693) algorithms performed on average the best in predicting polychronic-monochronic tendency of participants.

4 Discussion

The results of our study demonstrate the potential of eye-tracking technology and supervised ML algorithms to recognize users' polychronic-monochronic tendency. Our results show that Random Forest classifiers have the potential to support this hypothesis. Although the Random Forest classifiers have a higher accuracy than other tested classifiers, they perform not significantly better. Moreover, this supports our hypothesis that eye tracking data is valuable to predict polychronic-monochronic tendency.

However, the findings from this study are limited from several perspectives that can be improved in the future. First, the users' eye movements are highly relevant to the task. In this study, we focused on information processing with visualized information. Therefore, there is a need to test how the task and also the timeframe affect prediction accuracy. Furthermore, the information dashboard we used in this study does not include any color-coding; previous studies showed that different visual stimuli may have influenced users' eye movements; therefore, we need to test our model's reliability and the prediction accuracy with a more realistic information dashboard design. Besides, we tested users' eye movements for a short period (30 s) as we aimed to extract users' characteristics in a short time. However, the users' polychronic-monochronic tendency might be more accurately recognized by data of an extended period rather than a short period. Therefore, the ML algorithm performance is needed to be investigated based on longer task durations and the eye-movements. Moreover, the polychronic-monochronic tendency was not balanced across the sample of this study.

Furthermore, as the usage of eye tracking technology is not limited to the desktop, we assume that these findings can motivate the application of other types of eye trackers as well. For example, as extended reality (VR, AR, XR) researchers emphasized the usage of eye-movement data for designing intelligent services [24, 25], they can benefit from the findings of this study to design adaptive applications for these platforms based on users' polychronic-monochronic tendency.

5   Conclusion

Building neuro-adaptive systems by recognizing user characteristics and states via biosignal data is a research area with growing importance in the field of NeuroIS. In this paper, we investigated the feasibility of detecting the user's polychronic-monochronic tendency by leveraging eye tracking data and supervised machine learning. Our results provide the first insights about eye-based prediction of polychronic-monochronic tendency. Random Forest algorithms were the best performing supervised machine learning algorithms to recognize the user's polychronic-monochronic tendency.

References

1. vom Brocke, J., Hevner, A., Léger, P. M., Walla, P., & Riedl, R. (2020). Advancing a NeuroIS research agenda with four areas of societal contributions. *European Journal of Information Systems, 29*, 9–24. https://doi.org/10.1080/0960085X.2019.1708218
2. Dimoka, A., Davis, F. D., Pavlou, P. A., & Dennis, A. R. (2012). On the Use of neurophysiological tools in IS research: Developing a research agenda for NeuroIS. *MIS Quarterly, 36*, 679–702.
3. Rissler, R., Nadj, M., Li, M. X., Knierim, M. T., & Maedche, A. (2018). Got flow? Using machine learning on physiological data to classify flow. In *Conference on human factors computing systems—proceedings* 2018-April, 1–6. https://doi.org/10.1145/3170427.3188480
4. Hoppe, S., Loetscher, T., Morey, S. A., & Bulling, A. (2018). Eye movements during everyday behavior predict personality traits. *Frontiers in Human Neuroscience, 12*, 105. https://doi.org/10.3389/fnhum.2018.00105
5. Steichen, B., Carenini, G., & Conati, C. (2013). User-adaptive information visualization: Using eye gaze data to infer visualization tasks and user cognitive abilities. In *Proceedings of the 2013 international conference on intelligent user interfaces* (pp. 317–328). ACM. https://doi.org/10.1145/2449396.2449439
6. Guo, J. J., Zhou, R., Zhao, L. M., & Lu, B. L. (2019). Multimodal emotion recognition from eye image, eye movement and EEG using deep neural networks. In *Proceedings of the annual international conference of the IEEE engineering in medicine and biology society (EMBS)* (pp. 3071–3074). https://doi.org/10.1109/EMBC.2019.8856563

7. Lobo, J. L., Del Ser, J., De Simone, F., Presta, R., Collina, S., & Moravek, Z. (2016). Cognitive workload classification using eye-tracking and EEG data. In: *Proceedings of the international conference on human-computer interaction in aerospace*. Association for Computing Machinery. https://doi.org/10.1145/2950112.2964585
8. Lindquist, J. D., & Kaufman-Scarborough, C. (2007). The polychronic—monochronic tendency model. *Time Soc., 16*, 253–285. https://doi.org/10.1177/0961463X07080270
9. de Vasconcellos, V. C. (2017). Time pressure, pacing styles, and polychronicity: Implications for organizational management. In: *Organizational psychology and evidence-based management* (pp. 205–225). Springer International Publishing, Cham. https://doi.org/10.1007/978-3-319-64304-5_11
10. Kirchberg, D. M., Roe, R. A., & Van Eerde, W. (2015). Polychronicity and multitasking: A diary study at work. *Human Performance, 28*, 112–136. https://doi.org/10.1080/08959285.2014.976706
11. Berger, S., & Bruch, H. (2021). Role strain and role accumulation across multiple teams: The moderating role of employees' polychronic orientation. *Journal of Organizational Behavior, 42*, 835–850. https://doi.org/10.1002/job.2521
12. Lin, T. T. C. (2019). Why do people watch multiscreen videos and use dual screening? Investigating users' polychronicity, media multitasking motivation, and media repertoire. *International Journal of Human Computer Interaction, 35*, 1672–1680. https://doi.org/10.1080/10447318.2018.1561813
13. Kowler, E. (2011). Eye movements: The past 25 years. *Vision Research, 51*, 1457–1483.
14. Brasel, S.A., & Gips, J. (2011). Media multitasking behavior: Concurrent television and computer usage. *Cyberpsychology, Behavior and Social Network, 14*, 527–534. https://doi.org/10.1089/cyber.2010.0350
15. Just, M. A., & Carpenter, P. A. (1980). A theory of reading: From eye fixations to comprehension. *Psychological Review, 87*, 329–354.
16. Riedl, R., Davis, F. D., & Hevner, A. R. (2014). Towards a neuroIS research methodology: Intensifying the discussion on methods, tools, and measurement.
17. vom Brocke, J., Riedl, R., & Léger, P.-M. (2013). Application Strategies for Neuroscience in Information Systems Design Science Research. *The Journal of Computer Information Systems, 53*, 1–13.
18. Pfeiffer, J., Pfeiffer, T., Meißner, M., & Weiß, E. (2020). Eye-tracking-based classification of information search behavior using machine learning: Evidence from experiments in physical shops and virtual reality shopping environments. *Information Systems Research, 31*, 675–691.
19. Toreini, P., Langner, M., Maedche, A., Morana, S., & Vogel, T. (2022). Designing attentive information dashboards. *Journal of the Association for Information Systems, 22*, 521–552. https://doi.org/10.17705/1jais.00732
20. Toreini, P., & Langner, M. (2020). Designing user-adaptive information dashboards: Considering limited attention and working memory. In *27th European conference on information systems—information systems for a sharing society* (pp. 0–16). ECIS
21. Henderson, J. M., Shinkareva, S. V., Wang, J., Luke, S. G., & Olejarczyk, J. (2013). Predicting cognitive state from eye movements. *PLoS ONE, 8*, 1–6. https://doi.org/10.1371/journal.pone.0064937
22. 3. Model selection and evaluation—scikit-learn 0.21.3 documentation. https://sklearn.org/model_selection.html#model-selection-and-evaluation. Last accessed 16 May 2021.
23. 1. Supervised learning—scikit-learn 0.19.1 documentation. https://sklearn.org/supervised_learning.html#supervised-learning. Last accessed 16 May 2021.

24. Peukert, C., Lechner, J., Pfeiffer, J., & Weinhardt, C. (2020). Intelligent invocation: towards designing context-aware user assistance systems based on real-time eye tracking data analysis. In: *Lecture notes in information systems and organisation* (pp. 73–82). https://doi.org/10.1007/978-3-030-28144-1_8.
25. Peukert, C., Pfeiffer, J., Meißner, M., Pfeiffer, T., & Weinhardt, C. (2019). Shopping in virtual reality stores: The influence of immersion on system adoption. *Journal of Management Information Systems, 36*, 755–788. https://doi.org/10.1080/07421222.2019.1628889

The Effect of SVO Category on Theta/Alpha Ratio Distribution in Resource Allocation Tasks

Dor Mizrahi, Ilan Laufer, and Inon Zuckerman

Abstract Social Value Orientation (SVO) measures and quantifies the effect of players' social attitude regarding the divisions of resources and showed to be valuable in predicting the degree of cooperation across a wide variety of social dilemmas and resource allocation tasks (Prisoner's dilemma, Chicken game, Stag-hunt, Battle of the sexes). In this study, we would like to test whether people with different SVO values (specifically prosocial and individualistic) do not only exhibit different social preferences in these games, but whether they produce a different electrophysiological pattern. To this aim, we have utilized a specific electrophysiological index, the Theta/Alpha ratio, which is known to be correlated with the cognitive load. Our result demonstrates that there are indeed differences in the Theta/Alpha ratio of players with different SVO values.

Keywords EEG · Theta/alpha ratio · Social value orientation · Resource allocation

1 Introduction

The literature of the social and behavioral sciences have shown that human players take into account the consequences of their actions others when making a decision (e.g. [1–6]). This theory, that describes a person's preferences when making decisions, was later denoted as the Social Value Orientation (SVO) theory. There are several ways to measure a person's SVO value whether as a discrete measure [7, 8] or as a continuous measure [9], but in the end all measuring methods converge into the same four main human behavior profiles: (1) Individualistic—a player who is only concerned with their own outcomes (2) Competitive—a player who aspires to maximize their own outcome, but in addition also minimizes the outcome of others,

D. Mizrahi (✉) · I. Laufer · I. Zuckerman
Department of Industrial Engineering and Management, Ariel University, Ariel, Israel
e-mail: dor.mizrahi1@msmail.ariel.ac.il

I. Zuckerman
e-mail: inonzu@ariel.ac.il

© The Author(s), under exclusive license to Springer Nature Switzerland AG 2022   97
F. D. Davis et al. (eds.), *Information Systems and Neuroscience*,
Lecture Notes in Information Systems and Organisation 58,
https://doi.org/10.1007/978-3-031-13064-9_10

(3) Prosocial—a player who tends to maximize the joint outcome (4) Altruistic—who is motivated to help others at the expense of their own utility. Previous studies that have examined the distribution of electrophysiological indices in players while performing cognitive tasks showed that behavioral indices, such as the individual coordination ability [10, 11], has a significant effect on their brain activity (e.g. [12–15]). In this study we aimed to test whether there is a difference in cognitive load between players with different social value orientation category, specifically prosocial and individualistic, in the context of a resource allocation task. Cognitive load was measures by using an electrophysiological marker of cognitive load, namely, the Theta/Alpha Ratio (TAR) [16–18]. TAR is known to increase as cognitive load increases and vice-versa. Hence, in this study we utilize the TAR to find out whether the difference in SVO profile is accompanied by electrophysiological changes during execution of resource allocation tasks.

2   Materials and Methods

2.1   Experimental Setup

The study comprised the following stages. First, participants received an explanation regarding the overarching aim of the study and were given instructions regarding the experimental procedure and the interface of the application. Next, we have measured the SVO of the player using the "slider" method [9] and categorized it as "prosocial" or "individualistic". Then, each player was then presented with a resource allocation task that included 18 questions. In each question, the player was presented with two resource-sharing options which should be divided between himself and another participant unknown in the experiment (Fig. 1 shows the resource allocation application screen). The experiment is structured in such a way that one option appears on the right and the other option appears on the left. The number of points presented at the upper part of the screen is for the player and the number of points at the lower part is for the other player. The player must choose one of the options within 8 s otherwise none of the players will receive points. The position of the options on the screen (right or left) is randomly selected. The distribution of the two solutions was as follows, one option was prosocial, i.e., maximizing the joint profit of the two players, and another more individualistic option, i.e., maximizing the personal profit of the player. The prosocial option can be seen on the left side and the individualistic option on the right side. Finally, the output file with the experiment logs was uploaded also to a shared location for offline analysis.

Figure 2 portrays the outline of the experiment. The slide presenting the resource allocation was presented for a maximal duration of 8 s and the next slide appeared following a button press. The order of the 18 games was randomized in each session.

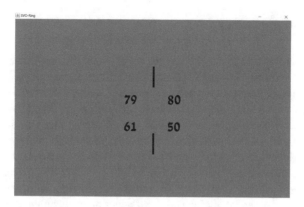

Fig. 1 Resource allocation application screen

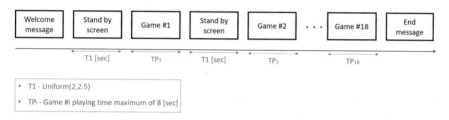

Fig. 2 Experimental paradigm with timeline

2.2 Participants

The participants were 4 students (2 prosocial and 2 individualistic) students from the university that were enrolled in one of the courses on campus (right-handed, mean age $= \sim 24$, $SD = 2$).

2.3 EEG Acquisition

The EEG Data acquisition process during the resource allocation tasks sessions was recorded by a 16-channel g.USBAMP bio signal amplifier (g.tec, Austria) at a sampling frequency of 512 Hz. 16 active electrodes were used for collecting EEG signals from the scalp based on the international 10–20 system. Recording was done by the OpenVibe [19] recording software. Impedance of all electrodes was kept below the threshold of 5 K [ohm] during all recording sessions.

2.4   Related Measures

Theta/Alpha Ratio (TAR)—In this study, we assessed the cognitive load in each
epoch using the Theta/Alpha ratio (TAR). The TAR measure of cognitive workload
is based on the hypothesis that an increase in workload is associated with an increase
in theta power with a simultaneous decrease in alpha power (e.g. [16–18]). In terms
of topographic distribution, previous studies have shown that workload decreased
alpha power at parietal regions and increased theta power at frontal regions (e.g. [12,
14, 15, 20]). Hence, in this study we have focused on the analysis of the of frontal
and prefrontal cluster electrodes (Fp1, F7, Fp2, F8, F3, and F4). For each epoch we
have calculated the accumulated cognitive load [21], by calculating the energy ratio
between theta and alpha bands for each participant on each single epoch.

3   Data Processing and Analysis

3.1   Preprocessing and Feature Extraction

To maximize the recorded EEG epochs signal-to-noise ratio the before extracting
the TAR value we have implemented data preprocessing pipeline. The pipeline used
a combined filter (band pass combined with a notch filter) for an artifact removal.
Then, the raw EEG signal was re-referenced to an average reference, and decomposed
using independent component analysis (ICA) [22] to remove additional artifacts. To
end the process the EEG signal than down sampled to 64 Hz following a baseline
correction. Data was analyzed on a 1-s epoch window from the onset of each game,
as presented in Fig. 3.

Next, we estimated the intensity of the cognitive workload in each epoch using
the TAR index. First, we have calculated the energy in the Theta and Alpha bands
by using the Discrete Wavelet Transform (DWT) [23, 24]. The DWT is based on a
multiscale feature representation. Every scale represents a unique thickness of the

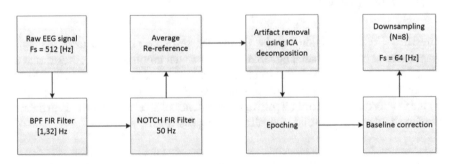

Fig. 3 Preprocess pipeline

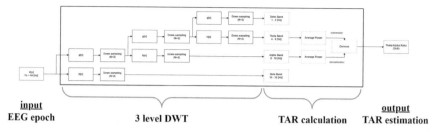

input   **output**
EEG epoch   **3 level DWT**   **TAR calculation**   **TAR estimation**

Fig. 4 Theta-Alpha ratio calculation based on 3 level DWT scheme

EEG signal [25]. Each filtering step contains two digital filters, a high pass filter, $g(n)$, and a low pass filter $h(n)$. After each filter, a down sampler with factor 2 is used to adjust time resolution. In this research, we used a 3-level DWT, which can extract the theta and alpha band (see Fig. 4) which are required to calculate TAR index. The output of the DWT (i.e., the power of theta and alpha bands) is a signal which variant over the epoch time. To calculate the TAR index for the entire segment we calculate the average power of the whole epoch, as described in Eq. 1, and divided them to calculate the TAR index of the current epoch.

$$P_x = \frac{1}{T} \sum_{t=1}^{T} x^2(t) \tag{1}$$

3.2 Theta/Alpha Ratio Distribution as Function of SVO categories

In this section we will analyze the distribution of 72 TAR values (36 of the prosocial players and 36 of the individualistic players) that were extracted from the EEG segments. To understand whether there is statistical significance in the data we performed a two-sample paired t-test. The test showed that players classified as prosocial players by the SVO theory had higher TAR values (M = 2.0654, SD = 1.1010) than those who classified as individualistic (M = 1.0519, SD = 0.4435), t(70) = 5.1234, p < 0.001. Visualization of the TAR samples distribution as a function of the SVO category using boxplot presented in Fig. 5.

The TAR ratio represents the workload that the player invests during the task. The greater the workload the higher the TAR index should be (simultaneous change in two indices—decreased alpha and increased theta). These results, which are based on electrophysiological measurements, show that humans who are prosocial according to the SVO theory invest a larger workload than their individualistic counterparts.

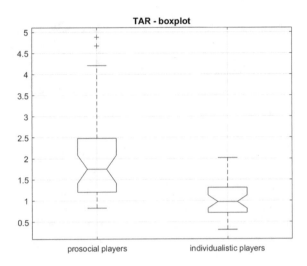

Fig. 5 TAR as function of SVO profiles

4 Conclusions and Future Work

To the best of our knowledge, this is the first time in which a difference is shown in an electrophysiological marker of cognitive load between individuals with different SVO values in the context of resource allocation tasks. Specifically, we have demonstrated this relationship by using a prefrontal and frontal cluster of electrodes. This result corroborates previous research showing that SVO profiles may affect the strategic behavior of players [5, 26, 27], and that different behavioral strategies and indices may be accompanied by electrophysiological changes [12–15]. Following the results obtained in this study, there are many avenues for future research. First, by increasing the number of subjects it will be possible to see whether a correlation can be found between the distribution of the TAR value and a continuous measurement of the SVO value, and not just a comparison with the categorical value as done in this study. Second, it will be possible to examine the effect of the structure of the questionnaire, such as the absolute number of resources or the difference between the various options, on the distribution of electrophysiological indices or on the activity of specific regions in the brain. Third, according to [28, 29] mental work load also affected parietal brain areas. It will be interesting to explore the effect of these areas on the TAR distribution and the correlation to the prefrontal and frontal areas. Finally, previous studies have shown that other measures culture [3, 30] and loss-aversion [31] may affect human behavior in decision making scenarios. It will be interesting to see if the TAR is also correlated with the above mentioned measures.

References

1. Balliet, D., Parks, C., & Joireman, J. (2009). Social value orientation and cooperation in social dilemmas: A meta-analysis. *Group Processes and Intergroup Relations, 12*, 533–547.
2. Mizrahi, D., Laufer, I., & Zuckerman, I. (2021). The effect of expected revenue proportion and social value orientation index on players' behavior in divergent interest tacit coordination games. In *International Conference on Brain Informatics* (pp. 25–34). Cham: Springer.
3. Mizrahi, D., Laufer, I., & Zuckerman, I. (2020). Collectivism-individualism: Strategic behavior in tacit coordination games. *PLoS One, 15*.
4. Mizrahi, D., Laufer, I., & Zuckerman, I. (2020). Individual strategic profiles in tacit coordination games. *Journal of Experimental & Theoretical Artificial Intelligence, 33*, 1–16.
5. Mizrahi, D., Laufer, I., & Zuckerman, I. (2021). Predicting focal point solution in divergent interest tacit coordination games. *Journal of Experimental and Theoretical Artificial Intelligence*, 1–21.
6. Mizrahi, D., Zuckerman, I., & Laufer, I. (2020). Using a stochastic agent model to optimize performance in divergent interest tacit coordination games. *Sensors., 20*, 7026.
7. Liebrand, W. B., & Mcllntock, C. G. (1988). The ring measure of social values: A computerized procedure for assessing individual differences in information processing and social value orientation. *European Journal of Personality, 2*, 217–230.
8. Van Lange, P. A. M., De Bruin, E. M. N., Otten, W., & Joireman, J. A. (1997). Development of prosocial, individualistic, and competitive orientations: Theory and preliminary evidence. *Journal of Personality and Social Psychology, 73*.
9. Murphy, R. O., Ackermann, K. A., & Handgraaf, M. J. J. (2011). Measuring social value orientation. *Judgment and Decision making, 6*, 771–781.
10. Mizrahi, D., Laufer, I., & Zuckerman, I. (2022). Modeling and predicting individual tacit coordination ability. *Brain Informatics*.
11. Mizrahi, D., Laufer, I., & Zuckerman, I. (2019). Modeling individual tacit coordination abilities. In *International Conference on Brain Informatics* (pp. 29–38). Cham, Haikou, China: Springer.
12. Laufer, I., Mizrahi, D., & Zuckerman, I. (2022). An electrophysiological model for assessing cognitive load in tacit coordination games. *Sensors, 22*, 477.
13. Mizrahi, D., Laufer, I., & Zuckerman, I. (2021). Level-K classification from EEG signals using transfer learning. *Sensors., 21*, 7908.
14. Mizrahi, D., Laufer, I., & Zuckerman, I. (2020). The effect of individual coordination ability on cognitive-load in tacit coordination games. In F. Davis, R. Riedl, J. vom Brocke, P.-M. Léger, A. Randolph, & T. Fischer (Eds.), *NeuroIS Retreat*. Vienna, Austria.
15. Mizrahi, D., Laufer, I., & Zuckerman, I. (2021). Topographic analysis of cognitive load in tacit coordination games based on electrophysiological measurements. In *NeuroIS Retreat*. Vienna, Austria.
16. Stipacek, A., Grabner, R. H., Neuper, C., Fink, A., & Neubauer, A. (2013). Sensitivity of human EEG alpha band desynchronization to different working memory components and increasing levels of memory load. *Neuroscience Letters, 353*, 193–196.
17. Fernandez Rojas, R., Debie, E., Fidock, J., Barlow, M., Kasmarik, K., Anavatti, S., Garratt, M., & Abbass, H. (2020). Electroencephalographic workload indicators during teleoperation of an unmanned aerial vehicle shepherding a swarm of unmanned ground vehicles in contested environments. *Frontiers in Neuroscience, 14*, 1–15.
18. Käthner, I., Wriessnegger, S. C., Müller-Putz, G. R., Kübler, A., & Halder, S. (2014). Effects of mental workload and fatigue on the P300, alpha and theta band power during operation of an ERP (P300) brain–computer interface. *Biological Psychology, 102*, 118–129.
19. Renard, Y., Lotte, F., Gibert, G., Congedo, M., Maby, E., Delannoy, V., Bertrand, O., & Le´cuyer, A. (2010). Openvibe: An open-source software platform to design, test, and use brain–computer interfaces in real and virtual environments. *Presence Teleoperators Virtual Environment, 19*, 35–53.
20. Fairclough, S. H., & Venables, L. (2004). Psychophysiological candidates for biocybernetic control of adaptive automation. *Human Factors in Design*, 177–189.

21. Bagyaraj, S., Ravindran, G., & Shenbaga Devi, S. (2014). Analysis of spectral features of EEG during four different cognitive tasks. *International Journal of Engineering and Technology, 6*, 725–734.
22. Hyvärinen, A., & Oja, E. (2000). Independent component analysis: Algorithms and applications. *Neural Networks, 13*, 411–430.
23. Shensa, M. J. (1992). The discrete wavelet transform: Wedding the a trous and Mallat algorithms. *IEEE Transactions on Signal Processing, 40*, 2464–2482.
24. Jensen, A., & la Cour-Harbo, A. (2001). Ripples in mathematics: The discrete wavelet transform. Springer Science & Business Media.
25. Hazarika, N., Chen, J. Z., Tsoi, A. C., & Sergejew, A. (1997). Classification of EEG signals using the wavelet transform. *Signal Processing, 59*, 61–72.
26. Mizrahi, D., Laufer, I., & Zuckerman, I. (2019). Optimizing performance in diverge interest tacit coordination games using an autonomous agent. In *The 21th Israeli Industrial Engineering and Management Conference.*
27. Krockow, E. M., Colman, A. M., & Pulford, B. D. (2016). Exploring cooperation and competition in the Centipede game through verbal protocol analysis. *European Journal of Social Psychology, 46*, 746–761.
28. Guan, K., Zhang, Z., Chai, X., Tian, Z., Liu, T., & Niu, H. (2022). EEG based dynamic functional connectivity analysis in mental workload tasks with different types of information. *IEEE Transactions on Neural Systems and Rehabilitation Engineering, 30*, 632–642.
29. Zhang, P., Wang, X., Chen, J., You, W., & Zhang, W. (2019). Spectral and temporal feature learning with two-stream neural networks for mental workload assessment. *IEEE Transactions on Neural Systems and Rehabilitation Engineering, 27*, 1149–1159.
30. Cox, T. H., Lobel, S. A., & Mcleod, P. L. (1991). Effects of ethnic group cultural differences on cooperative and competitive behavior on a group task. *Academy of Management Journal, 34*, 827–847.
31. Mizrahi, D., Laufer, I., & Zuckerman, I. (2020). The effect of loss-aversion on strategic behaviour of players in divergent interest tacit coordination games. In *International Conference on Brain Informatics* (pp. 41–49). Cham, Padova, Italy: Springer.

Is Our Ability to Detect Errors an Indicator of Mind Wandering? An Experiment Proposal

Colin Conrad, Michael Klesel, Kydra Mayhew, Kiera O'Neil,
Frederike Marie Oschinsky, and Francesco Usai

Abstract Mind wandering could have a variety of impacts on information systems phenomena, not least long monotonous tasks. Unfortunately, mind wandering states are difficult to measure objectively. In this paper, we describe work-in-progress to address this problem in a novel way. We describe two studies that will observe participants' ability to detect errors in a task as a correlate of mind wandering. Demonstrating the technique using a lecture paradigm, the studies employ previously investigated methods of measuring mind wandering as a baseline for the new technique. If successful, we will demonstrate a new method for measuring mind wandering that can be applicable to a broad range of information systems and psychological studies.

Keywords Mind wandering · Cognition · Attention · Vigilance · EEG

C. Conrad (✉) · K. Mayhew · K. O'Neil · F. Usai
Dalhousie University, Halifax, Canada
e-mail: colin.conrad@dal.ca

K. Mayhew
e-mail: kydra.mayhew@dal.ca

K. O'Neil
e-mail: kiera.oneil@dal.ca

F. Usai
e-mail: francesco.usai@dal.ca

M. Klesel
University of Twente, Enschede, The Netherlands
e-mail: michael@klesel.info

F. M. Oschinsky
Fraunhofer Institute for Computer Graphics Research IGD, Darmstadt, Germany
e-mail: frederike.marie.oschinsky@igd-r.fraunhofer.de

© The Author(s), under exclusive license to Springer Nature Switzerland AG 2022
F. D. Davis et al. (eds.), *Information Systems and Neuroscience*,
Lecture Notes in Information Systems and Organisation 58,
https://doi.org/10.1007/978-3-031-13064-9_11

1　Introduction

COVID-19 brought about a major shift in working environments and day to day life. In-person activities such as schooling, large gatherings, travelling, and work were all halted for months at a time, resulting in a need to radically change how they operate [1]. Prior to the pandemic, only 7.9% of the global population held a position where they worked from home [2]. Today, between 35 and 50% of workers in the United States and western European countries worked from home in some capacity [3]. As the importance of work from home has increased, so too has the need for making an individual's home environment productive [4].

However, the work-from-home environment comes with new challenges, both social and environmental. Individuals who work from home are likely to feel social isolation, which may negatively impact their performance. A study conducted by Toscano and Zappalà [5] found that there is a negative relationship between social isolation and remote work satisfaction, and a negative association between such isolation and stressful working conditions. Similarly, cognitive factors such as information overload [6] and work environment distraction [7] have been found to negatively impact productivity. We might wonder whether the persistence of self-generated thoughts or wandering minds throughout the workday could thus impact productive work-from-home spaces. This has motivated us to pursue research into the role that the presence of mind wandering can play in home workspace productivity and effective home workspace technology use.

It is currently difficult to measure mind wandering states in an ecologically valid for from home setting with objective measures. While questionnaires can give insight into the presence of mind wandering, even during technology use [8], they can tell us little about when mind wandering episodes occur. Alternatively, researchers have employed experience sampling probes [9] or electroencephalography [10] to measure mind wandering states. However, these are either distracting, as repeatedly prompting subjects to report on their mental state can be intrusive and take away from the task at hand, which is undesirable, or are difficult to employ remotely (i.e., EEG).

One approach that could overcome these limitations is to embed behavioural indicators of mind-wandering within the task itself. Instead of interrupting a person with questions about their mental state, we seek to infer their mental state by looking at how they perform the task with which they are engaged. Specifically, in this study we propose a method for detecting mind wandering episodes that relies on one of the most frequent tasks that anyone can encounter at work, regardless of their profession: attending to video-lecture wherein someone discusses a given topic. The only manipulation that we introduce is to insert, throughout the lecture's script, errors that render a given sentence meaningless relative to the preceding context. Our theory is that when mind wandering occurs, people would tune out the video stimuli and become less vigilant and will generate more errors. When attention drifts away from the main task people miss more task-related information than when focusing on the task, which has been corroborated in studies related to reading which showed that indeed episodes of mind-wandering increase the likelihood of missing errors [11, 12].

To further assess the validity of this approach and evaluate its generalizability, we employed a paradigm wherein subjects will be asked to listen to a video-lecture and indicate, by pressing a button, whenever they come across a sentence that contradicts the preceding ones. Using this paradigm we will conduct two studies: a proof-of-concept study employing only a behavioral task; and a more comprehensive EEG study. In the proof-of-concept study we will cross-validate the failure to detect errors with other well-established behavioral measures of mind wandering, such as experience sampling probes [9]. For this first study our research question and associated hypotheses are:

RQ1—How strongly is performance at detecting contradictory information in the video-lecture associated with mind wandering as recorded by sampling probes?

H1—There is a strong correlation between the extent to which subjects identify contradictory information and the extent to which they experience mind-wandering.

If H1 is supported by results of Study 1, we will further validate our approach using only EEG measures and no experience samples in Study 2. A second study will allow us to cross-validate these findings with past studies in the absence of confounding factors created by experience sample probes, would replicate our results, and would provide evidence for a truly passive measure of mind wandering [10, 13, 14]. Our main research question and associated hypotheses are:

RQ2—How strongly is performance at detecting contradictory information in the video-lecture associated with EEG correlates of mind wandering?

H2—There is a strong correlation between the extent to which subjects identify contradictory information and EEG markers of mind wandering (i.e. modulation of the amplitude of the P300 component elicited by auditory tones).

In the reminder of this paper we will describe our research methods for the studies before describing the potential contribution of the work.

2   Methods

2.1   Participants

For each of the two studies we are aiming to recruit 40 participants among the pool of undergraduate students. As the studies we are proposing are exploratory in nature—especially as there is little to no background literature documenting the size of the effects that we are aiming to detect—the sample size was not defined based on the result of an analysis of statistical power, rather on a number that is considered appropriate to conduct exploratory research involving the analysis of correlations [15]. To ensure a consistent sample, subjects from both studies will be excluded if they report uncorrected vision problems or physical impairments that would prevent them from using a computer keyboard or mouse or neurological conditions that could affect EEG (e.g., epilepsy or a recent concussion). For the second study the same criteria apply with subjects who have not taken part in Study 1.

2.2 Stimuli

We selected a lecture on machine learning based on freely available online courses through LinkedIn Learning as the material to be presented to participants. We first modified the transcript of the lecture to include 24 coherency errors. Then, one member of the research team recorded the lecture again based on the edited script. The following is an example of one of the errors that were inserted in the script of the lecture:

- Original text: "A machine might have an algorithm that says two types of data should be treated the same way. The machine will then use the algorithm to look for patterns."
- Edited text: "A machine might have an algorithm that says two types of data should be treated the same way. The machine will then use the algorithm to look for patterns, *based on a rule that states that two distinct types of data should be treated differently.*"

All participants will receive the same script including all 24 errors. Experience sampling probes [9] will be used to determine the extent to which participants were on-task or mind-wandering. To this end, each participant will receive 10 probes throughout the course of the entire lecture. Each probe will occur 5 s after the appearance of a coherency error. Three versions of the paradigm were created to ensure that, across participants, each coherency error was followed by a mind-wandering probe. Thus, each version of the paradigm contains 5 unique sampling probes (i.e., occur after coherency errors not probed in another version).

2.3 Procedure

Study 1. Upon the beginning of the experiment participants will be asked to enter a closed room equipped with only a computer and speakers. All participants will complete a quick questionnaire about their demographic information, and a multiple-choice test to assess their knowledge about the topic of the video-lecture. The latter test will be used to identify and exclude participants' with prior knowledge on machine learning, which will likely be a confounding factor in our analysis. Indeed, studies suggest that, while prior knowledge might have no effects on the extent to which people do experience mind-wandering, it does nonetheless facilitate information processing (i.e. text comprehension [16, 17]). Therefore, we expect that prior knowledge on the topic of machine learning will affect subjects' ability to detect incongruency errors independently from the extent to which their attention is on task. To prevent this from confounding our results, data from subjects scoring above chance level (25%) in the multiple-choice task will be excluded from data analysis.

The PsychoPy framework will be configured to record study start times, as well as the timing and response from the participants regarding embedded errors, and responses to the sampling probes. All participants will then be asked to sit through the 1-h pre-recorded lecture. Participants will be instructed to indicate with a button-press when they notice coherency errors within the lecture. To determine if missed-errors correspond to periods of self-identified mind wandering, participants will receive a mind-wandering probe 5 s after the onset of the coherency error. Participants' comprehension of the video-lecture will be tested before being debriefed on the nature of the contradictions that they encountered in the lecture. They will then be asked to leave upon completion of the task.

Study 2. Upon the beginning of the experiment participants will be asked to enter a closed room equipped with a computer, speakers and EEG device. Before being fitted with an EEG cap, participants will be asked to complete a quick questionnaire about their demographic information, and a multiple-choice test to assess their knowledge about the topic of the video-lecture. Participants will be fitted with horizontal and vertical electrooculograms (EoG) and 32 scalp electrodes (ActiCap, BrainProducts GmbH, Munich, Germany) positioned at standard locations according to the international 10–10 system and referenced to the midline frontal location (FCz). Electrode impedances will be kept below 15 kOhm at all channel locations throughout the experiment. EEG data will be recorded using a Refa8 amplifier (ANT, Enschende, The Netherlands) at a sampling rate of 512 Hz, bandpass filtered between 0.01 and 170 Hz, and saved using ANT ASAlab. While subjects are watching the pre-recorded lecture, single auditory tones will be presented in the background. Tone-presentation will occur 5 s after the onset of errors in the video-lecture. This is to ensure that, even if subjects infer the association between the presence of an error and the auditory tone, the tones cannot function as cues to the presence of an error. Button presses occurring after the onset of the auditory tones will be treated as missed errors, as they are likely attributable to the cueing effect, rather than on genuine error-detection. In total, 24 tones will be presented. Following the presentation of the lecture, the EEG will be removed.

2.4 Data Processing and Analysis

Behavioral Data. In both studies, subjects' responses will be of two types: *detected* or *undetected* error. Each of such responses will be assigned to one of two groups based on the mental state reported in the behavioral prompt that subjects receive 5 s after the onset of the error (i.e., *on-task* or *mind wandering*). To assess whether mental state predicts whether an error is detected or not, we will conduct a two-tailed t-test contrasting the *on-task* and *mind wandering* groups to test whether the difference in the type of response between them is statistically significant.

Neurophysiological Data. If results from Study 1 show a strong correlation between reported mental state and the ability to detect coherency errors in text, in Study 2 we will infer subjects' mental state based on their performance in the error detection task. Whenever participants will correctly detect an error in the lecture's script, we will assume that they were focused on the task at-hand.

To investigate the neural correlates of mind wandering, we will contrast these two categories of neural responses by looking at two distinct features: (1) neural oscillatory activity; and (2) event-related potentials (ERPs) elicited by auditory tones. The analysis of oscillatory activity will be carried out by selecting only the 10 s preceding the onset of an error. This way we parse out any potential confounding effect due to neural activity associated with the preparation and execution of a motor response that occurs when subjects detect an error. Instead, to analyze ERP responses we will look at changes in the amplitude of the P300 component associated with the two different mental states of interest. We will select neural activity occurring from -0.2 to 1 s after the onset of the auditory tone. The 200 ms preceding the onset of the auditory tone will be considered as the signal baseline.

EEG Statistical Modeling. Statistical analysis will be conducted through Generalized Additive Mixed Effects Modeling (GAMM), which extends the traditional Generalized Linear Mixed Model (GLMM) by modeling non-linear relationships between the dependent variable and the predictors [18]. Specifically, we will test three models that will differ only for the dependent variable: amplitude of the P300 component, defined as the mean amplitude in the time window going from 200 to 400 ms after the onset of the auditory tone; power in the Theta band of the neural oscillatory activity; power in the Delta band of the neural oscillatory activity. All models will include the following fixed and random effects. The only fixed effect that will be included is Mental State (2 levels: on-task, mind-wandering). Random effects will include the following variables: subject ID, electrode, age, gender.

3   Limitations, Contribution and Future Work

We anticipate some limitations to these findings. As a novel paradigm, it is entirely possible that we will find no relationship between the relevant measures. Furthermore, even if we find a relationship, it is possible that the correlation would be caused by a latent factor that underdetermines the observed relationships between mind wandering. It would also be important to replicate the findings, even if the relationships are observed in both versions of the study with both known behavioral and EEG correlates of mind wandering.

If successful however, we will identify a new measure of mind wandering which could be employed in future studies in a wide range of contexts. Again, we want to stress the importance of employing measures of mind wandering that can be obtained without interfering with a person's main task at-hand, which is the primary reason for us proposing this study. Nonetheless, this study is part of a wider initiative related to the impact of mind wandering in remote work and the applications of these findings could be wide-reaching, extending to domains such as human–computer interaction, ergonomics, education, and design science. The techniques might further complement the experience sample probe method that is frequently employed in psychological research [9] and may play a role in the fast-changing literature related to varieties of mind wandering and their family resemblances [8, 19]. Ultimately, this work is positioned to help bridge the gap between NeuroIS and the many interesting and similar ongoing conversations in its reference disciplines.

Acknowledgements This project is supported by the Social Sciences and Humanities Research Council of Canada with funding from the Insight Development program awarded to Colin Conrad.

References

1. Aliosi, A., & De Stefano, V. (2021). Essential jobs, remote work and digital surveillance: Addressing the COVID-19 pandemic panopticon. *International Labour Review.* https://doi.org/10.1111/ilr.12219
2. International Labour Organization. Working from home: Estimating the worldwide potential. https://www.ilo.org/
3. Morikawa, M. (2020). Productivity of working from home during the COVID-19 pandemic: Evidence from an employee survey. RIETI discussion paper series 20-E-073
4. Umishio, W., Kagi, N., Asaoka, R., Hayashi, M., Sawachi, T., & Ueno, T. (2022). Work productivity in the office and at home during the COVID-19 pandemic: A cross-sectional analysis of office workers in Japan. *Indoor Air, 32*(1).
5. Toscano, F., & Zappalà, S. (2020). Social isolation and stress as predictors of productivity perception and remote work satisfaction during the COVID-19 pandemic: The role of concern about the virus in a moderated double mediation. *Sustainability, 12*, 1–14.
6. Conrad, C., Deng, Q., Caron, I., Shkurska, O., Skerrett, P., & Sundararajan, B. (2022). How student perceptions about online learning difficulty influenced their satisfaction during Canada's Covid-19 response. *British Journal of Educational Technology.*
7. Galanti, T., Guidetti, G., Mazzei, E., Zappalà, S., & Toscano, F. (2021). Work from home during the COVID-19 outbreak: The impact on employees' remote work productivity, engagement, and stress. *Journal of Occupational and Environmental Medicine, 63*(7), e426.
8. Klesel, M., Oschinsky, F. M., Conrad, C., & Niehaves, B. (2021). Does the type of mind wandering matter? Extending the inquiry about the role of mind wandering in the IT use experience. *Internet Research, 31*(3), 1018–1039.
9. Wammes, J. D., Ralph, B. C., Mills, C., Bosch, N., Duncan, T. L., & Smilek, D. (2019). Disengagement during lectures: Media multitasking and mind wandering in university classrooms. *Computers & Education, 132*, 76–89.
10. Conrad, C., & Newman, A. (2021). Measuring mind wandering during online lectures assessed with EEG. *Frontiers in Human Neuroscience, 455.*
11. Schad, D. J., Nuthmann, A., & Engbert, R. (2012). Your mind wanders weakly, your mind wanders deeply: Objective measures reveal mindless reading at different levels. *Cognition, 125*, 179–194.
12. Zedelius, C. M., Broadway, J. M., & Schooler, J. W. (2015). Motivating meta-awareness of mind wandering: A way to catch the mind in flight? *Consciousness and Cognition, 36*, 44–53.
13. Conrad, C. A neurophysiological study of the impact of mind wandering during online lectures. Doctoral Thesis. http://hdl.handle.net/10222/76253
14. Baldwin, C. L., Roberts, D. M., Barragan, D., Lee, J. D., Lerner, N., & Higgins, J. S. (2017). Detecting and quantifying mind wandering during simulated driving. *Frontiers in Human Neuroscience, 406.*

15. Daniel, J. (2012). Choosing the size of the sample. In: *Sampling essentials: Practical guidelines for making sampling choices* (pp. 236–253). SAGE Publication Inc.
16. Unsworth, N., & McMillan, B. D. (2013). Mind wandering and reading comprehension: Examining the roles of working memory capacity, interest, motivation, and topic experience. *Journal of Experimental Psychology: Learning, Memory, and Cognition., 39*(3), 832–842.
17. Schurer, T., Opitz, B., & Schubert, T. (2020). Working memory capacity but not prior knowledge impact on readers' attention and text comprehension. *Frontiers in Education, 5,* 26. https://doi.org/10.3389/feduc.2020.00026
18. Baayen, R. H., Davidson, D. J., & Bates, D. M. (2008). Mixed-effects modeling with crossed random effects for subjects and items. *Journal of Memory and Language, 59*(4), 390–412.
19. Seli, P., Risko, E. F., Smilek, D., & Schacter, D. L. (2016). Mind-wandering with and without intention. *Trends in Cognitive Sciences, 20*(8), 605–617.

Resolving the Paradoxical Effect of Human-Like Typing Errors by Conversational Agents

R. Stefan Greulich and Alfred Benedikt Brendel

Abstract Conversational agents (CA) interact with users via natural language in a turn-taking communication. They have been shown to trigger anthropomorphism in users, leading to the perception of humanness. However, this effect is influenced by various aspects, such as the CA having a name and its choice of words. In this context, the expected effect of typing errors is paradoxical. On the one hand, making typing errors is human-like and should, therefore, increase the perception of humanness. On the other hand, users understand that CAs are not human, and typing errors can be perceived as unexpected. To investigate the effect of typing errors on perceived humanness, we employ the dual processing theory of human cognition and propose a NeuroIS experiment to test our hypothesis. In this research in progress paper, we present the experimental design, expected outcome, and reflect on possible implications for CA research.

Keywords Conversational agents · Dual processing · Electroencephalography · Typographical errors · NeuroIS

1 Introduction

The tendency of humans to anthropomorphize technical systems is well known [1] and Conversational Agents (CAs) have recently attracted extensive research in this regard [2]. The social response theory [3] and computers-are-social-actor (CASA) paradigm [4] explain that the degree of social cues (i.e., human-like characteristics, such as having a name and using emoticons) increases or decreases users' perceived

The original version of this chapter was revised. Chapter title has now been corrected. The correction to this chapter can be found at https://doi.org/10.1007/978-3-031-13064-9_36

R. S. Greulich (✉) · A. B. Brendel
Technische Universität Dresden, Dresden, Germany
e-mail: stefan.greulich@tu-dresden.de

A. B. Brendel
e-mail: alfred_benedikt.brendel@tu-dresden.de

F. D. Davis et al. (eds.), *Information Systems and Neuroscience*,
Lecture Notes in Information Systems and Organisation 58,
https://doi.org/10.1007/978-3-031-13064-9_12

113

humanness [5]. The importance of understanding the perception of humanness lies in its effect on users' affection, cognition, and behavior. For instance, perceived humanness has been shown to increase service satisfaction [6], enjoyment [7], and performance [8]. In this context, the social cue of typing errors is associated with contradicting explanations and predictions regarding their effects, leading to mixed results in the literature [9–11].

Through the lens of dual processing theory [12, 13], the reaction to typing errors of CAs can be conceptualized as follows. Dual processing theory states that humans have two distinct modes of thinking, termed system 1 and system 2. System 1 is subconscious, fast, and effortless, but is limited in its processing to simple and over-learned tasks based on heuristics and biases. If a conflict in the possible outputs of system 1 is detected, system 2 is engaged [14]. System 2 is more flexible, but it is also associated with a higher mental effort and a conscious examination of the situation [12–16]. There is increasing evidence that the perception of anthropomorphism is associated with system 1 processing [17, 18]. Regarding the occurrence of typing errors, users are aware that CAs are machines and are, therefore, expecting a machine-like level of precision [19, 20]. Thus, the occurrence of such an error can be expected to trigger the conflict detection of system 1, and the user switches to system 2, leading to a conscious evaluation of the situation. This brings the artificial nature of the interaction into the conscious reflection of the user and destroys the perception of a natural interaction. However, for system 1 to trigger system 2, making typing errors has to be unexpected. Hence, if the CA is perceived to be human-like, it should be less likely that system 1 triggers system 2 because making errors is human [21, 22]. In this case, making typing errors should lead to an increase in perceived humanness [23].

In conclusion, resolving this paradox of contradicting explanations and results [9–11] stands as an important and interesting area of future research. It could lead to the addition of a new social cue (typing errors) but will most certainly contribute to a deeper understanding of the interplay of CA design and users' perception and processing of the interaction.

Against this background, we extend current studies [9–11], which are based on self-reported measures (e.g., surveys), by planning to conduct an EEG study. The current approach of self-reported measures has the problem that asking about typing errors and their perception leads to retroactive processing of the situation [24]. To avoid this distortion, we propose to measure the humanness perception and involvement of system 1 and system 2 directly via electroencephalography (EEG) and pupillometry during the interaction with different CAs exhibiting typing errors. This leads to our research question:

RQ: *What is the influence of typing error on users' neural signals associated with the perception of humanness and dual processing?*

2 Theoretical Background

2.1 Dual Process Theory

The idea that human cognition is separated into two systems has existed for centuries in the fields of psychology and philosophy [15, 25]. Psychology research on this theory gained increased interest in the 90s [16], leading to the currently accepted formulation by Stanovich and West [12]. System 1 is the automatic subconscious processing of stimuli and relies on intuitive responses. On the other hand, system 2 involves higher cognitive processing, provides more flexibility, and general problem-solving skills, but at a higher cost than system 1. The switch from system 1 to system 2 is explained in the three-stage theory of dual processing [14]. This theory describes that the sensory input is processed by multiple instances of system 1. If the outputs of those instances are in conflict, system 2 is engaged [14]. Recent investigations into the dual process theory from a neuroscientific perspective give support to a separation between both in the form of separate brain networks (for review see [13, 26]), opening up the possibility of measuring the involvement of both systems during an interaction with a CA.

2.2 Human-Like Design of CA and Perceived Humanness

According to the social response theory [3], a CA with human-like features lets the user perceive and respond as if they are interacting with a social human partner [4, 27]. The intensity of this effect is increased when human-like features are used in the design of the CA [5, 28], such as gender [29], name [30], self-reference [31], dynamic response delays [32], or the use of emoticons [27]. This perception of humanness leads to an improved users experience, including higher enjoyment [33], social presence [34], and trustworthiness [35].

2.3 Typing Errors as Social Cues

There is a public perception of artificial systems, such as robots, to be precise and flawless [19, 20]. Humans, however, are expected to have flaws and produce errors, a notion which was codified about 2000 years ago in the quote "Errare humanum est [...]" (To err is human [...]) [21]. An artificial system is perceived to be more human if it exhibits human-like characteristics [23, 36, 37]. Textual conversation produced by humans is often flawed [38], including input errors on a keyboard, so-called typing errors [38]. Therefore, it would follow that an artificial system producing typing errors is perceived as more human than without [23]. However, investigations into text-based CAs producing typing errors showed that the perception of humanness is either reduced [11], shows no significant result [10], or is increased [9].

3 Hypotheses Development

We aim to investigate the interaction of typing errors and human-like design of CAs (in our case instantiated as chatbots). Based on the three-stage theory of dual processing, humans switch from system 1 to system 2, if the outputs of multiple instances of system 1 running in parallel are in conflict with each other [14]. Humans detect and correct for textual errors by detecting conflict between the input and heuristic understanding of the text, followed by reanalyzing the read text [39, 40]. Additionally, the production of errors by a system that is expected to be flawless [19, 20], can be expected to trigger the conflict detection of system 1 outputs. In essence, the system 1 outputs of a belief-based processing ("The CA is flawless and does not produce errors"), a heuristic processing ("The CA is talking about bikes"), and a semantic processing ("There is the word 'bkies' on the screen") are not reconcilable and system 2 is engaged. Therefore, we hypothesize:

H1: A chatbot's typing errors will increase the neural signals associated with system 2 and increased cognitive processing.

The interaction with a human like CA has been described in the IS literature as more effortless and easier to use than non-human CAs because of the transfer of human-to-human interaction patterns [2, 41]. Since humans utilize system 1 as much as possible to reduce the cognitive load of everyday tasks [42, 43], it stands to reason, that this ease of use is caused by the user staying in system 1 during the interaction with the CA. There is increasing evidence that the interaction is considered more natural and the partner more anthropomorphic if a switch to system 2 is avoided [17, 18]. Following from those premises, we hypothesis that:

H2: The human like design of a chatbot increases neural signals associated with system 1 and neural signals associated with decreased cognitive processing.

It is our hypothesis that remaining in system 1 eases the illusion of conversing with a human partner, while system 2 forces the recognition of the artificial nature of the situation. Therefore, the switch between both systems should be less prominent the stronger the CA is designed human like. Thus:

H3: The effect of typing errors on system 2 signals and cognitive processing is negatively moderated by the human like design.

Finally, we intent to test the validity of humanness questionaries, as we postulate, that the question of perceiving as human is manly based on staying in system 1, but questionnaires require a conscious reflection of the situation [24], which is a function of system 2 [44]. To test whether our postulation is true, we hypothesize:

H4: There is a correlation between the self-reported perceived humanness value and the neural signals associated with the perception of humanness.

4 Method

4.1 Participants

We plan to collect data from 20 to 30 participants (right-handed, native German speakers and no history of dyslexia), recruited from students at Technische Universität Dresden.

4.2 Task and Procedure

The participants will be instructed to rent an e-bike via each chatbot individually. The process will have a total of nine steps (e.g., stating type of e-bike, selecting day and time of rental) and take approximately three minutes to finish. The sequence of the exposed CAs will be pseudorandomized with a questionnaire in between (see Sect. 4.4 for details).

4.3 Treatments

We employed a 2 × 2 within-subject design with typing errors/no typing errors and humanlike/non-human like design as independent variables (see Table 1 and Fig. 1 for the experimental design). The human like chatbot utilizes the humanness increasing design elements of female gender [29], name [30], self-reference [31], dynamic response delays [32] and the use of emoticons [27]. For the typing errors, we included human like typing errors based on the categorization of Macneilage [38].

Table 1 Research design

	Typing errors	No typing errors
Human like	**Heloo**, I am Laura. 🤖 I am not a human but will do my best to help you. What can I do for you? ☺	Hello, I am Laura. 🤖 I am not a human but will do my best to help you. What can I do for you? ☺
Non-human like	**Welcomee** to the e-bike rental	Welcome to the e-bike rental

Typing errors are highlighted in bold

Fig. 1 Chatbot utilizing the human like design and no typing errors

4.4 Measures

Previous work showed, that the engagement or switch to system 2 is measurable via EEG and pupillometry [45–48]. While the pupil diameter also reacts to changes in illumination [49] and changes in gaze orientation [50], EEG allows for a much fine-grained examination of the neural signal [48]. We therefore plan to utilize both methods, with eye tracking and 16 channel EEG employed in the experiment.

To correct our model for possible confounds, we will collect prior exposure to CAs with a modified technology exposure questionnaire [51], as well as demographic data before the experiment. After each CA interaction, we will collect the perceived humanness [52], service satisfaction [53] and current emotional state [54].

5 Expected Results, Discussion and, Conclusion

In this research-in-progress paper, we present an experiment to resolve the contradicting effects of CA's typing errors on users' perception. Our study would provide valuable insight into the human processing of cues while interacting with CAs. Our research could direct the CA research away from a trial-and-error approach to finding design elements improving the perceived humanness and potential other aspects (e.g., service satisfaction, enjoyment, and trustworthiness).

The experiment is implemented, and we are awaiting approval from the ethics board of the Technische Universität Dresden. For the EEG results, we expect finding responses in line with the works of Williams et al. [48]. For the CAs exhibiting typing errors, we predict an increase in frontal theta power during the interaction signifying an increased involvement of system 2. For the CAs without typos, an increase in parietal alpha is expected, signifying an increased involvement of system 1. Similarly, the pupillometry should reflect the EEG recordings, with increase in pupil dilation between the interaction with a CA with and without typos in line with the results from [45, 47]. Additionally, for the comparison between human and non-human like design, we foresee an increase in theta power over the F3 and F4 electrode locations for non-human like designs as demonstrated in [55].

References

1. Howard, J. A., & Kunda, Z. (2000). Social cognition: Making sense of people. *Contemporary Sociology*. https://doi.org/10.2307/2654104
2. Diederich, S., Brendel, A., & Morana, S. (2022). On the design of and interaction with conversational agents: An organizing and assessing review of human-computer interaction research. *Journal of the Association for Information Systems*.
3. Nass, C., & Moon, Y. (2000). Machines and mindlessness: Social responses to computers. *Journal of Social Issues, 56*, 81–103. https://doi.org/10.1111/0022-4537.00153
4. Nass, C., Steuer, J., & Tauber, E. R. (1994). Computers are social actors. In *Proceedings of the ACM CHI Conference on Human Factors in Computing Systems* (p. 204).
5. Seeger, A.-M., Pfeiffer, J., & Heinzl, A. (2018). Designing anthropomorphic conversational agents: Development and empirical evaluation of a design framework. In *Proceedings of the International Conference on Information Systems (ICIS)* (pp. 1–17).
6. Gnewuch, U., Morana, S., & Maedche, A. (2017). Towards designing cooperative and social conversational agents for customer service. In *Proceedings of the International Conference on Information Systems (ICIS)* (pp. 1–13).
7. Hassanein, K., & Head, M. (2007). Manipulating perceived social presence through the web interface and its impact on attitude towards online shopping. *International Journal of Human-Computer Studies, 65*, 689–708. https://doi.org/10.1016/j.ijhcs.2006.11.018
8. Kalam Siddike, M. A., Spohrer, J., Demirkan, H., & Kohda, Y. (2018). People's interactions with cognitive assistants for enhanced performances. In *Proceedings of the Annual Hawaii International Conference on System Sciences* (pp. 1640–1648). IEEE Computer Society.
9. Bluvstein Netter, S., Zhao, X., Barasch, A., & Schroeder, J. (2021). "Hello! How May I Helo You?": How (corrected) errors humanize a communicator. *SSRN Electronic Journal*. https://doi.org/10.2139/ssrn.3894518

10. Bührke, J., Brendel, A. B., Lichtenberg, S., Greve, M., & Mirbabaie, M. (2021). Is making mistakes human? On the perception of typing errors in chatbot communication. In *Proceedings of the Annual Hawaii International Conference on System Sciences* (pp. 4456–4465). IEEE Computer Society.

11. Westerman, D., Cross, A. C., & Lindmark, P. G. (2019). I Believe in a thing called Bot: Perceptions of the humanness of "Chatbots." *Communication Studies, 70*, 295–312. https://doi.org/10.1080/10510974.2018.1557233

12. Stanovich, K. E., & West, R. F. (2000). Individual differences in reasoning: Implications for the rationality debate? *The Behavioral and Brain Sciences, 23*, 645–726. https://doi.org/10.1017/S0140525X00003435

13. Evans, J. S. B. T., & Stanovich, K. E. (2013). Dual-process theories of higher cognition: Advancing the debate. *Perspectives on Psychological Science, 8*, 223–241. https://doi.org/10.1177/1745691612460685

14. Pennycook, G., Fugelsang, J. A., & Koehler, D. J. (2015). What makes us think? A three-stage dual-process model of analytic engagement. *Cognitive Psychology, 80*, 34–72. https://doi.org/10.1016/j.cogpsych.2015.05.001

15. Evans, J. S. B. T. (2008). Dual-processing accounts of reasoning, judgment, and social cognition. *Annual Review of Psychology, 59*, 255–278. https://doi.org/10.1146/annurev.psych.59.103006.093629

16. Pennycook, G. (2018). A perspective on the theoretical foundation of dual process models. In *Dual process theory 2.0* (pp. 5–27). Routledge.

17. Jahn, K., & Nissen, A. (2021). Towards dual processing of social robots: Differences in the automatic and reflective system. In *International Conference on Information Systems, ICIS 2020—Making Digital Inclusive: Blending the Local and the Global*.

18. Spatola, N., & Chaminade, T. (2021). Cognitive load increases anthropomorphism of humanoid robot. The automatic path of anthropomorphism. *Journal of Experimental Psychology*. https://doi.org/10.31234/OSF.IO/KD4GE

19. Kriz, S., Ferro, T. D., Damera, P., & Porter, J. R. (2010). Fictional robots as a data source in HRI research: Exploring the link between science fiction and interactional expectations. In *Proceedings—IEEE International Workshop on Robot and Human Interactive Communication* (pp. 458–463).

20. Bruckenberger, U., Weiss, A., Mirnig, N., Strasser, E., Stadler, S., & Tscheligi, M. (2013). The good, the bad, the weird: Audience evaluation of a "real" robot in relation to science fiction and mass media. In *Lecture notes in computer science* (including subseries Lecture notes in artificial intelligence and lecture notes in bioinformatics) (pp. 301–310).

21. Seneca, L. A., Epistulae morales ad Lucilium (65) AD.

22. Rasmussen, J. (1990). Human error and the problem of causality in analysis of accidents. *Philosophical Transactions of the Royal Society B Biological Sciences, 327*. https://doi.org/10.1098/rstb.1990.0088

23. Turing, A. M. (1950). Computer machinery and intelligence. *Mind. LIX*, 433–460. https://doi.org/10.1093/MIND/LIX.236.433

24. Riedl, R., & Léger, P.-M. (2016). *Fundamentals of NeuroIS*. Springer.

25. Frankish, K., & Evans, J. S. B. T. (2012). The duality of mind: An historical perspective. In *In two minds: Dual processes and beyond*. Oxford University Press.

26. Lieberman, M. D. (2012). What zombies can't do: A social cognitive neuroscience approach to the irreducibility of reflective consciousness. In *In two minds: Dual processes and beyond*. Oxford University Press.

27. Feine, J., Gnewuch, U., Morana, S., & Maedche, A. (2019). A taxonomy of social cues for conversational agents. *International Journal of Human-Computer Studies, 132*, 138–161. https://doi.org/10.1016/J.IJHCS.2019.07.009

28. Diederich, S., & Benedikt Brendel, A. (2019). Towards a taxonomy of platforms for conversational agent design design science and design thinking view project. In *Proceedings of the Wirtschaftsinformatik* (pp. 1100–1114).

29. Nunamaker, J. F., Derrick, D. C., Elkins, A. C., Burgoon, J. K., & Patton, M. W. (2011). Embodied conversational agent-based Kiosk for automated interviewing. *Journal of Management Information Systems, 28*, 17–48.
30. Cowell, A. J., & Stanney, K. M. (2005). Manipulation of non-verbal interaction style and demographic embodiment to increase anthropomorphic computer character credibility. *International Journal of Human-Computer Studies*. https://doi.org/10.1016/j.ijhcs.2004.11.008
31. Schuetzler, R. M., Grimes, G. M., & Giboney, J. S. (2018). An investigation of conversational agent relevance, presence, and engagement. In *Proceedings of the Americas Conference on Information Systems (AMCIS)* (pp. 1–10).
32. Gnewuch, U., Morana, S., Adam, M. T. P., & Maedche, A. (2018). Faster is not always better: Understanding the effect of dynamic response delays in human-chatbot interaction. In *Proceedings of the European Conference on Information Systems (ECIS)* (pp. 1–17).
33. Lee, S. Y., & Choi, J. (2017). Enhancing user experience with conversational agent for movie recommendation: Effects of self-disclosure and reciprocity. *International Journal of Human-Computer Studies, 103*, 95–105.
34. Kim, Y., & Sundar, S. S. (2012). Anthropomorphism of computers: Is it mindful or mindless? *Computers in Human Behavior*. https://doi.org/10.1016/j.chb.2011.09.006
35. Araujo, T. (2018). Living up to the chatbot hype: The influence of anthropomorphic design cues and communicative agency framing on conversational agent and company perceptions. *Computers in Human Behavior, 85*, 183–189.
36. De Visser, E. J., Monfort, S. S., Goodyear, K., Lu, L., O'Hara, M., Lee, M. R., Parasuraman, R., & Krueger, F. (2017). A Little anthropomorphism goes a long way: Effects of oxytocin on trust, compliance, and team performance with automated agents. *Human Factors, 59*, 116–133. https://doi.org/10.1177/0018720816687205
37. Diederich, S., Brendel, A. B., & Kolbe, L. M. (2020). Designing anthropomorphic enterprise conversational agents. *Business & Information Systems Engineering, 62*, 193–209. https://doi.org/10.1007/s12599-020-00639-y
38. Macneilage, P. F. (1964). Typing errors as clues to serial ordering mechanisms in language behaviour. *Language and Speech, 7*, 144–159. https://doi.org/10.1177/002383096400700302
39. Van Herten, M., Chwilla, D. J., & Kolk, H. H. J. (2006). When heuristics clash with parsing routines: ERP evidence for conflict monitoring in sentence perception. *Journal of Cognitive Neuroscience, 18*, 1181–1197. https://doi.org/10.1162/jocn.2006.18.7.1181
40. Frazier, L., & Rayner, K. (1982). Making and correcting errors during sentence comprehension: Eye movements in the analysis of structurally ambiguous sentences. *Cognitive Psychology, 14*, 178–210. https://doi.org/10.1016/0010-0285(82)90008-1
41. Følstad, A., & Brandtzaeg, P. B. (2017). Chatbots and the new world of HCI. *Interactions, 24*, 38–42. https://doi.org/10.1145/3085558
42. Evans, J. S. B. T. (2010). Intuition and reasoning: A dual-process perspective. *Psychological Inquiry, 21*, 313–326. https://doi.org/10.1080/1047840X.2010.521057
43. Macrae, C. N., Milne, A. B., & Bodenhausen, G. V. (1994). Stereotypes as energy-saving devices: A peek inside the cognitive toolbox. *Journal of Personality and Social Psychology, 66*, 37–47. https://doi.org/10.1037/0022-3514.66.1.37
44. Kahneman, D. (2003). A perspective on judgment and choice: Mapping bounded rationality. *American Psychologist, 58*, 697–720. https://doi.org/10.1037/0003-066X.58.9.697
45. Kahneman, D., Peavler, W. S., & Onuska, L. (1968). Effects of verbalization and incentive on the pupil response to mental activity. *Canadian Journal of Psychology, 22*, 186–196. https://doi.org/10.1037/h0082759
46. Laeng, B., Sirois, S., & Gredebäck, G. (2012). Pupillometry: A window to the preconscious? *Perspectives on Psychological Science, 7*, 18–27. https://doi.org/10.1177/1745691611427305
47. Querino, E., Dos Santos, L., Ginani, G., Nicolau, E., Miranda, D., Romano-Silva, M., & Malloy-Diniz, L. (2015). Cognitive effort and pupil dilation in controlled and automatic processes. *Translational Neuroscience, 6*, 168–173. https://doi.org/10.1515/tnsci-2015-0017
48. Williams, C. C., Kappen, M., Hassall, C. D., Wright, B., & Krigolson, O. E. (2019). Thinking theta and alpha: Mechanisms of intuitive and analytical reasoning. *NeuroImage, 189*, 574–580. https://doi.org/10.1016/j.neuroimage.2019.01.048

49. Campbell, F. W., & Gregory, A. H. (1960). Effect of size of pupil on visual acuity. *Nature, 187*, 1121–1123. https://doi.org/10.1038/1871121c0
50. Sokolov, E. N. (1963). Higher nervous functions; the orienting reflex. *Annual Review of Physiology, 25*, 545–580. https://doi.org/10.1146/annurev.ph.25.030163.002553
51. Thompson, R. L., Higgins, C. A., & Howell, J. M. (1994). Influence of experience on personal computer utilization: Testing a conceptual model. *Journal of Management Information Systems, 11*, 167–187. https://doi.org/10.1080/07421222.1994.11518035
52. Seymour, M., Yuan, L., Dennis, A. R., & Riemer, K. (2021). Have we crossed the Uncanny valley? Understanding affinity, trustworthiness, and preference for realistic digital humans in immersive environments. *Journal of the Association for Information Systems, 22*(9). https://doi.org/10.17705/1jais.00674
53. Verhagen, T., van Nes, J., Feldberg, F., & van Dolen, W. (2014). Virtual customer service agents: Using social presence and personalization to shape online service encounters. *The Journal of Computing and communication, 19*, 529–545.
54. Steyer, R., Schwenkmezger, P., Notz, P., & Eid, M. (1994). Testtheoretische Analysen des Mehrdimensionalen Befindlichkeitsfragebogen (MDBF). *Diagnostica, 40*, 320–328.
55. Urgen, B. A., Plank, M., Ishiguro, H., Poizner, H., & Saygin, A. P. (2013). EEG theta and Mu oscillations during perception of human and robot actions. *Frontiers in Neurorobotics, 7*, 19. https://doi.org/10.3389/fnbot.2013.00019

The View of Participants on the Potential of Conducting NeuroIS Studies in the Wild

Anke Greif-Winzrieth, Christian Peukert, Peyman Toreini, and Christof Weinhardt

Abstract Traditionally, NeuroIS studies are performed in highly controlled laboratory environments. However, due the proliferation of mobile sensor equipment, it is nowadays also possible to also run experiments in less controlled environments, e.g., in home settings. Running NeuroIS experiments in these settings allows researchers to pursue novel research questions in non-artificial environments as well as conducting large scale studies. It is unclear, though, what drives participants' willingness to take part in such studies. Therefore, this article sets out to explore the anticipated opportunities and concerns of potential participants. Within a survey (n = 69), we captured the participants' view on running NeuroIS experiments outside controlled laboratory environments. Our preliminary results provide first insights on what aspects influence the participants' decision on taking part in NeuroIS@Home studies. We argue that it is of utmost importance to design NeuroIS@Home studies carefully, taking into account participant's concerns.

Keywords NeuroIS · Participant View · NeuroIS@Home · Home Environments

1 Introduction

During the pandemic, the way how research and teaching activities are usually performed has changed, e.g., due to researcher's limited access to infrastructures

A. Greif-Winzrieth (✉) · C. Peukert · P. Toreini · C. Weinhardt
Karlsruhe Institute of Technology (KIT), Karlsruhe, Germany
e-mail: anke.greif-winzrieth@kit.edu

C. Peukert
e-mail: christian.peukert@kit.edu

P. Toreini
e-mail: peyman.toreini@kit.edu

C. Weinhardt
e-mail: weinhardt@kit.edu

© The Author(s), under exclusive license to Springer Nature Switzerland AG 2022
F. D. Davis et al. (eds.), *Information Systems and Neuroscience*,
Lecture Notes in Information Systems and Organisation 58,
https://doi.org/10.1007/978-3-031-13064-9_13

or social distancing measures [1–4]. However, research institutions faced these challenges by adapting to the new circumstances through setting up new approaches for collecting data [1, 4]. One of the most common approaches was to switch from in situ experiments to remote experiments (e.g., from laboratory to online experiments). While conducting online experiments proved to still work quite well with remote arrangements [5] by including different measures to, e.g., control for users' attention, researchers who rely on neurophysiological techniques and associated sensor systems encountered severe difficulties [6]. Transferring the environment for these kinds of studies is even more challenging and, here, specifically trade-offs between the three kinds of validity, i.e., internal, external, and ecological validity, need to be carefully considered [7, 8]. For instance, aspects such as adequate participant preparation or calibration must not be neglected to obtain reliable data. To overcome the limitations, researchers set out to explore new approaches. In a recent study [9], we introduced the idea of providing participants with a box equipped with the necessary equipment and instructions to perform NeuroIS experiments remotely, which we refer to as NeuroIS@Home. Particularly, our approach builds upon the recent advances in sensor systems that facilitate to also conduct NeuroIS studies in the wild. First NeuroIS scholars have already conducted studies in comparable remote @home settings [4, 6, 10, 11]. Although these solutions pioneered around home experiments, it is still under investigation how to best conduct such studies outside controlled laboratory environments. Based on a systematic literature review on the use of wearable devices for the measurement of heart rate and heart rate variability, Stangl & Riedl [12] call for further methodological and theoretical research to advance the field.

Whereas we have already identified challenges and opportunities that experts see in such an approach [9], Demazure et al. [1] have proposed a methodology for conducting EEG studies in home environments, and Vasseur et al. [6] have identified some success factors of remote data collection, the perspective of participants on such studies has so far remained untouched. However, the success of NeuroIS@Home highly depends on the participants' willingness to participate. One important difference to ordinary lab settings is that the responsibility of the individual participant is higher since tasks that are normally done by the experimenters are transferred to the participant [1, 6]. Consequently, we set out to explore opportunities and concerns from a participant's point of view on this novel approach and pose the following research question: *What are opportunities and concerns of participants regarding participation in NeuroIS studies in the wild?*

To answer the research question, we conducted an online survey in which we invited potential NeuroIS@Home participants. We have selectively invited participants with different prior exposure to NeuroIS studies to capture a diverse picture. Overall, we try to complement the expert opinions captured in [9] with the view by participants to provide a comprehensive overview on the potential of running NeuroIS studies in the wild. Within this article, we report on the preliminary findings of parts of the survey and follow a rather exploratory approach. We therefore directly introduce the applied method of the student survey before we report on our preliminary findings and discuss potential avenues for future research.

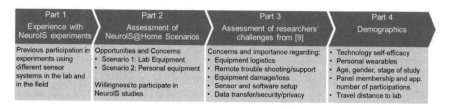

Part 1	Part 2	Part 3	Part 4
Experience with NeuroIS experiments	Assessment of NeuroIS@Home Scenarios	Assessment of researchers' challenges from [9]	Demographics
Previous participation in experiments using different sensor systems in the lab and in the field	Opportunities and Concerns • Scenario 1: Lab Equipment • Scenario 2: Personal equipment Willingness to participate in NeuroIS studies	Concerns and importance regarding: • Equipment logistics • Remote trouble shooting/support • Equipment damage/loss • Sensor and software setup • Data transfer/security/privacy	• Technology self-efficacy • Personal wearables • Age, gender, stage of study • Panel membership and app. number of participations • Travel distance to lab

Fig. 1 Overview survey structure

2 Method—Student Survey

To address our research question, we conducted an online survey consisting of four main parts that are summarized in Fig. 1.

In Part 1, participants were asked to indicate whether they have already taken part in NeuroIS studies employing different neurophysiological techniques and associated sensor systems (EEG, ECG, Eye-Tracking)[1] and, if so, to state for each sensor whether they participated on site, e.g., in a lab or remotely, e.g., at home (or both). To ease the participation of people, who are not familiar with the name of technologies, we provided images of the sensors for each technique. These images showed the following devices: Tobii 4C for eye tracking, cEEGGrids, Emotiv Flex, and Emotiv Epoc+ for EEG, polar H10 chest belt and Plux ECG sensor for ECG.

Part 2 started with an introduction to the NeuroIS@Home approach. We present Scenario 1, asking participants to imagine that they are "invited to an experiment that requires the use of neurophysiological techniques and associated sensor systems" and that the invitation states that they will "receive a box equipped with all necessary equipment and instructions for running the experiment remotely (e.g., at [their] home)." We asked the participants to provide at least three different opportunities and three different concerns about attending experiments with such a remote setup. In a second step, they were asked to select the most promising opportunity and the most severe concern. Furthermore, regarding the most severe concern, we were interested in whether participants may think of any solution to address the concern. Afterwards, we introduced a slightly adapted scenario in which instead of getting the required equipped in a box, participants are asked to use their own hardware (e.g., smartwatch, fitness wearables, webcam, VR goggles, etc.) in attending experiments (e.g., as suggested in [13])[2] and provide the recorded data. Again, at least three opportunities and concerns shall be listed. To also get an indication on how likely it is that they participate in NeuroIS studies, we surveyed different conceivable implementations of the approach (see Table 1) on a 5-point Likert scale (very unlikely | very likely) and captured the intention to take part in a regular laboratory NeuroIS study.

---

[1] We chose these techniques because they were used by researchers recruiting subjects from the panel we used for this survey.

[2] See Stangl & Riedl [12] for a comprehensive review on the usage of wearables (e.g., smartwatches or fitness trackers) to measure heart rate and heart rate variability.

Table 1 Sample demographics (N = 69)

Gender			Previous participation in experiments	
Female	39%		*1 to 5 experiments*	16%
Male	58%		*6 to 10 experiments*	38%
Diverse	3%		*More than 10 experiments*	46%
Age	M = 25.7, SD = 6.1		Previous participation in NeuroIS experiments	
Stage of study			*NeuroIS experiments*	77%
Undergraduate	42%		*Remote NeuroIS experiments*	35%
Graduate	48%		Travel time to the laboratory	
Ph.D.	7%		*Less than 15 min*	40%
Other	3%		*15 to less than 30 min*	35%
Duration of lab panel membership			*30 to less than 45 min*	12%
Less than 6 months	16%		*45 to 60 min*	4%
6 to less than 12 months	11%		*More than 60 min*	9%
12 to less than 24 months	9%			
24 months or more	64%			

Building on the results of our expert survey [9], in Part 3, we re-used the identified challenges that particularly apply to the participants (see Table 2) and asked the participants to evaluate whether they share the concerns (5-point Likert scale; not at all | a lot), how important this aspect is for their decision to participate (5-point Likert scale; not important at all | very important), and to describe how a potential solution might look like that would fit their expectations and needs. Finally, we elicited the participants' technology self-efficacy using three items adapted from Bandura [14], and asked whether they use smartwatches/fitness wearables in their everyday life. Finally, in Part 4, we collected some demographics (age, gender, duration of lab panel membership, number of participations, stage of study, travel time to the lab).

Participants were incentivized with a fixed payout of €3.00 and by taking part in a lottery in which one out of 20 participants will receive additional €20.00.

Table 2 Scales and results for study registration (M = Mean, SD = Standard Deviation)

Item	M	SD	> 3
How likely is it that you would register to participate...			
.. if you are invited to come to the lab to participate?	4.2	1.0	81%
...if you are asked to come to the lab to pick up a box equipped with all necessary equipment and instructions, run the experiment at home, and return the box to the lab?	3.1	1.4	64%
...if a box equipped with all necessary equipment and instructions is delivered to your home, you run the experiment at home, and send the box back?	4.1	1.0	75%
...if you are asked to use your own hardware (e.g., a smartwatch, fitness wearable, webcam, VR goggles, etc. that you own) and provide access to the recorded data?	3.7	1.2	67%

3 Results

Overall, 69 participants (39% female, 58% male, 3% diverse) recruited[3] from the participants panel of the Karlsruhe Decision & Design Lab (KD²Lab) took part in the study. The study took place in March 2022 and was performed via an online questionnaire. It took on average 22 min (SD = 8) to complete the questionnaire. The mean age of the participants was 25.7 years (SD = 6.1). All except one participant were students (42% undergraduate, 48% graduate, 7% Ph.D., 3% other). 16% have been registered in the panel for less than 6 months, 11% for 6 to less than 12 months, 9% for 12 to less than 24 months, and 64% for 24 months or more. 16% of participants had participated in 1 to 5 experiments before, 38% in 6 to 10 experiments and 46% in more than 10 experiments. Regarding prior NeuroIS experience, 77% stated that they had previously participated in experiments using neurophysiological tools and 35% stated that they had participated in such experiments in remote setups outside the lab. In terms of travel time to the laboratory, 41% stated that they can reach the lab in less than 15 min by their usual mode of transportation, for 35% it takes 15 to less than 30 min, for 12% 30 to less than 45 min, for 4% 45 to 60 min, and for 9% more than 60 min. Table 1 summarizes the sample demographics.

Participants submitted 240 (max. 7 per participant) concerns and 231 opportunities (max. 9 per participant) for Scenario 1 (i.e., use of equipment delivered from the lab). For Scenario 2 (i.e., use of personal equipment), they submitted 224 concerns (max. 6 per participant) and 216 opportunities (max. 8 per participant).

Although, a comprehensive labelling of the individual opportunities and concerns will be part of future research, we provide first insights in the following: Many opportunities and concerns articulated by the participants of this survey are in line with the findings from our experts survey [9] and support the importance of following

---

[3] The experiment was recruited with the software hroot [16].

Table 3 Scales and results for participants' concerns regarding challenges identified in [9]

Item	M	SD	> 3	Example of suggested solution
Do you have any concerns about...				
...equipment logistics?	3.5	1.0	64%	"Sent to a Packstation[4]"
...remote trouble shooting?	3.7	1.3	65%	"Live chat or telephone number"
...equipment damage/loss?	3.6	1.3	62%	"Insurance"
...sensor and software setup?	3.6	1.2	68%	"Video guides"
...data transfer/security/privacy?	3.3	1.4	52%	"Transfer with a USB Stick or sth similar"

the guidelines for conducting EEG experiments in the wild proposed by Demazure et al. [1]. Participants for example highlighted the importance of reliable solutions for remote trouble shooting and support (e.g., "no/no proper help if problems occur during experiment at home") and mentioned their concerns regarding bearing greater responsibility for correct setup, handling, and logistics of equipment (e.g., "You carry more responsibility."), which is also in line with previous findings [1, 6]. However, participants also came up with additional topics like health and safety ("[...] What if some wires short-cut? Give electrical schock?") as well as sustainability concerns (e.g., "Transport causes emissions") and cheating (e.g., "intentional not doing the study correct (save time)"). Among the most important opportunities from the participants' point of view are—similar to what we found in the expert survey [9]—convenience/ease of participation (e.g., "comfort of the home, less excitement etc."), scalability (e.g., "More participants through the remote possibility"), and external/ecological validity (e.g., "real life data not biased by the laboratory atmosphere"). Further, participants highlight flexibility in time and place (e.g., "Flexible schedule to participate in the study from home"). As presented in Table 2, more than 60% of participants state that it is rather or very likely (>3 on the Likert scale) that they would accept an invitation to participate for each of the scenarios. The data further suggests that travel time to the lab affects the assessment of the different scenarios (i.e., we find a trend that participants with longer travel times prefer the delivery of the box to their home).

Table 3 reports the participants' assessment of some of the challenges that we identified in [9] regarding their level of concern and lists some examples of suggested solutions for each category of challenges.

Figure 2 presents the participants' assessment of the importance of some of the challenges we identified in [9] regarding their decision about participating. Overall, the results suggest that the challenges are considered similarly important to the participants.

---

[4] A "Packstation" is a central place to pick up and send parcels powered by a well-known logistics service provider.

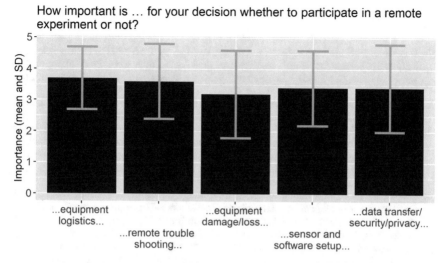

How important is ... for your decision whether to participate in a remote experiment or not?

Fig. 2 Participants' assessment of the importance of challenges identified in Greif-Winzrieth et al. [9] for their decision about participating

4 Concluding Note & Future Research

Within this article, we set out to explore the opportunities but also concerns of a novel approach of conducting NeuroIS experiments in the wild, especially from a participant's point of view. Thereby, we followed a rather exploratory approach and captured the participants' opinions by means of an online survey. We presented first promising results that need to be further explored in the course of our research project. Our next steps encompass further analyses of the collected data, e.g., by means of a qualitative content analysis [15] with the collected arguments, building categories, conduct in-depth interviews with both NeuroIS experts and participants, and triangulating the findings with previous results from other studies (e.g., [1, 6, 9]). Finally, based on our results, we want to derive a set of guidelines that aims at supporting researchers when planning to run NeuroIS experiments in the wild. Thereby, we will provide insights on the barriers to participation and propose possible steps researchers should take care of to ensure and increase participation. These guidelines will need to carefully consider further important issues we only briefly touched on in this paper such as validity, data quality, and ethical considerations of conducting NeuroIS studies in less controlled environments. Based on our preliminary results, we suggest that conducting NeuroIS studies in the wild bears a great potential for some, but for sure not all research contexts. Thus, we expect that the insights from our research will contribute to help researchers in carefully making design decisions and choosing appropriate methods to investigate their research questions in the NeuroIS field.

References

1. Demazure, T., Karran, A. J., Boasen, J., Léger, P. -M., Sénécal, S. (2021). Distributed remote EEG data collection for NeuroIS research: a methodological framework. In: Schmorrow, D. D. and Fidopiastis, C. M. (eds.) *Augmented Cognition 15th International Conference, AC 2021,* (pp. 3–22).
2. van der Aalst, W., Hinz, O., & Weinhardt, C. (2020). Impact of COVID-19 on BISE Research and Education. *Business & Information Systems Engineering, 62,* 463–466.
3. Demazure, T., Karran, A., Léger, P. (2021). Continuing doctoral student training for NeuroIS and EEG during a pandemic: a distance Hands-on learning syllabus. In: Davis, F.D., Riedl, R., vom Brocke, J., Léger, P., Randolph, A. B., & Müller-Putz, G. (eds.) *Information Systems and Neuroscience. NeuroIS 2021. Lecture Notes in Information Systems and Organisation, 52,* (pp. 204–211). Springer, Cham, Switzerland.
4. Labonté-lemoyne, E., Courtemanche, F., Coursaris, C., Hakim, A., Séneécal, S., Léger, P. -M.: Development of a new dynamic personalised emotional baselining protocol for Human-computer interaction. In: Davis, F. D., Riedl, R., vom Brocke, J., Léger, P., Randolph, A. B., & Müller-Putz, G. (eds.) *Information Systems and Neuroscience. NeuroIS 2021. Lecture Notes in Information Systems and Organisation, 52,* (pp. 237–242). Springer, Cham, Switzerland.
5. Gould, S. J. J., Cox, A. L., Brumby, D. P., & Wiseman, S. (2015). Home is where the lab is: a comparison of online and lab data from a time-sensitive study of interruption. *Human Computation, 2,* 45–67.
6. Vasseur, A., Léger, P., Courtemanche, F., Labonte-, E., Georges, V., Valiquette, A., Brieugne, D., Rucco, E., Coursaris, C., Fredette, M., Sénécal, S. (2021). Distributed remote psychophysiological data collection for UX evaluation: a pilot project. In: Kurosu, M. (ed.) *Human-Computer Interaction. Theory, Methods and Tools. HCII 2021. Lecture Notes in Computer Science, 12762,* (pp. 255–267). Springer.
7. Karahanna, E., Benbasat, I., Bapna, R., & Rai, A. (2018). Editor's comments: Opportunities and challenges for different types of online experiments. *MIS Quarterly, 42,* iii–x.
8. Riedl, R., Fischer, T., Léger, P. -M., Davis, F. (2020). A decade of NeuroIS research: Progress, challenges, and future directions. *ACM SIGMIS Database DATABASE Advances in Information Systems, 51,* 13–54.
9. Greif-Winzrieth, A., Peukert, C., Toreini, P., Adam, M. T. P. (2021). Exploring the Potential of NeuroIS in the Wild: Opportunities and Challenges of Home Environments. In: Davis, F. D., Riedl, R., vom Brocke, J., Léger, P., Randolph, A. B., and Müller-Putz, G. (eds.) *Information Systems and Neuroscience. NeuroIS 2021. Lecture Notes in Information Systems and Organisation,* 52, (pp. 38-46). Springer, Cham, Switzerland.
10. Berger, C., Knierim, M. T., Weinhardt, C. (2021). Detecting flow experiences in the field using video-based head and face activity recognition: a pilot study. In: Davis, F. D., Riedl, R., vom Brocke, J., Léger, P., Randolph, A. B., and Müller-Putz, G. (eds.) *Information Systems and Neuroscience. NeuroIS 2021. Lecture Notes in Information Systems and Organisation, 52,* (pp. 120–127). Springer, Cham, Switzerland.
11. Knierim, M. T., Pieper, V., Schemmer, M., Loewe, N., Reali, P. (2021) Predicting in-field flow experiences over two weeks from ECG data: A Case Study. In: Davis, F.D., Riedl, R., vom Brocke, J., Léger, P., Randolph, A.B., and Müller-Putz, G. (eds.) *Information Systems and Neuroscience. NeuroIS 2021. Lecture Notes in Information Systems and Organisation, 52,* (pp. 96–102). Springer, Cham, Switzerland.
12. Stangl, F. J., Riedl, R. (2022). Measurement of heart rate and heart rate variability with wearable devices: a systematic review. In: *17th International Conference on Wirtschaftsinformatik.* Nürnberg, Germany.
13. Fischer, T., & Riedl, R. (2017). Lifelogging as a viable data source for NeuroIS researchers: A review of neurophysiological data types collected in the lifelogging literature. *Lecture Notes in Information Systems and Organisation, 16,* 165–174.
14. Bandura, A. (1977). Self-efficacy: Toward a Unifying Theory of Behavioral Change. *Psychological Review, 84,* 191–215.

15. Mayring, P. (2015). Qualitative inhaltsanalyse : Grundlagen und Techniken. Beltz, Weinheim; Basel.
16. Bock, O., Baetge, I., & Nicklisch, A. (2014). hroot: Hamburg Registration and Organization Online Tool. *European Economic Review, 71*, 117–120.

Emoticons Elicit Similar Patterns of Brain Activity to Those Elicited by Faces: An EEG Study

Alessandra Flöck and Marc Mehu ⓘ

Abstract The present study investigated whether the patterns of brain activity elicited by emoticons are similar to those elicited by faces. In order to test this, participants were subliminally primed using either human faces, emoticons, or non-face control stimuli. Each prime group contained three levels of valence—positive, negative, or neutral. Brain activation was recorded via electroencephalography. Subsequently, three event related potential components of interest were identified, which are closely associated with the processing of faces, i.e., the P100, the N170, and the late positive potential. These ERPs were tested at two electrode sites. For each ERP component, peak amplitudes were calculated and used in repeated-measures ANOVAs. There were significant main effects of prime type across several ERP components, and several interaction effects prime type*prime valence on four out of six components. There was no statistically significant main effect of prime valence on any of the ERP components. While our results uncovered diverse patterns of brain reactivity to the different kinds of primes, there is sufficient evidence to suggest that, at some electrode locations, emoticons elicit similar brain reactions as faces do.

Keywords Emoticons,·Priming · P100 · N170 · LPP

1 Introduction

The idea that non-verbal communication is essential to human survival and reproduction has a long history in psychological and communication research [1, 2]. One of the newest challenges facing human beings nowadays is the correct interpretation and encoding of messages lacking any form of non-verbal signaling, as is the case for online communication [3]. This manner of communication has recently been referred to as Computer Mediated Communication (CMC), which implies communication to occur via the use of two or more electronic devices [3]. Within this framework, researchers have investigated the importance of emoticons as potential

A. Flöck · M. Mehu (✉)
Department of Psychology, Webster Vienna Private University, Vienna, Austria
e-mail: marc.mehu@webster.ac.at

© The Author(s), under exclusive license to Springer Nature Switzerland AG 2022   133
F. D. Davis et al. (eds.), *Information Systems and Neuroscience*,
Lecture Notes in Information Systems and Organisation 58,
https://doi.org/10.1007/978-3-031-13064-9_14

carriers and transmitters of messages usually taken up by non-verbal cues in face-to-face exchanges [4]. Research suggest that emoticons are not solely viable carriers of such messages, but they also bring about the advantage of a more controlled communication, as the use of emoticons is a conscious decision, thereby eliminating the possibility of transmitting unwanted emotions [5–7].

One way of investigating the neurobiological processes involved in the perception of facial expression using contemporary research equipment is by implementing electroencephalography (EEG) [4]. Here, analysis has focused mostly on event related potentials (ERPs), which reflect synchronized changes in neuron group activation across the human cortex. ERPs can be understood as brain signals which occur as a result of thought, perception, or in response to internal stimulation and involve a multiplicity of neurological processes, such as memory, expectations, variations in mental states, and attention modulation [8]. Because the mechanisms underlying emotion recognition and social signal processing have been postulated to be mostly automatic and unconscious, researchers have often paired EEG with subliminal priming techniques, so as to better understand the perceptual processes elicited by facial expressions [8–10]. This is possible due to the inherent quality of EEG to possess high temporal resolution, which makes possible the measurement of brain activation within milliseconds post stimulus onset [11].

In the context of face processing and recognition of facial expressions, three ERP components of interest have been defined: (1) the P100, (2) the N170, and (3), the late positive potential (LPP) [4, 12, 13]. These ERPs have often been deemed to be measurable in occipito-parietal and occipito-temporal areas of the human cortex. Both the P100 and the N170 component have been found to be modulated by emotional stimuli—such as happy or angry faces [4, 12, 14]. For the present study, three electrodes of interest were selected, namely electrode 6 (located frontally), and electrodes 35 and 39 (located occipitally, respectively on the left and right hemisphere) [8, 15]. While the electrode sites located occipitally have often been found to reflect emotion processing and priming effects, the role of the frontal lobe has solely been mentioned in the context of emotional processing [16]. In the present study, we investigate whether subliminal emotional priming may also evoke potentials in the frontal lobe.

Believing this to be an important step in demonstrating the utility of emoticons in online communication, the proposed project investigates whether emoticons elicit similar patterns of brain activation as faces do, by focusing on three electrode sites and ERP components formerly associated with face perception and emotion processing. In order to investigate this phenomenon, participants were subliminally primed with positively, negatively, and neutrally valenced human faces, emoticons, and control stimuli. We assume that this methodology allows us to access the affective realm of the participants' minds, as the data obtained is minimally influenced by conscious thought.

2 Materials and Method

2.1 Participants

Overall, 30 participants (17 women) partook in the present study. All of the participants reported to be above the age of 18, showed to be right-handed, and indicated to have normal or corrected-to-normal vision capacities. However, two participants had to be excluded from the experiment, given that they did not fulfil the inclusion criteria. The mean age of the participants—computed out of the remaining 28 partakers—was 21,7 years, with a standard deviation of 1,9 years.

2.2 Study Design

The present study used three different prime types, namely three actual human faces, three emoticons, and finally, three non-face control stimuli. Each of these stimuli contained a different valence, i.e., positive, negative, or neutral. Each trial included the sequential presentation of a fixation cross (500 ms), a mask (##/17 ms), one of the primes (17 ms), another mask (##/17 ms), a word (1000 ms), and finally a screen prompting the participant to give their judgment on perceived valence (see also Fig. 1.). The words were taken from the English Word Database of Emotional Terms (EMOTE) [17]. For the first part of the present study, 270 nouns with a valence ranging between 3.6 and 4.4 were selected and subsequently displayed in connection with one of the aforementioned primes. The human faces were extracted from the Radboud Faces Database (RaFD). In a similar manner as proposed by Gantiva et al. (2020), the present study used pictures portraying a Caucasian male, so as to ensure familiarity with the facial features of the model (see Fig. 2. for example stimuli) [4]. The stimuli were presented on a Dell E2214hb 21.5″ widescreen LED LCD monitor. The monitor brightness was constantly set to the medium setting for each participant. All of the stimuli were presented in a random order. The experiment was programmed using the E-Prime 2.0 software.

2.3 Data Collection

In order to collect data on brain activation, we used the Geodesic EEG™ System 400 with an embedded HydroCel Geodesic Sensor Net with 64 electrodes (silver chloride sensors). Potential changes in activation were continuously sampled at the rate of 1000 Hz, implementing the EGI Net Amps 400 amplifier, containing a built-in Intel chip. Further, an applied online lowpass filter of 60 Hz was implemented. Data was continuously recorded via the Net Station 5.4 software.

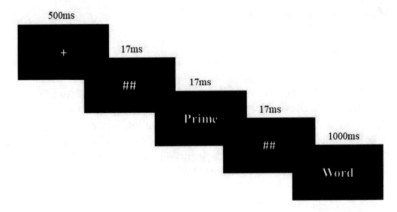

Fig. 1 Presentation sequence of a trial

Fig. 2 Example primes; all taken from the negative valence category

2.4   Procedure

Upon entering the laboratory, participants were asked to give informed consent about their willingness to partake in the present study. They were subsequently seated, and the EEG system was applied. After all electrodes were tested for connectivity and impedance, the participant received the instructions for the first part of the experiment, including how to diminish the chance of artefacts via reduced blinking and swallowing during the measuring periods.

2.5   Data Analysis

The processing of the acquired EEG signals was carried out using the EEG DISPLAY 6.4.9 software. Potentials were locked to the onset of the prime as pictured above. For each data set, an offline bandpass filter from 0.1, to 30 Hz was applied, before epochs were generated starting 100 ms pre-stimulus onset and ending 900 ms post stimulus. Visible artefacts, such as obvious blinking, and electrooculogram amplitudes over 75 mV were eliminated during the cleaning process. As a next step, averages for each condition and all electrodes were calculated and re-referenced to the common reference point. Lastly, grand averages were calculated to visualize all potential changes in brain activation.

For the statistical analysis we conducted 3 X 3 repeated-measures Analyses of Variance (ANOVA) with prime type (face, emoticon, and non-face), and valence (positive, negative, neutral) as within-subjects factors; and peak ERP amplitudes (P100, N170, and LPP) as dependent variables. When sphericity could not be assumed, the Greenhouse–Geisser correction was applied. Simple main effect analysis were conducted to follow-up on significant interaction effects. Peak amplitudes of the respective ERP components were used over a range of 20 ms before, to 50 ms after the assumed timepoint of the respective ERP components, to account for inter-individual variation in the timing of these components [14][14]. Further, post-hoc pairwise comparisons were computed to follow-up on the main effects uncovered by the ANOVAs. These analyses were conducted for three electrode locations: One located fronto-medially (electrode 6), and two located occipitally, on the left and right hemisphere (electrodes 35 and 39).

3   Results

In the following sections, results will be shown for all of the electrode locations at all timepoints of interest. The analysed electrodes were grouped so as to make more obvious the overall differences in activation between the locations.

3.1   P100 Electrode 6

The repeated-measures ANOVA revealed a significant effect of prime type [F (2,54) = 11.298, $p < 0.001$], a non-significant effect of prime valence [F (2,54) = 0.917, $p = 0.406$], and a significant interaction of prime type*prime valence [F (4,108) = 3.739, $p = 0.007$]. Pairwise comparisons revealed significant differences between emoticons and control stimuli ($p = 0.005$), and between faces and control stimuli ($p = 0.001$), both emoticon and faces elicited higher peak amplitudes than control stimuli. The effect of prime type was investigated separately for the different levels of prime valence. This analysis revealed a significant effect of prime type for the positive and neutral stimuli (all $ps < 0.001$). Post-hoc pairwise comparisons further revealed a significant difference between neutral faces and neutral controls ($MD = 2.642, p < 0.001$).

3.2 N170 Electrode 6

For the N170 component at electrode location 6, the repeated-measures ANOVA revealed a significant effect of prime type [F $(2,54) = 13.436$, $p < 0.001$), but a non-significant effect of prime valence [F $(2,54) = 0.406$, $p = 0.668$). Pairwise comparisons revealed significant differences between faces and control stimuli ($p = 0.017$), and between emoticons and control stimuli ($p < 0.001$). The interaction effect prime type*prime valence was also significant [F $(4,108) = 4,556, p = 0.002$]. The effect of prime type was evaluated separately for the different levels of valence. The main effect of prime type was observed for negative ($p < 0.001$), positive ($p = 0.044$), and neutral stimuli ($p < 0.001$). Post hoc comparison of prime type*prime valence revealed significant differences between negative emoticons and negative control stimuli ($MD = 2.444$, $p = 0.002$); and between negative faces and negative control stimuli ($MD = 2.139$, $p = 0.014$); between neutral emoticons and neutral controls ($MD = 2.004$, $p = 0.03$); and between neutral faces and neutral control stimuli ($MD = 2.917$, $p < 0.001$). In all these cases, control stimuli elicited greater N170 than face and emoticon stimuli at that location.

3.3 LPP Electrode 6

Analysis for the LPP at electrode location 6 revealed a significant effect of prime type [F $(2,54) = 8.864, p < 0.001$). There was no effect of prime valence [F $(2,54) = 0.67$, $p = 0.516$), and lastly no significant interaction effect of prime type*prime valence [F $(2.986,80.629) = 1.905$, $p = 0.136$]. Pairwise comparisons revealed significant differences between emoticons and control stimuli ($p = 0.013$), and between faces and control stimuli ($p < 0.001$). The LPP amplitude elicited by both emoticons and faces was larger than for control stimuli (Fig. 3).

3.4 P100 Electrode 35

The repeated-measures ANOVA revealed a significant effect of prime type [F $(2,54) = 9.650$, $p < 0.001$], a non-significant effect of prime valence [F $(2,54) = 0.107$, $p = 0.898$], and a slight trend for the interaction effect between prime type*prime valence displayed [F $(3.09,8.921) = 2.105$, $p = 0.098$]. Post hoc analyses revealed a significant difference between emoticons and faces ($p = 0.03$), and a significant difference between emoticons and control stimuli ($p < 0.001$). Emoticons appeared to elicit lower amplitudes than faces and control stimuli at that location.

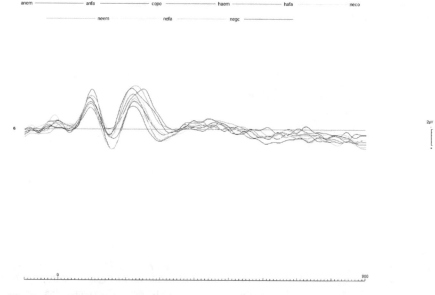

Fig. 3 Overall activation at electrode 6 from 100 ms pre- to 900 ms post-stimulus onset. Note. anem: negative emoticon; haem: positive emoticon; neem: neutral emoticon; hafa: positive face ; anfa: negative face; nefa: neutral face; copo: positive control; negc: negative control; neco: neutral control

3.5 N170 Electrode 35

Repeated-measures ANOVA revealed a significant effect of prime type [F (2,54) = 27.493, $p < 0.001$]. The effect of prime valence was not statistically significant [F (2,54) = 0.026, $p = 0.974$]. Still, the interaction effect between prime type*prime valence proved to be statistically significant [F (4,108) = 2.678, $p = 0.036$). Post hoc pairwise comparisons for the main effect of prime type revealed significant differences between emoticons and faces, and between faces and control stimuli (p respectively < 0.001). A follow-up analysis revealed that the effect of prime type was significant for all levels of valence (all $ps < 0.001$). Post hoc comparisons showed negative emoticons to be significantly different to negative faces ($MD = 2.435, p = 0.013$), and negative faces showed to be significantly different to negative controls ($MD = -3.041, p < 0.001$). Positive emoticons are significantly different to positive faces ($MD = 3.229, p < 0.001$). Positive faces were significantly different to positive controls ($MD = -2.299, p = 0.024$). Lastly, neutral faces were significantly different than neutral controls ($MD = -4.028, p < 0.001$). Faces appeared to elicit a stronger N170 at that location than emoticons and control stimuli.

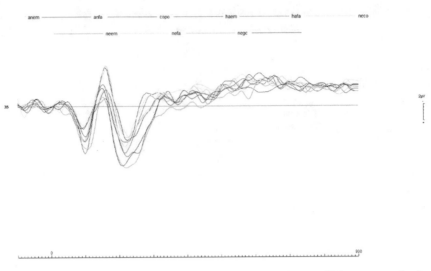

Fig. 4 Overall activation at electrode 35 from 100 ms pre- to 900 ms post-stimulus onset. Note. anem: negative emoticon; haem: positive emoticon; neem: neutral emoticon; hafa: positive face; anfa: negative face; nefa: neutral face; copo: positive control; negc: negative control; neco: neutral control

3.6 LPP Electrode 35

A repeated measures ANOVA was conducted. Here, again, a significant effect of prime type was found [F (2,54) = 22.871, $p < 0.001$). However, neither prime valence categories [F (2,54) = 0.256, $p = 0.775$), nor the interaction between prime type*prime valence [F (4,108) = 1.432, $p = 0.228$) were statistically significant. Post-hoc pairwise comparisons of prime type showed emoticons and control stimuli to be significantly different ($p < 0.001$). Furthermore, faces and control stimuli showed significant differences ($p < 0.001$). Control stimuli appeared to elicit higher LPP amplitudes than both emoticons and faces at this location (Fig. 4).

3.7 P100 Electrode 39

Repeated measures ANOVA revealed a significant effect of prime type [F (2,54) = 16.625, $p < 0.001$]. Prime valence was not significant [F (2,54) = 2.712, $p = 0.075$], and the interaction between prime type*prime valence revealed a trend [F (4,108) = 2.777, $p = 0.052$]. Post-hoc pairwise comparisons of prime type revealed both faces and emoticons to be significantly different to control stimuli (all $ps < 0.001$). Emoticons and faces appear to elicit greater amplitude than control stimuli at this timepoint and location.

3.8   N170 Electrode 39

Repeated measures ANOVA revealed a significant effect of prime type [F (2,54) = 38.767, $p < 0.001$]. Whereas prime valence showed to be non-significant [F (2,54) = 1.702, $p = 0.192$], the interaction between prime type*prime valence showed a slight trend [F (4,108) = 2.146, $p = 0.08$). Post-hoc pairwise comparisons of prime type revealed significant differences between faces and emoticons ($p < 0.001$), and faces and control stimuli ($p < 0.001$). Faces seem to elicit the most negative potentials across all valence categories.

3.9   LPP Electrode 39

Repeated measures ANOVA revealed a significant effect of prime type [F (2,54) = 15,912, $p < 0.001$], but non-significant effects of prime valence [F (2,54) = 2.437, $p = 0.097$], and of the interaction between prime type*prime valence [F (4,108) = 1.258, $p = 0.291$]. Post-hoc pairwise comparisons of prime type revealed both faces and emoticons to be significantly different to control stimuli (all $ps < 0.001$). Emoticons and faces appear to elicit more negative-going activation than the control stimuli (Fig. 5).

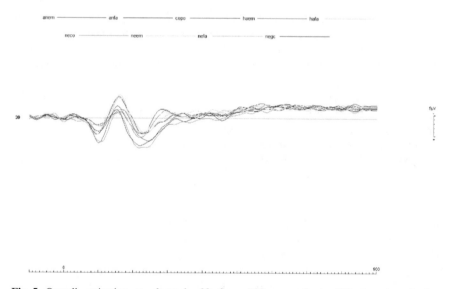

Fig. 5 Overall activation at electrode 39 from 100 ms pre- to 900 ms post-stimulus onset. Note. anem: negative emoticon; Hhaem: positive emoticon; neem: neutral emoticon; hafa: positive face; anfa: negative face; nefa: neutral face; copo: positive control; negc: negative control; neco: neutral control

4 Discussion

The present study set out to investigate whether the human brain processes emoticons in the same manner as actual human faces. This was done to establish whether emoticons could be considered equivalent to faces as viable transmitters of emotional messages. Implementing a priming paradigm, this study attempted to bypass subjective judgement given by the participants, and instead gather information directly from their affective processing system, the data obtained thereby only minimally influenced by conscious thought [19]. Three electrode locations were chosen for the present analysis, one located fronto-medially (electrode 6), and two residing just above the occipital cortex (electrodes 35/39). Additionally, three ERP components of interest had been identified, all frequently mentioned within the context of face processing, namely, the P100, the N170, and the LPP) [4, 12, 13].

Following the analysis of the ERP components of interest at three different electrode locations, the patterns of differences observed suggest that the type of prime more than its valence affected ERP components. Results suggest an overall trend of both emoticons and faces to elicit similar patterns of activation, showing a clear demarcation to the control stimuli in 6 out of 9 tests conducted. As aforementioned, the role of the frontal cortex in the processing of emotions and faces has not yet been fixed, the selection of a frontal electrode consequently of a purely explorative nature [20]. Interestingly, it appears as though even at frontal locations, we could observe a similarity of activation between emoticons and faces for the three ERP components. Concerning the measurement of the N170 component at this specific electrode location, one may be able to argue that the component actually measured was the well-known N200 component, which—although formerly hypothesized to be responsive to mismatches—is contemporarily also linked to cognitive control at early timepoints—specifically in response to visual stimuli [21].

Turning towards an investigation at the electrode locations 35 and 39, located by the occipital cortex of the brain, one comes to discover slight discrepancies to the existent literature. Although faces and emoticons seem to elicit rather similar activation in most of the tested cases, it appears as though it is not always the activation one may have expected. Looking at the P100 component at electrode location 39, as well as the LPP component at both electrode sites, one would have assumed for emoticons and faces to elicit the most positive going potentials, given that these components have long been observed in the given context—specifically when concerned with emotionally laden stimuli [12, 14]. Yet, the implemented control stimuli seem to elicit the most positive going potentials at the respective timepoints and sites. Concerning the N170 component at occipital sites, results suggest the potentials elicited by the human face primes to be the most negative going. These results seem to pose themselves as a conundrum, given that faces and emoticons elicit very similar patterns of activation across many testing conditions, but yet, rather different ones when concerned with the actual processing of faces as reflected most strongly by the N170. Consequently, our results show that while emoticons and faces share some activation patterns at certain location and for certain components, emoticons and faces nonetheless differ on a component that is important for the visual processing of faces.

Generally, the present study found a clear effect of prime type, which remained significant across all testing conditions. Prime valence, however, did not show such an effect. This appears specifically interesting given that prime valence had been assumed to be of importance in the modulation of both the P100 component and the N170 component [8]. As both the P100 and the N170 are closely correlated with the correct detection of facial emotion, and seem to solely be evoked by actual faces in within the proposed study, one may hypothesize the lack of modulation in the emoticons condition to possibly stem from a lack of correct emotion detection when faced with these stimuli [22, 23]. Further, one also has to take into consideration that although occipital sites have formerly been associated with both the reflection of priming effect as well as effects of emotion processing [15], several studies have reported P100 and N170 specific modulations to be primarily measurable on the right hemisphere—at occipito-parietal regions [22]. Consequently, one may turn to the analysis of further electrode locations as a next step, thereby comparing the measurable effects at multiple sites across both hemispheres.

Lastly, one may hold that a possible alternative explanation for the present findings may be the notion that face processing is highly susceptible to controlled processes [13]. Here, one may turn to the notion of reappraisal—a cognitive process influencing the emotion regulation of the human being—both implicitly and explicitly [24]. This is mostly the case when encountering biologically salient stimuli. Typically, when processing such stimuli, the amygdaloid complex is activated, simultaneously also orchestrating activation in other cortical and subcortical regions of interest in each processing stage [5, 13]. Moreover, the regions activated by the amygdaloid complex have been associated with areas involved in early visual processing, which have been postulated to possibly be the origin of both the P100 and N170 ERP components [13]. Applying this theory to the present study, it could be that the patterns of activation we observed result from cognitive processes intermingling with the affective responses stemming from the implicit processing of faces, or even face-approximating stimuli such as emoticons. Naturally, further research on the topic would be necessary to test the applicability of the concept of reappraisal to the present results, given that it seems as though it has neither been researched in the context of subliminal priming, as well as in connection with emoticons. Yet, the concept in and of itself seems to be a promising explanation.

Concluding, the present findings seem to present themselves quite divergently, as it seems that although faces and emoticons seem to be processed in a highly similar manner across most conditions, emoticons do not seem to elicit the face-processing-specific N170 component. Moreover, it seems as though the potentials elicited by emoticons are not modulated by emotionally laden content, which could be due to the fact that participants did not correctly recognize the emotion displayed by the emoticon primes. Further research may focus on the manner in which emoticons and faces seem to share similarities one a different plane—i.e. not directly reflected in the face processing sphere.

References

1. Boone, R. T., & Buck, R. (2003). Emotional expressivity and trustworthiness: the role of nonverbal behavior in the evolution of cooperation. *Journal Of Nonverbal Behavior, 27*(3), 20.
2. Schug, J., Matsumoto, D., Horita, Y., Yamagishi, T., & Bonnet, K. (2010). Emotional expressivity as a signal of cooperation. *Evolution and Human Behavior, 31*(2), 87–94. https://doi.org/10.1016/j.evolhumbehav.2009.09.006
3. Klopper, R. (2010). The evolution of human communication from nonverbal communication to electronic communications. *14.*
4. Gantiva, C., Sotaquirá, M., Araujo, A., & Cuervo, P. (2020). Cortical processing of human and emoji faces: An ERP analysis. *Behaviour & Information Technology, 39*(8), 935–943. https://doi.org/10.1080/0144929X.2019.1632933
5. Kim, K. W., Lee, S. W., Choi, J., Kim, T. M., & Jeong, B. (2016). Neural correlates of text-based emoticons: A preliminary fMRI study. *Brain and Behavior, 6*(8). https://doi.org/10.1002/brb3.500
6. McKenna, K. Y. A., Green, A. S., & Gleason, M. E. J. (2002). Relationship Formation on the Internet: What's the Big Attraction? *Journal of Social Issues, 58*(1), 9–31. https://doi.org/10.1111/1540-4560.00246
7. Derks, D., Bos, A. E. R., & von Grumbkow, J. (2008). Emoticons and Online Message Interpretation. *Social Science Computer Review, 26*(3), 379–388. https://doi.org/10.1177/0894439307311611
8. Elgendi, M., Kumar, P., Barbic, S., Howard, N., Abbott, D., & Cichocki, A. (2018). Subliminal Priming—State of the Art and Future Perspectives. *Behavioral Sciences, 8*(6), 54. https://doi.org/10.3390/bs8060054
9. Doyen, S., Klein, O., Simons, D. J., & Cleeremans, A. (2014). On the other side of the mirror: priming in cognitive and social psychology. *Social Cognition, 32*(Supplement), 12–32. https://doi.org/10.1521/soco.2014.32.supp.12
10. Pell, M. D. (2005). Nonverbal emotion priming: evidence from the facial affect decision task. *Journal of Nonverbal Behavior, 29*(1), 45–73. https://doi.org/10.1007/s10919-004-0889-8
11. Koenig, T., Prichep, L., Lehmann, D., Sosa, P. V., Braeker, E., Kleinlogel, H., Isenhart, R., & John, E. R. (2002). Millisecond by millisecond, year by Year: normative eeg microstates and developmental stages. *NeuroImage, 16*(1), 41–48. https://doi.org/10.1006/nimg.2002.1070
12. Zheng, X., Mondloch, C. J., & Segalowitz, S. J. (2012). The timing of individual face recognition in the brain. *Neuropsychologia, 50*(7), 1451–1461. https://doi.org/10.1016/j.neuropsychologia.2012.02.030
13. Blechert, J., Sheppes, G., Di Tella, C., Williams, H., & Gross, J. J. (2012). See what you think: reappraisal modulates behavioral and neural responses to social stimuli. *Psychological Science, 23*(4), 346–353. https://doi.org/10.1177/0956797612438559
14. Tanaka, J. W., & Pierce, L. J. (2009). The neural plasticity of other-race face recognition. *Cognitive, Affective, & Behavioral Neuroscience, 9*(1), 122–131. https://doi.org/10.3758/CABN.9.1.122
15. Tanaka, M., Yamada, E., Maekawa, T., Ogata, K., Takamiya, N., Nakazono, H., & Tobimatsu, S. (2021). Gender differences in subliminal affective face priming: A high-density ERP study. *Brain and Behavior, 11*(4). https://doi.org/10.1002/brb3.2060
16. Kragel, P. A., Kano, M., Van Oudenhove, L., et al. (2018). Generalizable representations of pain, cognitive control, and negative emotion in medial frontal cortex. *Nature Neuroscience, 21*, 283–289. https://doi.org/10.1038/s41593-017-0051-7
17. Grühn, D. (2016). An english word database of emotional terms (EMOTE). *Psychological Reports, 119*, 290–308.
18. Turano, M. T., Marzi, T., & Viggiano, M. P. (2016). Individual differences in face processing captured by ERPs. *International Journal of Psychophysiology, 101*, 1–8. https://doi.org/10.1016/j.ijpsycho.2015.12.009

19. Ferreira, H. A., Saraiva, M. (2019). Subjective and Objective Measures. In: Ayanoğlu, H., Duarte, E. (eds) Emotional Design in Human-Robot Interaction. *Human–Computer Interaction Series. Springer, Cham.* https://doi.org/10.1007/978-3-319-96722-6_9

20. Zhen, Z., Fang, H., & Liu, J. (2013). The hierarchical brain network for face recognition. *PLoS ONE, 8*(3), e59886. https://doi.org/10.1371/journal.pone.0059886

21. Folstein, J. R., & Van Petten, C. (2007). Influence of cognitive control and mismatch on the N2 component of the ERP: A review. *Psychophysiology, 0*(0), 070915195953001–???. https://doi.org/10.1111/j.1469-8986.2007.00602.x.

22. Utama, N. P., Takemoto, A., Koike, Y., & Nakamura, K. (2009). Phased processing of facial emotion: An ERP study. *Neuroscience Research, 64*(1), 30–40. https://doi.org/10.1016/j.neures.2009.01.009

23. Hinojosa, J. A., Mercado, F., & Carretié, L. (2015). N170 sensitivity to facial expression: A meta-analysis. *Neuroscience & Biobehavioral Reviews, 55*, 498–509. https://doi.org/10.1016/j.neubiorev.2015.06.002

24. Ochsner, K. N., & Gross, J. J. (2008). Cognitive emotion regulation: Insights from social cognitive and affective neuroscience. *Current Directions in Psychological Science, 17*, 153–158.

Facilitating NeuroIS Research Using Natural Language Processing: Towards Automated Recommendations

Nevena Nikolajevic, Michael T. Knierim, and Christof Weinhardt

Abstract Facing rapidly growing numbers of scientific publications, attempts to structure and interconnect scientific articles are becoming increasingly relevant. Various approaches to scientific article mapping are discussed in academic literature. However, these are mainly developed upon citation-based similarity measures. We propose an extension to these concepts of literature mapping by introducing a knowledge graph-based approach. Our contribution is twofold. We introduce a holistic concept to systematically capture literature in the field of NeuroIS research based on a relational mapping of main research concepts in the field. Additionally, we implement a content-based recommender system for literature in the NeuroIS domain that serves us as a proof-of-concept and a performance baseline for further developments. By providing a relational overview of the research domain and a dedicated recommender system, we hope to facilitate the entry to the field for new researchers and enhance further research.

Keywords Recommender systems · Knowledge graphs · Natural language processing · Literature mapping

1 Introduction

As scientific output is growing ever faster, making a comprehensive overview of related work becomes increasingly difficult. Acknowledging prior work, however, provides the necessary basis for fruitful further research. The field of NeuroIS

N. Nikolajevic (✉) · M. T. Knierim · C. Weinhardt
Institute for Information Systems and Marketing, Karlsruhe Institute of Technology, Karlsruhe, Germany
e-mail: nevena.nikolajevic@kit.edu

M. T. Knierim
e-mail: michael.knierim@kit.edu

C. Weinhardt
e-mail: christof.weinhardt@kit.edu

© The Author(s), under exclusive license to Springer Nature Switzerland AG 2022
F. D. Davis et al. (eds.), *Information Systems and Neuroscience*,
Lecture Notes in Information Systems and Organisation 58,
https://doi.org/10.1007/978-3-031-13064-9_15

147

research is no exception with regard to fast-growing numbers of scientific publications. The term was coined 15 years ago, and interest has increased ever since [1]. As it is an overlapping field of different research areas—information systems research, neurosciences, and psychology—targeted topics are combined with methods borrowing from different scientific areas, making it sometimes difficult to navigate in the academic field [2].

Knowledge graphs are an emerging technique for data and knowledge management that is finding application in different fields, such as recommendation systems [3]. It is our idea to apply this technique to the above-mentioned issue to systematically capture the research field of NeuroIS. As a result, we anticipate to get a better overview and thus a better understanding of the research domain. We also hope to uncover existing synergies, research gaps, and facilitate the entry to the research field for new researchers by linking studied constructs to existing measurement methods. Thus, we aim to facilitate the search for relevant literature for researchers that do not have a comprehensive understanding of terminology, yet. As the field of NeuroIS is set up to be multidisciplinary, enhancing knowledge transfer across disciplines is particularly important. By mapping key aspects of the research domain and interlinking previous research work, we anticipate to make it easier for researchers to acquire broader expertise in the field.

The proposed approach is developed in an iterative way. To obtain an early proof-of-concept and basis for further developments, we conduct initial testing and evaluation cycles with domain experts. In our work-in-progress contribution we discuss our implemented content-based recommender system which we will use as a necessary performance baseline for further developments of our long-term goals, especially for our knowledge-graph based approach.

The rest of our research paper is structured as follows. In Sect. 2 we discuss related research contributions. In Sect. 3 we give a comprehensive overview of our developed overall concept, and illustrate our current approach and implementation. Section 4 is dedicated to the discussion of our results. Finally, in Sect. 5, we conclude our work and give a brief outlook on future work.

2   Related Work

In this section, we discuss two main concepts that are relevant to our work. We introduce current trends in the field of recommender systems research and subsequently depict existing solutions regarding literature mapping.

In recent years recommender systems research has been focused on the development of knowledge graph-based recommenders [3, 4]. The advantage of this concept is that relational information feeds into the creation of recommendations [3]. Especially in the case of recommenders designed for textual data this feature can be conductive as to emphasize specific features in addition to a plain content-based approach. Further, explainability of recommendations increases due to the transparent relationships of items within a knowledge graph [3].

Scientific literature mapping tools intend to structure the vast amount of available scientific publications and thus facilitate knowledge exchange and further research. In this regard many different approaches and tools exist as e.g. [5–7]. However, they mostly all have in common that similarities and relations between scientific literature are captured based on citations and co-citations. In the course of our work, we combine these two approaches.

3 Approach

In the scope of this paper, we discuss the implementation of a content-based recommender system [8] that we specifically develop for NeuroIS-related literature. Additionally, it serves as proof-of-concept for the outlined long-term research goals and as a performance baseline. To create a bigger picture, however, we start off by introducing our overall concept, then we illustrate our current implementations.

With our work, we aim to enhance the exchange and knowledge transfer within the NeuroIS domain. To do so, we address two objectives. Firstly, we develop a dedicated recommendation approach that supports researchers in finding related work specific to their needs. Secondly, we strive to map the NeuroIS discourse in a knowledge graph-based approach, to allow for a comprehensive overview of the field and enable explorative searches. Both objectives address the abovementioned issue of information explosion as well as a missing clear-cut delimitation of the research field and resulting challenges as e.g., inconsistent terminology [2]. To achieve our goals, we develop an ontology [9] that links the studied aspects of NeuroIS research. In exchange with domain experts, we identified the following concepts as the starting points to the ontology: (1) researched construct [2], as e.g., flow, technostress or mental workload, (2) applied measurement method [2], as e.g., eye-tracking, electroencephalography (EEG) or functional magnetic resonance imaging (fMRI), (3) used measurement tool [2], as e.g., EEG-tools as Emotiv Epoc+ and mBrain Train Smartfones, and (4) conducted type of research [2], as e.g., experimental study or literature review. Further classification characteristics as e.g., the investigation level [2], i.e., group or individual treatment as part of an experimental study, are possible. At this point, we focus on the four above-mentioned concepts. As ontologies can be flexibly extended with further concepts and apt relations between these concepts, we will consider further aspects in the future [10]. Figure 1 illustrates our current ontology-framework.

By assigning specific individual items to the concepts of an ontology a knowledge graph is obtained [11]. To instantiate our proposed ontology with concepts and corresponding relations, we consider articles that have been published in the NeuroIS domain. The developed knowledge graph is the basis to all further considerations. As it is a relational mapping, we anticipate obtaining a comprehensive overview of the NeuroIS research field, illustrating present dynamics, i.e., research streams, gaps, and future trends. Apart from this encompassing approach, more specific explorations are possible by setting a seed in form of a single or multiple research articles.

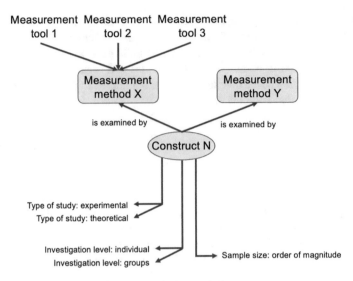

Fig. 1 Proposed ontology-framework (author's illustration)

Thus, research articles related to the set focal point can be explored. By setting a fictive starting seed, research topics that have been unconsidered in the past are brought into display. Building on the comprehensive knowledge graph approach, we aim at designing a more dedicated recommendation system for research articles that additionally considers content-based similarities between articles [8]. By combining knowledge graph-based and content-based recommendations to a hybrid recommender system, we hope to enhance recommendation quality and thus facilitate the search for related work and improve the exchange of knowledge [12]. In the following, we discuss the implementation of the content-based recommender system.

To keep our work as efficient as possible, we start off by implementing our content-based recommender system using term-frequency inverse-document-frequency (tf-idf) scores to calculate similarities across research articles [13]. Tf-idf is a weighting scheme that helps to identify important terms in a document given a corpus of documents [13]. As a result, it is possible to sort articles by their content similarity. In the first run, we consider titles and abstracts of research articles to train our model. We build up our data corpus using proceedings from the NeuroIS-retreat in the period between 2015 and 2021 as articles from before do not include a separate abstract. Additionally, we add articles that are found on arXiv and match the following search string: "physiological signal*" OR "psychophysiology" OR "neurophysiology" OR "neuroIS" OR "neuro-information system". We decide for these two sources especially due to their accessibility. As the NeuroIS-retreat is built deliberately in support of the NeuroIS research, we consider the published proceedings to be an adequate baseline for our data corpus. In case of arXiv there is no quality assurance through a peer-review process, however, due to a fast-publishing approach, we consider it to be a good indicator of current research streams and interests. We compose the search string based on search strings we found in literature [14]. To evaluate our implemented recommender and set a baseline to our future work, we choose a qualitative approach, i.e., we consulted three domain experts. We asked each of them to send us five articles they include into their NeuroIS-research. These articles simulate the respective reading history. Based on these input papers, we generated recommendations for further literature, i.e., we proposed seven articles—including title and abstract—each. To obtain recommendations, we calculate a cosine similarity matrix for the data corpus, based on the articles' respective tf-idf values and average the similarity scores given the input papers. Thus, we obtain only one score per article. We rank all papers by descending similarity values and return those that are most similar to the reading history. Further, we asked each of the domain experts to rate the recommendations following a structured evaluation questionnaire that we enclosed in our response. The evaluation is structured into two parts. First, we ask for an overall assessment regarding the perceived similarity of input articles, i.e., articles that were selected by the respective domain expert, output articles, i.e., articles that are selected by the recommender, and finally, of all articles. In the second part, the focus is set on each of the recommended articles individually. Here, we ask the experts to rate each article regarding the perceived similarity with respect to their reading history. Finally, we assess if the experts are interested in reading the rest of each respective article. Apart from questions that target the overall similarity of articles, we include questions asking specifically about the perceived similarity of discussed constructs, measurement methods and regarding the type of article. Thereby, we hope to obtain a more differentiated evaluation. Table 1 depicts all questions asked in the second part of our evaluation. The evaluation scale used is the same in the first part. We discuss our results in the following section.

4 (Preliminary) Results

We will refer to the interviewed domain experts as (participant) A, B and C. Regarding the overall evaluation, we aggregate the results from A and B due to their similarity. Both participants rated the articles in their reading history as "partly" to "very similar" across all categories. A rated the recommendations as "partly" to "very similar", whereas B indicated all categories with "partly similar". Regarding the pool of all articles, they indicated that articles where "partly" to "very similar" based on discussed construct/s, measuring method/s and type of article. However, both participants rated the articles to be "very similar" on an overall impression level. Participant C rated the read articles as "very similar" and the recommendations as "partly similar" across all categories. The overall article pool was rated as "not at all similar", however. On the individual article level results were not as clear. This is shown in Table 1.

However, in slightly more than half of the cases articles were rated to be "very similar" with respect to the reading history on an overall impression level. In even more

Table 1 Accumulated results of the evaluation on individual article level

	Very similar	Partly similar	Not at all similar	Does not apply to any article
Article is similar to the articles I read (overall impression) (%)	52	15	33	
Article represents an intersection of the articles I read (%)	57	14	29	
	Applies to all articles	*Applies to 3–4 articles*	*Applies to 1–2 articles*	*Does not apply to any article*
Article discusses constructs covered in the articles I read (%)	24	24	24	28
Article discusses measurement methods that are used in the articles I read (%)	24	29	14	33
Article corresponds by the type of article, to the articles I read (%)	33	62	5	0
	Yes	*No*		
After reading the title and abstract of the article, are you interested in reading the rest of the article? (%)	67	33		

cases (57%) the participants indicated the respective article to represent a themat-
ical intersection of the read articles. The participants expressed their interest to read
the rest of the respective article in two-thirds of all cases. Further, participants were
asked to indicate in how many of the articles they read, overlaps could be identified
regarding the discussed construct/s, measurement method/s and the type of article
given a specific recommendation. With respect to the construct/s and measurement
method/s discussed results were quite evenly distributed over all four defined levels
("applies to all articles", "applies to 3–4 articles", "applies to 1–2 articles", "does not
apply to any article"). In case of corresponding article type nearly all recommended
articles were rated to match at least 3 of 5 articles from the reading history.

5 Conclusion and Outlook

In the scope of our work-in-progress paper, we have outlined the long-term goals to
our research project and implemented a first component that simultaneously serves
us as proof-of-concept. The results obtained in the expert evaluation predominantly
support the functionality and usefulness of our content-based recommender system.
We will thus use these results to set a baseline to our future work.

To improve the content-based recommender, we will evaluate the observed defi-
ciencies—recommendations given to participant C—in a second cycle. We assume
that these shortcomings can be explained by two reasons. Firstly, we decided for a
relatively simple natural language encoding approach in the first cycle to make our
work only as complex as necessary to avoid overfitting and long computation times.
Hence, it is possible that our currently used model is at this stage limited to capture
more complex dependencies and contexts.

Also, recommendations made by a recommendation system can only be as good
as the data corpus it is built upon. Thus, we will extend our data corpus by further
articles and train our model on entire articles in the future. Further, we will implement
our proposed knowledge-graph and combine it with the content-based recommender
into a hybrid recommendation system. Thereby, we anticipate refinements in our
recommendation results. We understand our work as an iterative process in close
exchange with domain experts to ensure that our approach reflects the actual needs
of researchers in the field of NeuroIS.

References

1. Riedl, R., Fischer, T., Pierre-Majorique, L., Davis, F. (2020). A decade of NeuroIS research:
Progress, challenges, and future directions. In *SIGMIS database* (pp. 13–54). ACM.
2. Riedl, R., & Pierre-Majorique, L. (2016). *Fundamentals of NeuroIS: Information systems and
the brain.* Springer.

3. Wang, H., Zhang, F., Wang, J., Zhao, M., Li, W., Xie, X., Guo, M. (2018). RippleNet: Propagating user preferences on the knowledge graph for recommender systems. In *Proceedings of the 27th ACM International Conference on Information and Knowledge Management* (pp. 417–426). ACM.
4. Guo, Q., Zhuang, F., Qin, C., Zhu, H., Xie, X., Xiong, H., He, Q. (2021). A survey on knowledge graph-based recommender systems. In *Transactions and knowledge and data engineering.* IEEE. https://doi.org/10.1109/TKDE.2020.30.28705.
5. Connected Papers: https://www.connectedpapers.com
6. Inciteful: https://inciteful.xyz
7. Litmaps: https://www.litmaps.co
8. Lops, P., Gemmis, M., Semararo, G. (2011). Content-based recommender systems: State of the art and trends. In *Recommender systems handbook* (pp. 73–105). Springer.
9. Gruber, T. (1993). A translation approach to portable ontology specifications. *Knowledge Acquisition, 5*, 199–220.
10. Petasis, G., Karkaletsis, V. Paliouras, G., Krithara, A., Zavitsanos, E. (2011). Ontology population and enrichment: State of the art. In *Knowledge-driven multimedia information extraction and ontology evaluation* (pp. 134–166). Springer.
11. Ehrlinger, L. Wöß, W. (2016). Towards a definition of knowledge graphs. In *SEMANTICS* (Posters, Demos, SuCCESS).
12. Cano, E., Morisio, M. (2017). Hybrid recommender systems: A systematic literature review. In *Intelligent data analysis* (pp. 1487–1524). IOS Press.
13. Salton, G., Wong, A., & Yang, C. (1975). A vector space model for automatic indexing. *Communications of the ACM, 18*, 613–620.
14. Knierim, M. T., Rissler, R., Dorner, V., Maedche, A., Weinhardt, C. (2018). The psychology of flow: A systematic review of peripheral nervous system features. In *Information Systems and Neuroscience* (pp. 109–120). Springer.

CareCam: An Intelligent, Camera-Based Health Companion at the Workplace

Dimitri Kraft, Angelina Schmidt, Frederike Marie Oschinsky, Lea Büttner, Fabienne Lambusch, Kristof Van Laerhoven, Gerald Bieber, and Michael Fellmann

Abstract Health assistant tools at the workplace may contribute to preventing work-related absenteeism, increasing overall employee satisfaction, and reducing the costs of sickness or presenteeism in the long term. The tools may be integrated into a digital corporate health management strategy. Despite their huge potential, a major drawback of common tools (e.g., wearables, dedicated cameras) are that they require direct interaction, skin contact, or come with a high acquisition cost. A concept for unobtrusive and software-based monitoring to increase long-term health, improve working conditions or show the necessary adjustments to the new work situation can help to solve these problems. This paper presents a concept that shows how a simple webcam can be utilized to record vital signs, posture, and behavior during working hours. It offers individual and intelligent interventions and recommendations based on these data to reduce psychological and physical stress. Our approach demonstrates

D. Kraft (✉) · F. M. Oschinsky · L. Büttner · G. Bieber
Fraunhofer IGD, Rostock, Germany
e-mail: Dimitri.Kraft@igd-r.fraunhofer.de

F. M. Oschinsky
e-mail: FrederikeMarie.Oschinsky@igd-r.fraunhofer.de

L. Büttner
e-mail: Lea.Buttner@igd-r.fraunhofer.de

G. Bieber
e-mail: Gerald.Bieber@igd-r.fraunhofer.de

A. Schmidt · F. Lambusch · M. Fellmann
University of Rostock, Rostock, Germany
e-mail: Angelina.Schmidt@uni-rostock.de

F. Lambusch
e-mail: Fabienne.Lambusch@uni-rostock.de

M. Fellmann
e-mail: Michael.Fellmann@uni-rostock.de

K. Van Laerhoven
University of Siegen, Siegen, Germany
e-mail: kvl@eti.uni-siegen.de

© The Author(s), under exclusive license to Springer Nature Switzerland AG 2022
F. D. Davis et al. (eds.), *Information Systems and Neuroscience*,
Lecture Notes in Information Systems and Organisation 58,
https://doi.org/10.1007/978-3-031-13064-9_16

155

that the required parameters can be used to offer user-tailored interventions based on simple rules. We present a prototypical implementation of an intelligent health companion and show avenues for future research.

Keywords Imaging photoplethysmography · Pose estimation · Machine learning · Computer vision · Facial expression recognition · Multi-modal approach · Workplace health · Workplace assistance

1   Introduction

Working conditions have a substantial impact on the overall health of workers. Tools, such as smartwatches and fitness trackers, are designed to encourage a less sedentary lifestyle during leisure time. They are flooding the market and gain popularity. For instance, they are designed to reduce obesity and diabetes by motivating the wearer to lead an active lifestyle, e.g., to achieve 10.000 steps per day, exercise more, and monitor eating and drinking habits. Regular exercise in everyday life is proven to be beneficial to health. Extending their added value to the professional life seems obvious.

In addition to promoting physical health, exercise can also increase mental vitality [18]. As a result, it can reduce depression and anxiety, which can improve the social climate in the workplace [9]. In everyday working life, sedentary activities in particular result in restrictions on natural movement. Sitting for long periods of time leads to one-sided postures caused by repeatedly adopting certain postural patterns and a lack of changes in position [14]. Adult individuals in the United States, measured between 2003 and 2004, spend between 55 to 69% of the day sitting [17]. In the workday, this equates to 3/4 of working hours. The lack of physical activity that occurs with excessive sitting has been shown to lead to various chronic diseases, as well as cardiovascular and metabolic diseases [12]. Likewise, insufficient exercise is reflected in reduced productivity and increased illness and absenteeism. To support office workers, various programs have been developed in recent years that work in posture correction and movement motivation ranging from moving breaks to company sports to the promotion of sports programs outside the work environment [9]. Health-related physical and psychological risk factors can be reduced through physical activity breaks [20]. It has been shown that a change in posture while sitting at least every 30 min has a positive effect on health [20]. In addition, health programs are more successfully adopted if they are supported by the employer [9]. However, these approaches do not yet exploit the potential of digital technologies and visual assistance. We want to address these two shortcomings. Our approach to support the office worker consists of a contactless technology, which relies solely on a webcam, measuring vital (heart rate, respiratory rate, heart rate variability), behavioral (detects prolonged monotonic sitting, posture, blinking frequency), ergonomic (distance to camera, head tilt to screen) and various environmental parameters (illumination). We utilize these features and propose user-tailored and context-depended recommendations and suggestions. The CareCam approach enable a novel view on workplace health and stress research in general since the only requirement is a low-cost webcam. In this research-in-progress paper, we briefly illustrate the state of knowledge and discuss the extracted parameters using an intelligent camera-based health companion at the workplace.

2 Literature Review

Interrupting the sitting posture with short standing sequences provides a quick and easy simple posture change. It gets the body moving as it activates the sympathetic nervous system, interrupting parasympathetically controlled sitting [13]. Sit-stand workstations are a way to easily switch between sitting and standing during work hours. Because of its ease of use, the height-adjustable desk can be adapted to the workplace and reduces the amount of time spent sitting [1]. However, studies have found that in these expensive sit-stand workstations, only 20% of workers use a change in desk position per day [23]. Only regular reminders improved usage behavior [19]. In the comparison to free fitness memberships offered by employers, health trackers show better success in usage behavior and thus in improved health. The continuous recording of various health parameters and quick accessibility results in a promoted perception of one's activity, fitness, and thus well-being [8, 10]. In the office work environment with predominantly sedentary jobs, fitness wristbands with short moving interruptions are said to promote performance. Camera-based health monitoring systems are still the subject of research activities today. Vildjiounaite propose in their work a person-specific stress monitoring system that captures users' motion trajectories using depth cameras in the office [21]. In [16] the concept of COSMOS is outlined. This system captures physiological information e.g. vital signs such as heart rate, facial expressions, and eye blinks (Fig. 1).

3 Methods

Our approach to an intelligent health assistant at work utilizes various computer vision techniques to measure vital signs, physical activity, posture, behavior, and environmental influences. We measure heart rate and heart rate variability via imaging photoplethysmography (iPPG), a technique which relies on slight variations in facial skin color with each heartbeat. To quantify the heartrate by using a camera, we employ a face detection algorithm to detect the face in an image and extract the red, green, and blue pixel components of the facial region. These components are used to calculate the blood volume pulse signal. The post processing consists of applying a bandpass Butterworth filter in [0.75, 4.0] Hz band to filter out noise and other artefacts. After filtering we estimate the predominant frequency component using

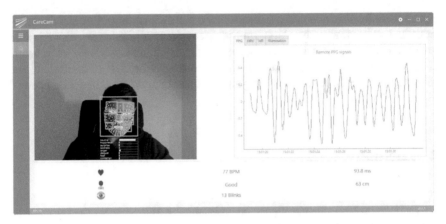

Fig. 1 Screenshot from CareCam. Left Panel shows the current camera image with corresponding overlay (face detection, keypoint detection, facial expression recognition). Right panel shows the pulse extracted from video. Lower panel provides aggregated information (e.g., pulse rate, pulse rate variability)

the FFT Welch [22] method. Because iPPG is susceptible to motion and illumination artifacts, we use a signal quality estimation algorithm to measure heart rate only when conditions are sufficient for valid measurements. Pose estimation techniques are used to track the sitting pattern, calculate head, and body tilt and sitting posture classification. We utilize BlazePose [3] to detect up to 33 keypoints. We trained a posture classification algorithm on top, expecting at least 9 keypoints from the upper region of the body (shoulders, nose, eyes, ears), calculate Euclidian distance features between these keypoints and use them to classify the current sitting posture during work using a random forest classifier. The sitting pattern is also inferred from these keypoints. Facial landmark detection is used to measure distance to camera (based on iris diameter), blinking frequency and blinking variability (the pattern of blinking) and to classify facial expression. The distance between face and camera may be estimated using the camera matrix and an object (iris) found in the image. Since the iris is relatively stable across population (11 mm+−0.5 mm), we can estimate the distance to camera. Blinking frequency is calculated using the eye aspect ratio approach proposed in [5], with a slight adaption to work with 3D coordinates (orientation of the face is considered). For facial expression classification we utilize the FER+ dataset [2] and train our algorithm on the extracted landmarks. To estimate the respiratory rate, we train a convolutional autoencoder to estimate the respiratory signal from the iPPG. This is work and progress but already showed acceptable performance. The signal transformation approach was already used in other domains e.g., PPG to ECG [6], SCG to ECG [11] signal transformation. Figure 2 provides an overview of the extracted parameters and their corresponding interventions. The concept may be extended using other different environmental sensors, e.g., carbon dioxide, temperature, and humidity sensors.

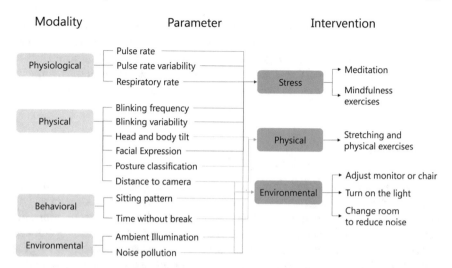

Fig. 2 Extracted parameters using the camera, optionally with a microphone to measure noise pollution. We use specific parameter for targeted interventions throughout the workday. These example interventions may be also categorized in soft interventions (reminder and notifications) and hard interventions (modal window)

4 Discussion

Research showed that targeted mindfulness interventions during the workday may improve mental and physical stress if they are followed over at least six weeks [6]. Reminder and interventions should consider context (e.g., time, calendar entries, location), general behavior (break schedule, sitting pattern) and history (past interventions) to improve adherence. We suspect that reminder should be subtle, unobtrusive, comprehensible, and easy to follow, while interventions should be well timed and should not disturb the user during a productive flow session. Camera-based systems, combined with contextual information, e.g., calendar entries, current running applications, can provide such intelligent mechanisms, since computer vision algorithms enable a wide variety of extracted information, compared to traditional wearable technologies. The downside of using camera-based systems at the workplace is the general rejection of such systems due to privacy concerns [4, 15]. However, during the covid-19 pandemic the need for video conferencing devices, such as webcam, dramatically increased and the general availability of such devices is no longer a concern. The privacy aspect on the other hand needs to be carefully addressed. While video-based systems show reasonable performance in measuring stress, they also are far less likely to be adopted and accepted. We suspect that acceptance can be increased if total transparency and data sovereignty is granted (what, how and when data is recorded, who have access and where it is stored). While the technology can be adopted voluntarily, persons (e.g., in the background) may be also monitored, who are not willing to do so. This concern is reasonable but may be addressed (e.g., segmentation of the background, user face recognition, and image encryption). Our approach to occupational health can also be applied to other fields, such as neuroscience, where we can measure affect or physiological responses to stressors. Furthermore, the CareCam can be used as an open-source toolkit to provide researchers with easy access to various physiological and behavioral parameters.

5 Conclusion

Workplace health is gaining more and more popularity since stress and musculoskeletal disorders are steadily increasing among office workers. Our approach to provide help is an intelligent assistant, operating an on variety of data and learning the general physiological, behavioral, and environmental patterns of a user. Reminder and Interventions help to cope with stressful states and unhealthy posture during the day. We outline a basic concept consisting of measure, extract and intervene. We show which parameters can be extracted from different modalities and how interventions can be used to support the user during the workday. However, camera-based solutions still suffer from low adoption in automatic stress detection, as privacy and data sovereignty concerns are often not addressed.

References

1. Alkhajah, T. A., Reeves, M. M., Eakin, E. G., Winkler, E. A. H., Owen, N., & Healy, G. N. (2012). Sit-stand workstations: a pilot intervention to reduce office sitting time. *American Journal of Preventive Medicine, 43* (3), S. 298–303.
2. Barsoum, E., Zhang, C., Canton Ferrer, C., & Zhang, Z. (2016). Training deep networks for facial expression recognition with crowd-sourced label distribution. In *ACM International Conference on Multimodal Interaction (ICMI)*.
3. Bazarevsky, V., Grishchenko, I., Raveendran, K., Zhu, T., Zhang, F., & Grundmann, M. (2020). Blazepose: on-device real-time body pose tracking. arXiv:2006.10204
4. Carneiro, D., Novais, P., Augusto, J. C., & Payne, N. (2019). New methods for stress assessment and monitoring at the workplace. *IEEE Transactions on Affective Computing, 10*(2), S. 237–254. https://doi.org/10.1109/TAFFC.2017.2699633
5. Cech, J., & Soukupova, T. (2016). Real-time eye blink detection using facial landmarks. In *Center for Machine Perception, Department of Cybernetics, Faculty of Electrical Engineering, Czech Technical University, Prague*, S. 1–8.
6. Chin, B., Slutsky, J., Raye, J., & Creswell, J. D. (2019). Mindfulness training reduces stress at work: A randomized controlled trial. *Mindfulness, 10*(4), S. 627–638.
7. Chiu, H.-Y., Shuai, H.-H., & Chao, P. C.-P. (2020). Reconstructing QRS complex from PPG by transformed attentional neural networks. *IEEE Sensors Journal, 20*(20), S. 12374–12383. https://doi.org/10.1109/JSEN.2020.3000344
8. Chung, C.-F., Gorm, N., Shklovski, I. A., & Munson, S. (2017). Finding the right fit: understanding health tracking in workplace wellness programs. In *Proceedings of the 2017 CHI Conference on Human Factors in Computing Systems*, S. 4875–4886.
9. Dratva, J., Etzer-Hofer, I., Feer, S., & Sommer, B. (2020). Auswirkungen von Sport und Bewegung während der Arbeitszeit auf die Gesundheit und die Produktivität am Arbeitsplatz.

10. Guitar, N. A., MacDougall, A., Connelly, D. M., & Knight, E. (2018). Fitbit activity trackers interrupt workplace sedentary behavior: A new application. *Workplace Health & Safety, 66*(5), S. 218–222.

11. Haescher, M., Höpfner, F., Chodan, W., Kraft, D., Aehnelt, M., & Urban, B. (2020). Transforming seismocardiograms into electrocardiograms by applying convolutional autoencoders. In *ICASSP 2020–2020 IEEE International Conference on Acoustics, Speech and Signal Processing (ICASSP)* (pp. S. 4122–4126). IEEE.

12. Healy, G. N., Wijndaele, K., Dunstan, D. W., Shaw, J. E., Salmon, J., Zimmet, P. Z., & Owen, N. (2008). Objectively measured sedentary time, physical activity, and metabolic risk. *Diabetes Care, 31*(2), 369–371. https://doi.org/10.2337/dc07-1795

13. Healy, G. N., Winkler, E. A. H., Owen, N., Anuradha, S., & Dunstan, D. W. (2015). Replacing sitting time with standing or stepping: associations with cardio-metabolic risk biomarkers. *European Heart Journal, 36*(39), S. 2643–2649.

14. Juul-Kristensen, B., Kadefors, R., Hansen, K., Byström, P., Sandsjö L., & Sjøgaard, G. (2006). Clinical signs and physical function in neck and upper extremities among elderly female computer users: the NEW study. *European Journal of Applied Physiology, 96*(2), S. 136–145.

15. Kallio, J., Vildjiounaite, E., Kantorovitch, J., Kinnula, A., & Bordallo López, M. (2021). Unobtrusive continuous stress detection in knowledge work—statistical analysis on user acceptance. *Sustainability, 13*(4), S. 2003. https://doi.org/10.3390/su13042003

16. Maeda, N., Hirabe, Y., Arakawa, Y., & Yasumoto, K. (2016). COSMS: Unconscious stress monitoring system for office worker. In: P. Lukowicz, A. Krüger, A. Bulling, Y.-K. Lim & S. N. Patel (Hg.), *Proceedings of the 2016 ACM International Joint Conference on Pervasive and Ubiquitous Computing: Adjunct.* ACM, S. 329–332.

17. Matthews, C. E., Chen, K. Y., Freedson, P. S., Buchowski, M. S., Beech, B. M., Pate, R. R., & Troiano, R. P. (2008). Amount of time spent in sedentary behaviors in the United States, 2003–2004. *American Journal of Epidemiology, 167*(7), 875–881. https://doi.org/10.1093/aje/kwm390

18. Michishita, R., Jiang, Y., Ariyoshi, D., Yoshida, M., Moriyama, H., & Yamato, H. (2016). The practice of active rest by workplace units improves personal relationships mental health and physical activity among workers. *Journal of occupational health*, S. 16–182.

19. Robertson, M.M., Ciriello, V.M., & Garabet, A.M. (2013). Office ergonomics training and a sit-stand workstation: Effects on musculoskeletal and visual symptoms and performance of office workers. *Applied Ergonomics, 44*(1), S. 73–85.

20. Ryan, C. G., Dall, P. M., Granat, M. H., & Grant, P. M. (2011). Sitting patterns at work: objective measurement of adherence to current recommendations. *Ergonomics, 54*(6), 531–538. https://doi.org/10.1080/00140139.2011.570458

21. Vildjiounaite, E., Huotari, V., Kallio, J., Kyllönen, V., Mäkelä, S.-M., & Gimel'farb, G. (2018). Detection of prolonged stress in smart office. In K. Arai (Hg.), *Intelligent Computing* (Bd. 857, pp. S. 1253–1261). Springer (Advances in Intelligent Systems and Computing Ser).

22. Welch, P. (1967). The use of fast Fourier transform for the estimation of power spectra: A method based on time averaging over short, modified periodograms. *IEEE Transactions on Audio and Electroacoustics, 15*(2), S. 70–73. https://doi.org/10.1109/TAU.1967.1161901.

23. Wilks, S., Mortimer, M., & Nylén, P. (2006). The introduction of sit-stand worktables; aspects of attitudes compliance and satisfaction. *Applied Ergonomics, 37*(3), S. 359–365.

Decision Delegation and Intelligent Agents in the Context of Human Resources Management: The Influence of Agency and Trust. A Research Proposal

Marion Korosec-Serfaty, Sylvain Sénécal, and Pierre-Majorique Léger

Abstract The integration of artificially intelligent decision support system agents as part of the external resourcing function of human resources management raises the issue of the effectiveness of the collaborative process between these agents and human resources managers and the cognitive and perceptual factors underlying the willingness to delegate such decisions. However, little is known about the neuropsychophysiological factors leading to the willingness to delegate decision-making to artificially intelligent decision support system agents in collaborative decision-making within human resources management. This research proposal explores how the perception of agency and trust affects the willingness to delegate personnel selection decisions to such agents. A single-factor within-subject design will be developed, where information provisioning as a proxy for situation awareness will be manipulated. Neuropsychophysiological and perceptual data will be collected to identify the neuropsychophysiological correlates of the perceptions of agency and trust and determine how they affect this delegation process.

Keywords Intelligent agents · Decision delegation · Agency · Trust · Human resources management · Human–Machine teams

1 Introduction

Investing in human resources leads to enhanced organizational agility and collaborative processes, wherein human resources management (HRM) plays a central bridging role between business and information technology personnel [1]. To achieve

M. Korosec-Serfaty (✉) · S. Sénécal · P.-M. Léger
Tech3Lab, HEC, Montréal, Québec, Canada
e-mail: marion.korosec-serfaty@hec.ca

S. Sénécal
e-mail: sylvain.senecal@hec.ca

P.-M. Léger
e-mail: pierre-majorique.leger@hec.ca

F. D. Davis et al. (eds.), *Information Systems and Neuroscience*,
Lecture Notes in Information Systems and Organisation 58,
https://doi.org/10.1007/978-3-031-13064-9_17
163

such a goal, the HRM function must become a unifying element by integrating artificially intelligent decision support system (DSS) agents as part of its resourcing function towards selecting external personnel who contribute to the organization's overarching strategic goal. Yet, this new integration raises the question of the effectiveness of the collaborative decision process between these agents and human resources (HR) managers and the cognitive and perceptual factors underlying the willingness to delegate such decisions.

While decision delegation and intelligent agents are not new topics in the information system (IS) and management fields [2], there has been limited research and investigation of the neuropsychophysiological factors [3], such as the perception of agency and trust, that drive the willingness to delegate decision-making to DSS agents in the context of collaborative decision-making within HRM. This study will thus propose to investigate the extent to which the perception of agency and trust affects the willingness to delegate personnel selection decisions to intelligent DSS agents.

This research will draw upon the situation awareness model [4] as a theoretical lens to understand the decision-making process of HR managers' willingness to delegate personnel selection to intelligent DSS agents. This model characterizes situation awareness as having three levels (i.e., perceptions of elements in the current situation, comprehension of the current situation and projection of future status), each dealing with the human factors associated with the current and future state of information within a dynamic environment, all of which are critical for allowing one's ability to take effective decisions in a timely manner.

A task consisting of recruiting external personnel for a fictitious organization with or without the collaboration of intelligent DSS agents will be developed to reproduce an ecologically valid HRM recruitment context. Using a scenario-based single-factor within-subject design and the Wizard-of-Oz methodology, the proposed experiment will manipulate information provisioning as a proxy for situation awareness and assess participants' responses using neurophysiological and perceptual measures to identify the neuropsychophysiological correlates of the perceptions of agency and trust and determine how these perceptions affect this delegation process.

In response to calls for research to investigate the cognitive and affective effects of information systems use in NeuroIS [5] and to explore the psychological factors driving managerial decision delegation to artificial intelligence in a strategic context [3], this proposal will foster knowledge building on the human factors underlying the willingness to delegate decisions to intelligent DSS agents. Furthermore, it will furnish practitioners with recommendations to aid in developing these agents for HRM by utilizing the situation awareness model and information provision design.

2 Methods

2.1 Experimental Design

To this aim, a scenario-based randomized within-subject experimental design will be developed to assess the effect of the three levels of reliable information provision as a proxy for situation awareness [6] on the sense of agency and the perception of trust: (1) none, (2) partial and (3) complete information provisioning (Fig. 1).

Experimental task. The experimental task will consist of 15 trials where participants will have to pre-select candidates for fictitious interviews, divided into three blocks, with five selections per block. At the beginning of the task, participants will be provided with a single scenario about a fictitious organization looking to hire. This scenario will include descriptions of the post to be filled and the organizational recruitment criteria. Participants will be further instructed that they will have to decide whether to pre-select potential candidates based on these criteria and that, in Blocks 2 and 3, this decision will have to be based on an intelligent DSS agent's recommendations (see Fig. 2).

Experimental conditions. In all conditions, participants will have to decide whether to pre-select a candidate based on these recruitment criteria: without the input of the intelligent DSS agent in Condition 1, with partial information provisioning from the intelligent DSS agent in Condition 2 (decision suggestion and each

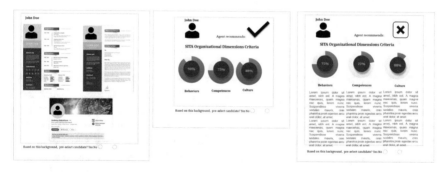

Fig. 1 Interface and experimental conditions

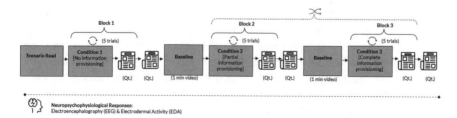

Fig. 2 Experimental design

criterion scores) and with complete information provisioning in Condition 3 (decision suggestion, each criterion scores and details on how the decision was made). The experiment will always start with Condition 1, while the other two will be randomized per participant.

Stimuli presentation. Participants will be given 15 min to read the scenario before the experimental task. In Block 1, they will have 5 min to read each candidate profile and 1 min to indicate whether to pre-select them by indicating yes or no on the screen. In Blocks 2 and 3, a first message from the intelligent DSS agent will appear via a chat window informing participants that profile N is currently being evaluated. The report, including the agent's recommendation, will be displayed 10 s later. The message and the report will be customized with the candidate's name to make it more realistic. Participants will have 1 min to read the report and decide whether they agree with the intelligent DSS agent's decision by selecting yes or no. After each trial, a 3-second fixation cross will be displayed, and a 1-minute video will be shown between each block to return participants to a baseline state. The experiment will last one hour.

Experimental protocol. The task will be administered using the Wizard-of-Oz methodology [7], a paradigm that stimulates interactions and investigates users' responses with hypothetical systems [8]. The methodology's critical success factor is that participants must assume that they interact with a fully operational intelligent agent [7]. The experimenter will thus perform online chat practice sessions with the team member acting as the Wizard to make the interactions convincing.

2.2   Measures

Neuropsychophysiological measures of the sense of agency. Electroencephalography (EEG) data will be recorded continuously at 512 Hz using a 32-channel EEG (Brain Products GmbH). Theta (θ) band activity in the premotor and parietal cortices will be measured as an index of the modulations of the sense of agency. Studies have found stronger activation of the theta brain wave frequency associated with learning, memory, intuition, and concentration over the brain areas linked with decision making (i.e., left frontal areas of the premotor cortex), illustrating that the more an individual feels in control of their own actions, the more these areas are activated [9]. By contrast, during the experience of a low sense of agency, previous research found stronger activation of the theta wave over the regions of the brain associated with intuition and social interactions (i.e., temporo-parietal areas of the right posterior-parietal cortex). These responses indicate that the more an individual tends to attribute specific actions to another, the more these brain areas are activated [10]. This will allow inferring the effect of the different levels of information provisioning on participants' feelings of being in control, potentially reflecting various levels of the feeling of agency [9]. Moreover, a previously validated nine-point scale will be used to assess perceptions of the sense of agency following each block [11].

Psychophysiological measures of trust. Research has shown that the experience of emotions is associated with the activation of the central nervous system [12], which manifests as variations in electrodermal activity (EDA). EDA is considered as the consequence of the direct mediation by the sympathetic branch of the autonomic nervous system, especially during social interactions, as in the case of human interaction with avatars [13], which can be inferred to transfer over to intelligent agents as autonomous avatars. In the context of collaborative decision-making, lower EDA activity is an indication of lower activation of the central nervous system, which can be inferred as indicating a higher level of trust towards the opposite team member, and therefore, the willingness to delegate decisions. EDA activity will thus be recorded continuously to extract Skin Conductance Levels as autonomic nervous system measures associated with emotional arousal and, thus, a change in trust or the onset of trust-based decision [12]. Post-hoc synchronization of the physiological data will be applied via the COBALT software [14–16]. Additionally, a previously validated visual analog scale will be used to record subjective measures of trust after each trial within Blocks 2 and 3 of the experiment [17].

2.3 Anticipated Analysis

EEG Data. Spectral power analysis will be performed in the θ (4–8 Hz) frequency band [9]. For Block 1, trials will be separated into one epoch: decision confirmation (1000 ms before and 1000 ms after participants confirmed their decision by pressing the corresponding keyboard key). For Blocks 2 and 3, trials will be separated into three epochs: (1) trial beginning (500 ms before and 9000 ms after message #1), (2) feedback processing (500 ms before and 9000 ms after the DSS agent's reports display) and (3) decision (1000 ms before and 1000 ms after participants' decisions confirmation). Analyses will be performed using a permutation test statistical analysis, and post-hoc tests will be applied using False Discovery Rate (FDR) correction for multiple comparisons. Differences between conditions will be assessed with paired-sample t-tests.

EDA Data. Skin Conductance Levels raw data will be standardized into participant-level z-scores and statistically analyzed using a general linear model regression with random intercept after outlier removal (mean $\pm 3 \times$ SD). Differences between experimental conditions will be assessed with paired-sample t-tests.

Perceptual Data. The visual analog scale used to assess perceptions of trust will be first converted to 0–100 and then averaged per participant and condition. The sense of agency scores will be averaged per participant and condition. Differences between experimental conditions will be assessed with pair-sampled t-tests for both measures.

3 Preliminary Suppositions, Expected Contributions and Future Directions

3.1 Preliminary Suppositions

It is postulated that, in the specific context of human–machine collaborative decision-making and delegation, and when provided with complete information by a intelligent DSS agent, there will be a stronger activation of the theta frequency band over the left frontal areas of the premotor cortex, signifying a stronger sense of agency. Moreover, there will be a lower activation of EDA activity, indicating a lower activation of the sympathetic branch of the central nervous system and, therefore, a lower level of emotional arousal, which will be inferred as illustrating a higher level of trust towards the intelligent DSS agent. Similarly, the self-reports upon the perception of agency and trust will be higher. Upon triangulation and verification of these findings, if a positive correlation is found, it will be inferred that the participants have a high likelihood towards a willingness to delegate the selection of personnel to intelligent DSS agents in the context of collaborative decision-making.

Reporting these results would allow a more profound discussion pertaining to information provisioning methods and ways in which the concept of situation awareness can be applied to inform the design of intelligent DSS agents to produce the optimal human–machine team circumstance. Moreover, they will demonstrate that the development of a shared situation awareness model between humans and machines can contribute to combating bias through the reduction in information uncertainty.

Positive findings would further demonstrate that it is possible to integrate intelligent DSS agents into the HRM hierarchy as potential trusted team members with the ability to select candidates. Thus, allowing to infer that a balanced collaboration between human and machine is possible, beyond the common fear of complete replacement and into the spheres where HRM is augmented through these technologies.

3.2 Implications for Research and Practice

This proposal will first contribute to the field by developing a more holistic view of HR managers and their interactions with intelligent DSS agents. The findings will further add to the literature on decision delegation by analyzing managerial willingness to delegate decisions to artificially intelligent technologies. In addition, they will bring a new perspective to HRM by utilizing the situation awareness model as a proposed approach to reduce information uncertainty in human–machine teaming. It will demonstrate the importance of information transparency on agency and trust in human–machine teaming and collaborative decision-making with respect to this proposition.

Leveraging the potential provided by DSS is a key value driver for HRM efficiency and effectiveness. Based on these findings, this research proposal will suggest guidelines for developing intelligent DSS agents for HR, using the situation awareness-information provision design to allow for greater transparency of information with regard to personnel recruitment.

3.3 Future Directions

While this study was developed with EEG and frequency analysis in mind, a potentially stronger route to testing the influence of the perception of agency and trust on the willingness to delegate personnel selection decisions to intelligent DSS agents would be through the use of functional magnetic resonance imaging (fMRI). The use of such technology would allow pinpointing with exactitude the brain areas directly responsible for decision-making and decision delegation, potentially leading to more fruitful EEG studies investigating the temporal aspects of these factors. With minor alterations to the experimental design, fMRI would further allow identifying the brain areas associated with trust in relation to decision-making and decision delegation in the context of unreliable information provision (i.e., when the intelligent DSS agent gets it wrong).

References

1. Oehlhorn, C. E., Maier, C., Laumer, S., & Weitzel, T. (2020). Human resource management and its impact on strategic business-IT alignment: A literature review and avenues for future research. *The Journal of Strategic Information Systems, 29*, 101641. https://doi.org/10.1016/j.jsis.2020.101641
2. Schneider, S., & Leyer, M. (2019). Me or information technology? Adoption of artificial intelligence in the delegation of personal strategic decisions. *Managerial and Decision Economics, 40*, 223–231.
3. Keding, C. (2021). *Understanding the interplay of artificial intelligence and strategic management: four decades of research in review.* Springer International Publishing.
4. Endsley, M. R. (1995). Toward a theory of situation awareness in dynamic systems. *Human Factors, 37*, 32–64.
5. vom Brocke, J., Hevner, A., Léger, P. M., et al. (2020). Advancing a NeuroIS research agenda with four areas of societal contributions. *European Journal of Information Systems, 29*, 9–24. https://doi.org/10.1080/0960085X.2019.1708218
6. Demir, M., McNeese, N. J., & Cooke, N. J. (2017). Team situation awareness within the context of human-autonomy teaming. *Cognitive Systems Research, 46*, 3–12. https://doi.org/10.1016/j.cogsys.2016.11.003
7. Riek, L. (2012). Wizard of Oz studies in HRI: A systematic review and new reporting guidelines. *J Human-Robot Interact, 1*, 119–136. https://doi.org/10.5898/jhri.1.1.riek
8. Hinds, P. J., Roberts, T. L., & Jones, H. (2004). Whose job is it anyway? A study of human-robot interaction in a collaborative task. *Human-Computer Interact, 19*, 151–181. https://doi.org/10.1207/s15327051hci1901&2_7

9. Jeunet, C., Albert, L., Argelaguet, F., & Lécuyer, A. (2018). Do you feel in control?: Towards novel approaches to characterise, manipulate and measure the sense of agency in virtual environments. *IEEE Transactions on Visualization and Computer Graphics, 24*, 1486–1495. https://doi.org/10.1109/TVCG.2018.2794598

10. Farrer, C., & Frith, C. D. (2002). Experiencing oneself vs another person as being the cause of an action: The neural correlates of the experience of agency. *NeuroImage, 15*, 596–603. https://doi.org/10.1006/nimg.2001.1009

11. Dewey, J. A., Knoblich, G. (2014) Do implicit and explicit measures of the sense of agency measure the same thing? *PLoS One, 9*

12. Riedl, R., & Javor, A. (2012). The biology of trust: Integrating evidence from genetics, endocrinology, and functional brain imaging. *Journal of Neuroscience, Psychology, and Economics, 5*, 63–91. https://doi.org/10.1037/a0026318

13. Garau, M., Slater, M., Pertaub, D. P., & Razzaque, S. (2005). The responses of people to virtual humans in an immersive virtual environment. *Presence Teleoperators Virtual Environ, 14*, 104–116. https://doi.org/10.1162/1054746053890242

14. Courtemanche, F., Léger, P.-M., Fredette, M., Sénécal, S. (2022). COBALT—Photobooth: Integrated UX Data System

15. Courtemanche, F., Labonté-LeMoyne, E., Léger, P. M., et al. (2019). Texting while walking: An expensive switch cost. *Accident Analysis and Prevention, 127*, 1–8. https://doi.org/10.1016/j.aap.2019.02.022

16. Courtemanche, F., Léger, P. M., Dufresne, A., et al. (2018). Physiological heatmaps: A tool for visualizing users' emotional reactions. *Multimed Tools Appl, 77*, 11547–11574. https://doi.org/10.1007/s11042-017-5091-1

17. Yang, X. J., Unhelkar, V. V., Li, K., & Shah, J. A. (2017). Evaluating effects of user experience and system transparency on trust in automation. *ACM/IEEE Int. Conf. Human-Robot Interact. Part, F1271*, 408–416.

A Stress-Based Smart Retail Service in Shopping Environments: An Adoption Study

Nurten Öksüz

Abstract Since many years, retailers have tried to leverage the benefits of e-commerce by transferring concepts of online shops into retailing. However, shopping experience in physical stores is different from the one in online shops due to potential stress factors arising in-store. Two previous scientific works focused on unobtrusively detected customers' perceived stress in order to provide tailored mobile services in real-time. The purpose of this paper is to assess the adoption of the envisioned stress-based smart retail service amongst potential customers. The results show that tailored services are perceived positively and thus, have the potential to contribute to an enhanced shopping experience.

Keywords Retailing · Stress-based services · Technology adoption study

1 Introduction

For years, retailers have been driven by the trend towards digitization, especially since the growing omnipresence of mobile devices in our daily lives [1, 2]. However, the shopping experience in physical stores differs from the one in online shops. This is due to the fact of arising stress factors such as long queues at the checkout, distances to walk to the products or limited time to shop [3]. Thus, customer's receptiveness for various services during shopping fluctuates. Additionally, studies have shown that perceived stress is one of the most influential factors in causing customers to abandon the purchasing process [4–6]. Nonetheless, to measure customers' stress during shopping in real-time and providing tailored mobile services is still an ongoing challenge. Previous work presents a potential methodological basis for a stress-based smart retail service (SBRS). The aforementioned basis shows that perceived stress during shopping can be identified in real-time through the combination of machine learning (ML) and neuroscientific methods [7]. The results revealed that ML is a valuable tool to predict perceived stress with an accuracy of 79.5% by classifying

N. Öksüz (✉)
Deutsches Forschungszentrum Für Künstliche Intelligenz (DFKI), Saarbrücken, Germany
e-mail: Nurten.oeksuez@dfki.de

© The Author(s), under exclusive license to Springer Nature Switzerland AG 2022
F. D. Davis et al. (eds.), *Information Systems and Neuroscience*,
Lecture Notes in Information Systems and Organisation 58,
https://doi.org/10.1007/978-3-031-13064-9_18

customers' stress level, analyzing unobtrusively measured heart rate (commercially available health sensor) and movement data (smartphone sensor). Follow-up work introduced a prototype of the envisioned SBRS providing two different services based on the individual's stress level [8]. However, to provide more value to individual customers in retailing, we need to have a better understanding on how customer behavior is affected by such new digital technologies and how smart retail services need to be adapted in response. Prior to this, there is a need for analyzing how customers would adopt such a service. This paper aims to do so by conducting a study to assess the adoption of the aforementioned SBRS amongst potential users. The paper addresses the following research question: *How would in-store customers adopt a smart retail service providing tailored mobile services in real-time based on individual stress level in shopping environments?*

2 State-of-the-Art

Radhakrishnan et al. [9] emphasize that real-time mining methods using sensors from personal mobile and wearable devices for data analysis purposes can enhance in-store shopping experience. Sensor- and vision-based technologies such as camera and robots can be individually combined in order to personally identify individual customer characteristics (e.g. gestures, speech) [10–12]. Furthermore, sensor-equipped robots can interact with customers and adapt to their behavior such as providing assistance [10]. Kowatsch and Maass [3] have presented that mobile recommendation agents (MRAs) increase the value of product information in physical stores. There exist publications introducing recommendation systems for supermarkets, fashion and electronic stores [13–15]. Further technologies, such as store navigation and product locators add value in the context of retailing [16]. However, there are only few approaches making emotions a subject of discussion. A previous work of the authors presents a new and unique approach, i.e. a stress-based smart retail service (SBRS) focusing on customers' perceived stress level during shopping. Perceived stress is proven to have a negative effect on customer purchase behavior, perceived shopping experience, and consequently on customer satisfaction and the success of the store.

3 Stress-Based Smart Retail Service

The introduced SBRS consists of a mobile app designed for collecting acceleration data from a mobile phone and unobtrusively measured heart rate data from a smartwatch during shopping [7]. The app recognizes when the customer enters the store (e.g. through visible light communication), the heart rate and acceleration data is constantly sent to the backend in an encrypted and anonymized format to guarantee an adequate level of privacy protection [7]. In the scope of the backend, the heart rate stream as well as acceleration data is analyzed with the help of ML to find patterns in the data and classify the customer into the two classes *stressed* and *relaxed* [7]. For the classification, the integrated ML model extracts mean acceleration and mean heart rate values of 60-s windows as well as significant time series characteristics from the heart rate curve to then analyses the data doing binary classification. Various ML models have been used for classification task to then focus on the best-performing model. Based on the results, the SBRS provides individually tailored services in real-time. If the customer is classified as *stressed*, additional help during shopping by finding products is provided. If the customer is classified as *relaxed*, the service introduces recommend products fitting to the customer's preferences and shopping list [8].

4   Method

The experiment consists of two parts: (1) playing through pre-defined shopping scenarios (equipped with an unobtrusive heart rate sensor and mobile phone), and (2) two paper-and-pencil questionnaires regarding the perception of the shopping scenario and the adoption of the additionally presented SBRS. The SBRS that is tailored to the pre-defined shopping scenarios is described in form of short stories as part of conceptual models. The study consisted of 100 participants in total aged between 18 and 30 years (female = 63, male = 37). Both groups received a short description about the envisioned situation-specific smart retail service and its functionalities that is mapped to the shopping situation they just played through. Furthermore, the participants have been informed that using the service would involve having to wear a sensor (such as smartwatch).

Conceptual models are means by which a designer expresses his or her own understanding of an envisioned information system such as the SBRS [17]. Studies have shown that the more structured a conceptual language is, the better the mental representations are [18]. For evaluation of the shopping scenarios, the subjects played through the pre-defined as-is narratives in a laboratory supermarket setting to then answer questions about the perceived stress, and confusion caused by the store layout. The difference between the relaxed shopping situation (group 1) and the stressful shopping situation (group 2) is that in group 2, participants were given two shopping stressors which were identified by [19], namely a product from the list not available on the shelves and a time restriction. To assess the participants' perceived stress during the played-through shopping scenarios, we designed the first survey adopting constructs from shopping stress [5, 20, 21], shopping excitement [22], and confusion about the store layout [23]. In a second part, we aimed to assess individuals' perceptions of the envisioned SBRS with respect to its potential adoption into their shopping routines. For this, each participant received a to-be narrative describing the SBRS service tailored to their shopping scenario they played through based on the group they belong to (*relaxed* or *stressed*, see Table 1). For the second survey, we adopted constructs from situation-service-fit (SSF) [25], behavior-service-fit (BSF)

Table 1 To-be Narratives for Group 1 (N = 50) and Group 2 (N = 50)

To-Be Narrative Group 1: Relaxed shopping scenario

It is Friday afternoon and you want to enjoy your evening by having a dinner. You decide to go to the supermarket to buy some food. You have a shopping list for your purchase. After arriving at the supermarket, you put all the products on your list into the shopping cart and look for additional products that might attract your attention. While strolling through the store, the situation-specific smart retail service on your mobile phone analyses your stress level and recognizes that you are relaxed and strolling around. The situation-specific smart retail service on your mobile phone gives you a notification: "Check your recommendations". Based on your shopping list and your preferences, the service recommends you a bar of chocolate "fine dark"

To-Be Narrative Group 2: Stressed sopping scenario

It is Friday evening. Some friends are going to visit you at 8 pm and you totally forgot to prepare something. You do not have anything to eat at home. You decide to go to the supermarket to buy some food. You prepared a shopping list before going to the store. When you arrive at the supermarket, you see that you have only 3 min left to make your purchase before the supermarket closes. You run into the supermarket and start putting the products from your shopping list into the shopping cart. You try to find canned food from the brand "Sonnen Bassermann", but you are not able to find it. The situation-specific smart retail service analysis your stress level and recognizes that you are stressed. It sends a notification to your smartphone: "Do you need help?". You checkmark "Sonnen Bassermann" on the list and the service provides the information on where to find the product

[24], perceived usefulness (PU) [25], flexibility of the service (FoS) [26], intention to use (ItU) [26], attitude towards usage (AtU) [26], as well as general technological affinity of the user [27]. All responses were measured on 7-point Likert scales (1 = strongly disagree, 7 = strongly agree).

5 Results

The results show that group 1 significantly perceived the shopping situation less stressful than group 2. Furthermore, the groups did not differ concerning shopping excitement with group 2 having a higher (but still insignificant) rating value than group 1. Moreover, the groups differed concerning confusion about the store layout with group 1 having a lower rating value than group 2 (Table 2).

Second, each subject had to assess the envisioned SBRS. All constructs for the two groups were rated with mean values ≥ 4 and negative skew values, which both are indicators for an overall positive rating of the SBRS. In group 1, participants rated BSF the highest with a mean of 4.98. Furthermore, SSF, PU, FoS, ItU, and

Table 2 Item Analysis of the Questionnaire Survey of Group 1 and Group 2 (Welch's t-test)

Construct		# Of items	Internal consistency		Mean		St.Dev (SD)		Skew	
			G1	G2	G1	G2	G1	G2	G1	G2
Part 1	Shopping stress	6	0.88	0.86	2.52	4.42	1.07	1.20	0.83	– 0.82
	Shopping excitement	3	0.77	0.78	4.44	4.59	1.18	1.10	–0.57	– 0.73
	Confusion in store layout	2	0.81	0.96	2.84	3.93	1.47	1.77	1.09	– 0.14
Part 2	Situation-service Fit	1	–	–	4.70	4.96	1.46	1.37	–0.56	– 1.32
	Behavior-Service fit	1	–	–	4.98	4.92	1.56	1.60	–0.17	– 0.14
	Perceived usefulness	2	0.81	0.81	4.64	4.66	1.29	1.15	–0.87	– 0.57
	Flexibility of service	1	–	–	4.86	4.86	1.03	0.95	–0.41	– 0.76
	Intention to use	3	0.93	0.96	4.35	4.05	1.60	1.59	–0.51	– 0.02
	Attitude towards usage	3	0.95	0.90	4.63	4.48	1.43	1.22	–0.74	– 0.51
	Technical enthusiasm	2	0.87	0.83	5.06	5.00	1.33	1.26	–0.42	– 0.45

AtU are rated positively with ItU rated the lowest with a mean of 4.35. For group 2, a similar picture emerges with highest mean for SSF (4.96) and BSF (4.92) and slightly lower mean values for BSF and ItU compared to group 1. The mean values for SSF, PU, and AtU are slightly higher compared to group 2.

6 Discussion and Implications

The goal of this paper was to analyze how users adopt the envisioned stress-based smart retail service (SBRS) tailored to the user's stress level while shopping in-store (RQ). We first applied a laboratory experiment where we designed two shopping scenarios—relaxed and stressful—to simulate pre-defined as-is situations based on literature in the scope of in-store shopping. Then, the participants of the laboratory experiment had to complete two paper-and-pencil questionnaires so that (a) the perceived stress level of the participants during the shopping scenarios and (b) the adoption of an envisioned SBRS providing tailored services in real-time. The analysis results of the study and the questionnaire have shown the success of the intention of creating relaxed and stressful shopping scenarios. Regarding the questions for adoption of the SBRS, we see that both groups answered all questions with a mean value > 4, indicating that the SBRS was assessed positively and that customers would welcome different services tailored to different shopping situations. This is supported by the fact that SSF, BSF and FoS were rated the highest. Even though the participants of the *relaxed* group rated the service more positive than the *stressed* one, the difference is not significant. Thus, we can assume that customers being stressed might need services specifically tailored to their sopping situation. The analysis results also show that the SBRS reacting to the stress level of customers in real-time is a hot topic in the NeuroIS community [28–32]. By classifying customers with respect to their stress levels, retailers have the potential to create an interactive shopping experience and improve customer loyalty by providing suitable services. The envisioned SBRS could enable applications such as: (a) targeted and customized advertisement: e.g. new product launch; (b) proactive retail help: a shop assistant to help customers finding products or recommendations for additional products suitable for purchases in the shopping list. However, it has to be mention that the use of information and communication technologies itself can constitute a source of stress, namely digital stress. Thus, it is possible that the SBRS might cause stress [32, 34, 35].

7 Conclusion and Future Work

Despite the fact that the stress-based smart retail service (SBRS) was rated positively and thus, provides multiple advantages not only for retailers but also for customers, they need to be motivated to use the SBRS, i.e. download the app and use it. For instance, retailers have the option to provide incentives (discounts or vouchers) for the initial sign-in. Moreover, the fact that the SBRS is supposed to be privacy-friendly can attract users. In this context, taking adequate measures to ensure data security as well as privacy is crucial. This paper serves as a first evaluation of the SBRS. Due to the exploratory nature of this work, several limitations need to be overcome by future research. We used narratives to present the envisioned SBRS. In a next step, the service is planned to be assessed in practice to compare the results to the ones presented in this paper. For this, we plan to conduct a laboratory study where the ML-based service will be integrated into a mobile phone. The recruited subjects wearing unobtrusive heart rate sensors and carrying the mobile phone will be able to use the installed service in the pre-defined shopping scenarios to then evaluate the service. Thus, additional constructs such as perceived ease of use [26] can be assessed too. Furthermore, focusing on additional shopping scenarios and generating new services based on perceived stress is also a topic of future work. In addition, it should not be neglected that the SBRS itself might be a source of stress (technostress) and thus, constitutes a potential limitation and deserves future investigation.

References

1. Li, Y. M., Lin, L. F., & Ho, C. C. (2007). A social route recommender mechanism for store shopping support. *Decision Support Systems, 94*, 97–108.
2. Liang, T. P., Lai, H. J., & Ku, Y. C. (2006). Personalized content recommendation and user satisfaction: Theoretical synthesis and empirical findings. *Journal of Management Information Systems, 23*(3), 45–70.
3. Kowatsch, T., & Maass, W. (2010). In-store consumer behavior: How mobile recommendation agents influence usage intentions, product purchases, and store preferences. *Computers in Human Behavior, 26*(4), 697–704.
4. Albrecht, C. M., Hattula, S., & Lehmann, D. R. (2017). The relationship between consumer shopping stress and purchase abandonment in Task-oriented and Recreation-oriented consumers. *Journal of the Academy of Marketing Science., 4*(5), 720–740.
5. Baker, J., & Wakefield, K. L. (2012). How consumer shopping orientation influences perceived crowding, excitement, and stress at the mall. *Journal of the Academy of Marketing Science, 5*(40), 791–806.
6. Zhu, W., Timmermans, H. (2008). Cut-off models for the 'Go-Home'decision Of pedestrians. In: Shopping Streets. Environment and Planning B: Planning and Design, *35*(2), 248–260.
7. Öksüz, N., Maass, W. (2020). A Situation-specific smart retail service based on vital signs. In: *Internation Conference on Information Systems (ICIS)*.
8. Öksüz, N., Manzoor, H. M., Harig, A., Maaß, W. (2020). A ML-based smart retail service prototype using biosignals. In: *Workshop on Information Technology and Systems. Workshop on Information Technology and Systems* (WITS).
9. Radhakrishnan, M. Misra, A. (2019). Can earables support effective user engagement during weight-based gym exercises?.In: *Proceedings of the 1st International Workshop on Earable Computing*, 42–47.
10. Bertacchini, F., Bilotta, E., & Pantano, P. (2017). Shopping with a robotic companion. *Computer Human Behavior, 77*, 382–395.
11. Chu, T. H. S., Hui, F. C. P., & Chan, H. C. B. (2013). Smart shopping system using social vectors and RFID. *International Multiconference on Engineering and Computer Science, 1*, 239–244.
12. Kamei, K., Ikeda, T., & Shiomi, M. (2012). Cooperative customer navigation between robots out-side and inside a retail Shop—an implementation on the ubiquitous market platform. *Ann. Telecommunication., 7*(8), 329–340.
13. Christidis, K., & Mentzas, G. (2013). A Topic-based recommender system for electronic marketplace platforms. *Expert Systems with Applications, 40*(11), 4370–4379.
14. Hwangbo, H., Kim, Y. S., & Cha, K. J. (2018). Recommendation system development for fashion retail ecommerce. *Electronic Commerce Research and Applications, 28*, 94–101.
15. Keller, T., & Raffelsieper, M. (2014) An E-commerce like platform enabling bricks-and-mortar stores to use sophisticated product recommender systems. In: *Proceedings of the 8th ACM Conference on Recommender system*, 367–368.
16. Linzbach, P., Inman, J. J., & Nikolova, H. (2019). E-commerce in a physical store: which retailing technologies add real value? *NIM Marketing Intelligence Review, 11*(1), 42–47.
17. Kuechler, W. L., & Vaishnavi, V. (2016). So, talk to me: The effect of explicit goals on the comprehension of business process narratives. *Mis Quarterly, 30*, 961–979.
18. Maaß, W., Storey, V. C. (2015). Logical design patterns for information system development problems. In: Johannesson, P., Lee, M. -L., Liddle, S. W., Opdahl, A. L., López, O. P., (eds.) *Conceptual Modeling—34th International Conference, ER 2015, Lecture Notes in Computer Science, 9381*, (pp. 134–147), Springer.
19. Aylott, R., & Mitchell, V. W. (1998). An exploratory study of grocery shopping stresors. *International Journal of Retail & Distribution Management, 26*(9), 362–373.
20. Miller, E. G., Kahn, B. E., & Luce, M. F. (2008). Consumer wait management strategies for negative service events: a coping approach. *Journal of Consumer Research, 34*, 635–648.

21. Russell, J. A., & Pratt, G. (1980). A description of the affective quality attributed to environments. *Journal of Personality and Social Psychology, 38*(2), 311–322.
22. Wakefield, K. L., & Baker, J. (1998). Excitement at the mall: determinants and effects on shopping response. *Journal of Retailing, 74*, 515–539.
23. Garaus, M., Wagner, U. (2013). Retail shopper confusion: An explanation of avoidance behavior at the point-of-sale. ACR North American Advances
24. Maass, W., Kowatsch, T., Janzen, S., & Filler, A. (2012). Applying situation-service fit to physical environments enhanced by ubiquitous information systems. In: *20th European Conference on Information Systems (ECIS). 221*, 1–12.
25. Venkatesh, V., & Davis, F. D. (2000). A theoretical extension of the technology acceptance model: Four longitudinal field studies. *Management Science, 46*(2), 186–204.
26. Wixom, B. H., & Todd, P. A. (2005). A theoretical integration of user satisfaction and technology acceptance. *Information Systems Research, 16*(1), 85–102.
27. Edison, S. W., & Geissler, G. L. (2003). Measuring attitudes towards general technology: Antecedents, hypotheses and scale development. *Journal of targeting, Measurement and Analysis for Marketing, 12*(2), 137–156.
28. Riedl, R., Kindermann, H., Auinger, A., & Javor, A. (2012). Technostress from a neurobiological perspective: System breakdown increases the stress hormone cortisol in computer users. *Business & Information Systems Engineering, 4*(2), 61–69.
29. Adam, M., et al. (2017). Design blueprint for stress-sensitive adaptive enterprise systems. *Business & Information Systems Engineering, 59*, 277–291.
30. Riedl, R.; Léger, P.-M.: Fundamentals of NeuroIS – Information Systems and the Brain. Springer (2016)
31. Vom Brocke, J., Hevner, A., Léger, P.-M., Walla, P., & Riedl, R. (2020). Advancing a neurois research agenda with four areas of society contributions. *European Journal of Information Systems, 29*(1), 9–24.
32. Riedl, R. (2013). On the biology of technostress: Literature review and research agenda. *DATA BASE for Advances in Information Systems, 44*(1), 18–55.
33. Fischer, T.; Riedl, R.: Lifelogging for Organizational Stress Measurement: Theory and Applications. SpringerBriefs in Information Systems. Springer, (2019)
34. Fischer, T., & Riedl, R. (2019). Technostress research: A nurturing ground for measurement pluralism? *Communications of the Association for Information Systems, 40*, 375–401.
35. Fischer, T.; Riedl, R. (2020): On the stress potential of an organizational climate of innovation: A survey study in Germany. *Behaviour & Information Technology*, Published online: 10 Nov 2020.

Collecting Longitudinal Psychophysiological Data in Remote Settings: A Feasibility Study

Sara-Maude Poirier, Félix Giroux, Pierre-Majorique Léger, Frédérique Bouvier, David Brieugne, Shang-Lin Chen, and Sylvain Sénécal

Abstract Research methods to better understand the habituation process of individuals' repeated task performance are constantly improving. However, data collection methods from a longitudinal perspective have been overlooked. Thus, the aim of this study is to explore the feasibility of collecting psychophysiological data remotely over several days. Through a five-days longitudinal study, behavioral data, facial emotions, and electrodermal activity were collected remotely. Behavioral results revealed that consumers tend to improve their performance in executing the tasks over time, regardless of their difficulty levels. Psychophysiological data showed that negative emotions were experienced by participants on Day 3 and tend to decrease on Day 5. Thus, it is possible that the novelty of the first remote session on Day 1 prevented participants from expressing negatives emotions even if they found the tasks difficult. This study highlights potential limitations of cross-sectional studies investigating the habituation process and validates the feasibility of conducting longitudinal psychophysiological data collection remotely in a naturalistic setting.

S.-M. Poirier (✉) · F. Giroux · P.-M. Léger · F. Bouvier · D. Brieugne · S.-L. Chen · S. Sénécal
HEC Montréal, Montréal, QC 3T 2A7, Canada
e-mail: sara-maude.poirier@hec.ca

F. Giroux
e-mail: felix.giroux@hec.ca

P.-M. Léger
e-mail: pierre-majorique.leger@hec.ca

F. Bouvier
e-mail: frederique.bouvier@hec.ca

D. Brieugne
e-mail: david.brieugne@hec.ca

S.-L. Chen
e-mail: shan-lin.chen@hec.ca

S. Sénécal
e-mail: sylvain.senecal@hec.ca

© The Author(s), under exclusive license to Springer Nature Switzerland AG 2022   179
F. D. Davis et al. (eds.), *Information Systems and Neuroscience*,
Lecture Notes in Information Systems and Organisation 58,
https://doi.org/10.1007/978-3-031-13064-9_19

Keywords Longitudinal study · Remote data collection · NeuroIS ·
Psychophysiological · Habituation · Facial expressions · Electrodermal activity

1 Introduction

There is a growing literature in behavioral sciences and human–computer interaction
(HCI) research investigating how individuals learn through the execution of repetitive
tasks [1, 2]. It emerges that continuous experience with a task or a technological artifact leads to a habituation effect defined as "well-learned action sequences, originally
intentional, that may be activated by environmental cues and then repeated without
conscious intention" [3]. To better capture unconscious states of individuals' minds
that characterized the habituation process, authors have started to integrate neurophysiological tools such as fMRI and eye tracking [4, 5]. Although the methods for
evaluating the habituation process are constantly being refined, the context in which
these studies seem to be conducted still limits our understanding of the phenomenon.
Indeed, most research in the domain has used cross-sectional studies, not allowing
observing the evolution of the habituation process over time. Interestingly, Vance
et al. [6] were the first to measure habituation through neural activity by adopting a
longitudinal perspective. Nevertheless, the authors were only interested in the cognitive aspect of habituation. The affective state of individuals is also important to inform
habituation formation since it reveals information about their satisfaction during the
task and tell us about its effectiveness [3, 7]. Moreover, to counteract the lack of
ecological validity often criticized in laboratory studies, new methods for remote
data collection must be developed [8, 9]. Indeed, collecting data in more natural
settings could counteract some of the challenges faced by longitudinal studies. For
instance, it could reduce the systematic attrition bias, especially for research focusing
on participants for whom travelling to lab can be more challenging (e.g., people with
disabilities) [10]. Therefore, more research is needed in the development of longitudinal data collection methods evaluating the behavioral and affective dimension of
the habituation process.

Thus, the aim of this study is to explore the feasibility of collecting psychophysiological data remotely over several experimental sessions. Precisely, over five days,
we have investigated if it was possible to collect behavioral and affective data related
to individual participants' habituation process. To perform this study, psychophysiological data were collected during the execution of an N-back task (i.e., memory task
where participants have to detect green squares that appeared in N-positions before)
with four levels of complexity [11]. Participants' behavior (i.e., task performance)
was used to infer their habituation process over the experiment sessions. Moreover,
psychophysiological emotional valence and arousal were measured via Automatic
Facial Expression Analysis (AFEA) and electrodermal activity (EDA), respectively,
and synchronized with a cloud-based software analysis platform. Psychophysiological data is an interesting complement to behavioral data since it can overcome social
desirability bias. This study is timely and relevant since the Covid-19 pandemic and

work-from-home policies have created great challenges in the pursuit of laboratory studies. One of the main insights from this health crisis was the urgent need for science to adapt to various situations [8, 12]. Thus, this research makes methodological contributions for researchers in conducting longitudinal studies with psychophysiological data collected remotely. Finally, this research allows the collection of data in naturalistic settings. This is particularly important for accessibility-related research, where bringing participants into the lab may be more difficult [13]. Therefore, the development of remote methods allows for more inclusiveness in the conduct of research.

2 Methodology

2.1 Design, Participants, and Context

This longitudinal study used a 5 (Time: 1 vs. 2 vs. 3 vs. 4 vs. 5) $\times$ 4 (N-back task complexity levels) within-subjects design. Nineteen participants (8 women, age M = 53, SD = 13.7) took part in this five-day study conducted remotely at their home. On Days 1, 3 and 5, physiological and behavioral data was collected with the help of a moderator. On Days 2 and 4, only behavioral data was collected without a moderator.

This study was approved by the Research and Ethics Board of our institution and was part of a larger project investigating the impact of habituation in an online transactional banking experience. This paper focuses on the N-back tasks performed before and after the banking task.

2.2 Apparatus and Instruments

Participants' facial expressions were captured by their personal computer webcam via the Remote Liveshare Moderated Testing features of the Lookback.io (Montreal, Quebec) platform. EDA was measured using the COBALT Bluebox, a custom low-cost psychophysiological sensing device based on the BITalino hardware [14–17]. To capture the changes of skin conductance response (SCR), two sticky sensors were self-placed by the participant on his/her non-dominant hand.

2.3 Measures and Data Post-processing

Behavioral data was measured as a percentage of successful hits in the first and second daily sessions of the N-back task. Data was aggregated by task difficulty level and by the task day by combining the performance of the two daily sessions together. The same was done for psychophysiological data measured during the first and second daily sessions of N-back tasks. Psychophysiological emotional valence was measured using AFEA, an unobtrusive method used in research and industry to assess humans' facial emotions via low-cost camera with minimal video resolution of 640×480 pixels. Participants' face video recordings were post-processed by the scientifically validated AFEA software program Noldus FaceReader [18, 19]. This software detects and classifies 6 basic emotions (i.e., happiness, anger, disgust, sadness, fear, and surprise) according to the Facial Action Coding System [20]. Noldus FaceReader also computes an emotional valence metric based on the difference of the intensity of the positive emotion of happiness and the most prominent negative emotion detected.

Psychophysiological emotional arousal was measured via the COBALT Bluebox [15–17]. We analyzed the phasic changes of SCR and we obtained z-score from the EDA data [14]. Finally, psychophysiological emotional valence and arousal were post-processed and synchronized using the COBALT Photobooth cloud-based software analysis platform [21–23]. This software manages the latency between the two measures and determines a combined metric with one second precision in relation to the state of the stimulus.

2.4 Experimental Task and Protocol

One day before the start of the data collection, participants were asked to complete a consent form. All participants received a COBALT Bluebox at their home and had to use their personal computer. Moreover, a moderator explained to the participants how to install the tools (e.g., Bluebox sensors) and how to open the online platform Lookback. To ensure that the self-installation went properly, we followed the guidelines proposed by Vasseur et al. [12] and Giroux et al. [8]. These guidelines have been used in several studies that have demonstrated its validity [14, 24, 25]. Every day, participants were asked to complete the N-back task with 4 increasing levels of difficulty. In these tasks, participants had to perform a memory task where they had to detect green squares that appeared in N-positions before. They had to click on [M] when the green square appears in the same position and on [X] when the position was different. After the fifth day of the experiment, all participants received a financial compensation and were instructed to send back the COBALT Bluebox to our lab.

2.5 Analysis

To analyze behavioral data, we performed four cumulative logistic regressions (i.e., one per task difficulty level) modeling the probability of the *Number of successful hits* having higher ordered value over the five *Task days*. We also modeled the interaction between *Task difficulty levels* and *Task days*. To analyze the psychophysiological data, we performed two linear regressions with random intercept model on the dependent variables of *Psychophysiological emotional valence* and *Psychophysiological emotional arousal*, with *Task difficulty levels* and *Task days* as independent variables. Finally, p-values were adjusted for multiple comparisons using the method of Holm-Bonferroni [26].

3 Results

3.1 Behavioral Results

The number of participants who did the second daily session, as contrasted with the first one, tended to decrease over days (see Fig. 1). As observed in Fig. 1, behavioral results demonstrate that higher task difficulty level (i.e., Level 4) led to less successful hits than lower task difficulty (i.e., Levels 1, 2, and 3). Moreover, our cumulative logistic regression on the *Number of successful hits* revealed a significant effect of *Task days* in *Task difficulty levels* 1 ($\beta = 0.3865$, p = 0.0001), 2 ($\beta = 0.5183$, p = 0.0001), 3 ($\beta = 0.4138$, p = 0.0004), and 4 ($\beta = 0.3702$, p = 0.0009). The effect of *Task days* on the *Number of successful hits* did not differ significantly among the *Task difficulty levels*.

3.2 Psychophysiological Results

Linear regression on *Psychophysiological emotional valence* revealed a significant effect of *Task days*. Participants' psychophysiological emotional valence in response to the N-back task decreased between Days 1 and 3 for *Task difficulty level 1* ($\beta = 0.1396$, p = 0.0344) and *Difficulty level 3* ($\beta = 0.1396$, p = 0.0199). Although not significant, psychophysiological emotional valence seems to have increased between task days 3 and 5. As for *Psychophysiological emotional arousal*, our regression showed no significant effect of *Task days*, although there was a trend indicating lower EDA throughout the task Days 1, 3, and 5, for each task difficulty level (see Fig. 1).

4 Discussion

The behavioral data of our feasibility showed a significant positive linear relationship between task days and task difficulty levels, which indicates that participants constantly improve their performance over days. Therefore, it is possible to infer that as the days go by, participants do develop a certain level of habituation with the

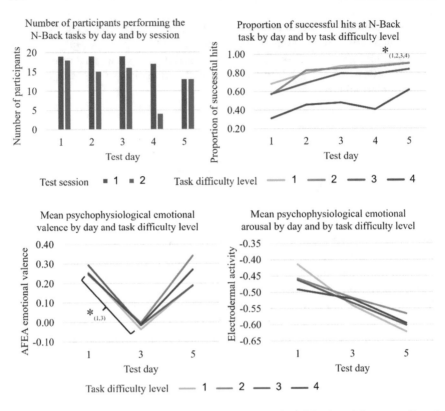

Fig. 1 Histogram showing the number of participants from which behavioural data was collected for each of the five test days and daily sessions (top left). Line graphs showing the increased proportion of successful hits over the five test days for all four task difficulty levels (top right), the mean psychophysiological emotional valence indexed with AFEA (bottom left) and mean psychophysiological emotional arousal indexed with EDA (bottom right) over the three moderated test days

task. Conversely, our psychophysiological data revealed that the negative emotions appeared only on Day 3 and tend to decrease on Day 5. The fact that participants had more difficulty with the task on Day 1 but without expressing negative facial emotions may be explained by the habituation process of the remote experiment procedures themselves. Indeed, it is possible that the novelty of the first remote session prevented participants from expressing the emotions that they were experiencing during the task, which can be a threat to cross-sectional studies.

As a first contribution, our study shows the feasibility of our methodology for collecting longitudinal psychophysiological data remotely. However, not all participants completed the second task session of the day in the unmoderated test days. Therefore, the moderated tests may have contributed to the achievement of the second session of the day [26]. This means that researchers should provide more detailed explanations for participants when the sessions are not moderated to maximize the completion of tasks. This contribution may lead to more flexible psychophysiological data collection in a naturalistic environment without compromising validity and reliability [9]. As a second contribution, we provide some preliminary guidelines for researchers who want to study habituation over several days. We suggest that researchers should be careful with the interpretation of psychophysiological data in the first session due to participants' habituation to the experiment procedures. Researchers may also want to conduct one or few remote practice sessions in the beginning of a longitudinal study to familiarize the participants with the protocol and better capture their psychophysiological reactions to the tasks.

Our study is not without limitations. The duration of our longitudinal study was limited to five days, with two of them being unmoderated, which may have decreased participation rate during these sessions. More research should apply our methodology in longer time perspectives, especially for more complex tasks. Furthermore, since the data collection was done at home, we did not control for the time-of-day participants performed the task. Therefore, their energy level could vary, which could potentially have affected the results. Finally, we hope that our methodological innovation will encourage multidisciplinary research domains to conduct longitudinal studies on habituation, learning and rehabilitation, which can have the benefit from being experienced in naturalistic settings like users, students, workers, or patients' home [13].

References

1. Chein, J., & Schneifer, W. (2012). The brain's learning and control architecture. *Current Directions in Psychological Science, 21*, 78–84.
2. Gardner, B., & Rebar, A. L. (2019). *Habit formation and behavior change.* Oxford University Press.
3. Guinea, A., & Markus, M. (2009). Why break the habit of a lifetime? Rethinking the roles of intention, habit, and emotion in continuing information technology use. *MIS Quarterly, 33*, 433–444.
4. Anderson, B. B., Vance, A., Kirwan, C. B., Jenkins, L. J., & Eargle, D. (2016). From warning to wallpaper: Why the brain habituates to security warnings and what can be done about it. *Journal of Management Information Systems, 33*(3), 713–743.
5. Anderson, B. B., Jenkins, L. J., Vance, A., Kirwan, C. B., & Eargle, D. (2016). Your memory is working against you: How eye tracking and memory explain habituation to security warnings. *Decision Support Systems, 92*.
6. Vance, A., Jenkins, L. J., & Anderson, B. B. (2018). Tuning out security warnings: A longitudinal examination of habituation through fMRI, eye tracking, and field experiments. *MIS Quarterly: Management Information Systems, 42*, 355–380.
7. Lankton, N., Wilson, E., & Mao, E. (2010). Antecedents and determinants of information technology habit. *Information & Management, 47*, 300–307.
8. Giroux, F., Léger, P.-M., Brieugne, D., Courtemanche, F., Bouvier, F., Chen, S.-L., Tazi, S., Rucco, E., Fredette, M., Coursaris, C., & Sénécal, S. (2021). Guidelines for collecting automatic facial expression detection data synchronized with a dynamic stimulus in remote moderated user tests. In *International Conference on Human Computer Interaction.*

9. Markman, A. (2017). Combining the strengths of naturalistic and laboratory decision-making research to create integrative theories of choice. *Journal of Applied Research in Memory and Cognition, 7.*

10. Coen, A. S., Patrick, D., & Shern, D. L. (1996). Minimizing attrition in longitudinal studies of special populations: An integrated management approach. *Evaluation and Program Planning, 19,* 309–319.

11. Jaeggi, S., Buschkuehl, M., Perrig, W. J., & Meier, B. (2010). The concurrent validity of the N-back task as a working memory measure. *Memory (Hove, England), 18,* 394–412.

12. Vasseur, A., Léger, P.-M., Courtemanche, F., Labonte-Lemoyne, E., Georges, V., Valiquette, A., Brieugne, D., Rucco, D., Coursaris, C., Fredette, M., & Sénécal, S. (2021). Distributed remote psychophysiological data collection for UX evaluation: A pilot project. In *International Conference on Human-Computer Interaction.*

13. Giroux, F., Couture, L., Lasbareilles, C., Boasen, J., Stagg, C. J., Flemming, M., Sénécal, S., & Léger, P.-M. (2022). Usability evaluation of assistive technology for ICT accessibility: Lessons learned with stroke patients and able-bodied participants experiencing a motor dysfunction simulation (In preparation).

14. Swoboda, D., Boasen, J., Léger, P.-M., & Pourchon, R. (2022). Comparing the effectiveness of speech and physiological features in explaining emotional responses during voice user interface interactions. *Applied Sciences, 12,* 1269.

15. Courtemanche, F., Sénécal, S., Fredette, M., & Léger, P.-M. (2022). COBALT-bluebox: Multi-modal user data wireless synchronization and acquisition system. Declaration of invention No. AXE-0045, HEC Montréal, Montréal, Canada.

16. Courtemanche, F., Sénécal, S., Léger, P.-M., & Fredrette, M. (2022). COBALT-capture: Real-time bilateral exchange system for multimodal user data via the internet. Declaration of invention No. AXE-0044, HEC Montréal, Montréal, Canada.

17. Léger, P.-M., Courtemanche, F., Karran, A. J., Fredette, M., Dupuis, M., Zeyad, H., Fernandez-Shaw, J., Côté, M., Del Aguila, L., Chandler, C., Snow, P., Vilone, D., & Sénécal, S. (2022). Caption and observation based on the algorithm for triangulation (COBALT): Preliminary results from a beta trial (In preparation).

18. Tanja, S., Andres, G. R., & Olivier, C. S. (2019). Assessing the convergent validity between the automated emotion recognition software Noldus FaceReader 7 and facial action coding system scoring. *PLOS ONE, 14.*

19. Loijens, L., & Krips, O. (2022). FaceReader méthodology note [cited 2022 March 15]. https://www.noldus.com/resources/pdf/noldus-white-paper-facereader-methodology.pdf

20. Ekman, P., & Friesen, W. V. (1978). Facial action coding system: A technique for the measurement of facial movement.

21. Léger, P.-M., Courtemanche, F., Fredette, M., & Sénécal, S. (2019). A cloud-based lab management and analytics software for triangulated human-centred research. In *Information systems and neuroscience* (pp. 93–99). Springer.

22. Courtemanche, F., Fredette, M., Sénécal, S., Léger, P.-M., Dufresnes, A., Georges, V., & Labonte-Lemoyne, E. (2019). U.S. Patent No. 10,368,741. Washington, DC: U.S. Patent and Trademark Office.

23. Courtemanche, Léger, P.-M., Fredrette, M., & Sénécal, S. (2022). COBALT-photobooth: Integrated UX data system. Declaration of invention No. VAL-0045, HEC Montréal, Montréal, Canada.

24. Labonté-LeMoyne, E., Courtemanche, F., Coursaris, C., Sénécal, S., & Léger, P.-M. (2021). Development of a new dynamic personalised emotional baselining protocol for human-computer interaction. In *Proceedings of the NeuroIS 2021*, Vienna, Austria, 1–3 June.

25. Carmichael, L., Poirier, S.-M., Coursaris, C., Léger, P.-M., & Sénécal, S. (2021). Does media richness influence the user experience of chatbots: A pilot study (pp. 204–213).

26. Keppel, G., & Wicken, T. D. (2004). *Design and analysis: A researcher's handbook*, 4th edn. Pearson/Prentice Hall.

Eye-Gaze and Mouse-Movements on Web Search as Indicators of Cognitive Impairment

Jacek Gwizdka, Rachel Tessmer, Yao-Cheng Chan, Kavita Radhakrishnan, and Maya L. Henry

Abstract The aging population brings a drastic increase in age-related conditions that affect cognition. Early detection of cognitive changes is important but challenging. Standard paper-and-pencil neuropsychological assessments of cognition require trained personnel and are costly and time-consuming to administer. Moreover, they may not capture subtle cognitive impairments associated with aging and age-related disorders that impact activities of daily living. Eye-tracking during cognitive tasks has been demonstrated to provide sensitive metrics of cognitive changes in mild cognitive impairment (MCI) and dementia. However, the eye-tracking studies with older adults and individuals with cognitive impairment to date have utilized specialized and decontextualized cognitive tasks rather than everyday activities. We propose to examine the feasibility and utility of web page interactions, search query characteristics, eye-tracking, and mouse-movement-derived measures collected during everyday web search tasks as a metric for cognitive changes associated with healthy and pathological aging.

J. Gwizdka (✉) · Y.-C. Chan
Information eXperience Lab, School of Information, The University of Texas at Austin, Austin, U.S.
e-mail: neurois2022@gwizdka.com

Y.-C. Chan
e-mail: ycchan@utexas.edu

R. Tessmer · M. L. Henry
Department of Speech, Language, and Hearing Sciences, The University of Texas at Austin, Austin, U.S.
e-mail: rachel.tessmer@austin.utexas.edu

M. L. Henry
e-mail: maya.henry@austin.utexas.edu

K. Radhakrishnan
School of Nursing, The University of Texas at Austin, Austin, U.S.
e-mail: kradhakrishnan@mail.nur.utexas.edu

M. L. Henry
Department of Neurology, Dell Medical School, The University of Texas at Austin, Austin, U.S.

© The Author(s), under exclusive license to Springer Nature Switzerland AG 2022
F. D. Davis et al. (eds.), *Information Systems and Neuroscience*,
Lecture Notes in Information Systems and Organisation 58,
https://doi.org/10.1007/978-3-031-13064-9_20

Keywords Cognitive impairment · Eye-tracking · Web search

1 Introduction

Human longevity has increased dramatically over the past century, reshaping the population in fundamental ways. In the United States, this aging of the population, resulting from reduced birth rate and increased longevity, has significant social and economic consequences [1]. With population aging comes a drastic increase in the incidence and prevalence of age-related conditions that affect cognition such as stroke and other vascular pathologies, mild cognitive impairment (MCI), and dementia. In 2020, the prevalence of Alzheimer's disease in the United States was over six million individuals, whereas the prevalence of MCI was over twelve million individuals [2]. It is estimated that 15% of individuals with MCI develop dementia after only two years [3], and that a third of individuals with MCI develop Alzheimer's dementia within five years [4]. Projected increases in the number of individuals living with dementia and MCI will impact disease burden in the United States, placing stress on individuals, families, healthcare resources, and social and economic structures [2, 5].

Clinical researchers agree on the importance of early detection of functional cognitive impairments in older adults; however methods to capture and meaningfully characterize these impairments are lacking [5]. The majority of studies examining neuropsychological functions in healthy aging and age-related disorders use standard assessments that may fail to capture subtle cognitive impairments in everyday activities. Eye-tracking (including pupillometry) during cognitive tasks has been used as a sensitive metric for cognitive changes in MCI and Alzheimer's dementia [6]. However, the eye-tracking studies with older adults and individuals with cognitive impairment to date have typically utilized decontextualized cognitive tasks rather than everyday, functional activities. There is a great need for research evaluating how age-related and pathological changes in cognitive function manifest in more ecologically-valid tasks and contexts such as technology use. In addition, studies have shown that measures derived from computer usage, specifically mouse movement patterns, hold potential as biomarkers of MCI and dementia [7, 8].

Nearly 70% of all seniors in the United States have internet access, with the highest rate (over 80%) among older adults aged 65–69 [9]. Given the pervasiveness of web use in the everyday lives of older adults, human interactions with websites represent an important functional context for examining behavior patterns associated with aging and age-related disorders. We propose a study examining the feasibility and utility of collecting web use behavior data (web search task performance, mouse movements, and eye-tracking measurements during web searching) as markers of cognitive status in healthy and pathological aging.

2   Related Work

2.1   Computer Use, Mouse Movements, and Dementia

General computer usage has been investigated as a potential indicator of dementia. For example, interaction data collected on four representative computer tasks (file navigation, document editing, email, and web browsing) was shown to differentiate individuals with and without cognitive impairment [8]. Individuals with cognitive impairment were generally slower, entered less text, and had more frequent mouse clicks, though the distance of mouse movements resembled individuals without cognitive impairment. The last finding is in partial contrast with another study that found participants with MCI made fewer but more variable mouse movements, were less efficient, and took longer pauses [7].

Web searching is one of the common everyday activities of Internet users. Verbal expression, including the expression of information needs in textual queries, is considered a differentiator between cognitively healthy adults and those with cognitive impairments. A longitudinal study (conducted over a six-month period) with 42 participants found that older adults with lower cognitive function repeated more terms and used more typical English words in their queries [10].

2.2   Eye-Tracking Measures and Dementia

Eye-tracking (including pupillometry) is an attractive technology for detecting cognitive impairment, and it has been used as a sensitive metric for cognitive changes in MCI and Alzheimer's dementia. Thus far, most eye-tracking studies have employed cognitive tasks that, by design, tested specific cognitive functions, most often memory. For example, visual paired comparison tasks [11, 12] and attention/memory for objects in visual stimuli [6] were shown to be correlated with neuropsychological measures. Other research has demonstrated differences between individuals with amnesic and non-amnesic MCI on an anti-saccade task [13]. There is a lack of studies that demonstrate the feasibility of eye-tracking measures as a diagnostic marker for MCI and/or dementia. Our proposed study aims to bridge this gap in the research.

2.3   Neuropsychological Assessments in Aging & Age-Related Disorders

Distinguishing between cognitive changes associated with typical aging and those that are indicative of age-related disorders can be challenging [14]. In typical aging, changes in cognition can be mediated by a variety of factors, such as level of education [14, 15], and researchers have documented increased heterogeneity in performance on various cognitive tasks, such as those assessing attention, executive functioning, and some non-verbal abilities [16]. This heterogeneity makes the distinction between early pathological and non-pathological cognitive changes in older adults particularly challenging.

Early and accurate detection and characterization of cognitive changes is essential for providing individuals with appropriate clinical care [15]. Many studies have employed neuropsychological assessments to establish distinct profiles of performance for typically aging individuals and individuals with age-related disorders. For instance, constellations of cognitive deficits have been established in Alzheimer's dementia, including impaired performance on assessments of semantic knowledge, episodic memory, and executive functioning [17]. Although neuropsychological assessments are well suited to detect frank cognitive deficits, they may not be sensitive enough to capture the subtle changes in abilities that impact activities of daily living in individuals with MCI. Thus, direct observation of performance on daily tasks is needed to identify functional deficits in these individuals [18]. Furthermore, neuropsychological assessments can be costly and require an extended administration time [19], necessitating consideration of other tools that can help identify cognitive changes in older adults. Our proposed study aims to provide an alternate tool using eye-tracking and other measures derived from ecologically-valid internet search tasks.

3   Method

A controlled within- and between-subject experiment will be conducted in the Information eXperience (IX) lab at the University of Texas at Austin, with additional data being collected through tele-assessment. In order to examine health information search behavior in older adults with and without cognitive impairment, we will collect eye-tracking data during ecologically-relevant web search tasks. In addition, participants will complete a battery of well-established neuropsychological assessments. The study has been aproved by the University of Texas at Austin's Institutional Review Board.

3.1   Participants

For the pilot study, we plan to recruit 10 healthy older adults (>age 65) without cognitive impairment (group HOA). Additionally, we plan to recruit 10 older adults diagnosed with MCI or dementia (N = 10 MCI/dementia; group CI). Scores from the neuropsychological assessment battery described in Sect. 3.6 will be used to corroborate diagnoses in line with current consensus criteria for MCI and dementia [22–24]. In a pre-screening procedure, HOA participants must score >26 on the

Montreal Cognitive Assessment [25]. All participants must have normal or corrected-to-normal vision. Participants will be compensated with a $40 gift card.

3.2 Experimental Design

Participants from HOA and CI groups will participate in two experimental sessions. In the first session, participants will conduct four web search tasks on health-related topics in person in the IX lab. In the second session, participants will complete a neuropsychological assessment battery remotely over Zoom. In addition, half of the group of HOA participants will be invited for an additional session where we will test data collection with a "home-level" eye-tracker (i.e., an inexpensive instrument solution). This will allow us to compare metrics obtained on the same user tasks from a high-end eye-tracker with those from a low-end eye-tracker to determine feasibility of in-home data collection. Comparable data quality between eye-tracking devices could enable using our naturalistic paradigm approach as a remote assessment tool in future studies. The experimental design is shown in Fig. 1. Blue arrows indicate within- and between-group comparisons for statistical significance testing (inferential statistics) and predictive classification performance (machine learning algorithms).

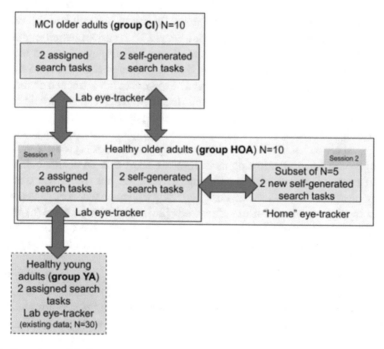

Fig. 1 Experimental design

Fig. 2 Our custom google search engine results page (SERP)

3.3 Web Search Tasks

In the first experimental session, participants will be asked to perform four health-related search tasks on Google search engine. Search results will be retrieved from Google in real-time in the background using SerpAPI.[1] A custom search engine result page (SERP) will be used to achieve control over the display of search results by including only organic search results and limiting the number of results to seven per page in order to ensure that eye fixations can be more accurately tracked on individual search results (Fig. 2). Before performing the four search tasks, participants will perform a training task to get accustomed to the experimental setup.

Two assigned search tasks. Participants will complete two assigned search tasks that were employed in a prior experiment with younger adults (group YA; N = 30, mean age = 24.5 years), allowing us to compare HOA and YA search behaviors in order to assess the effects of non-pathological aging on task performance and eye-tracking measures. The tasks were designed using the simulated work-task approach in order to trigger realistic information needs [26]. The task topics are (a) vitamins and (b) low blood pressure (Table 1). Each task will have sub-questions that will require participants to save webpages which contain answers to the sub-questions. Task order will be randomized.

Two self-generated tasks. Additionally, participants will complete two search tasks on health topics related to their own needs. Comparability between self-generated tasks will be achieved by providing participants with common task types: (a) information about dietary supplements and (b) dietary restrictions related to

---

[1] https://serpapi.com/.

Table 1 Two assigned search tasks

Task 1: Vitamin A

A family member has asked your advice in regard to taking vitamin A for health improvement purposes. You have heard conflicting reports about the effects of vitamin A, and you want to explore this topic in order to help your cousin. Specifically, you want to know:

(1) What is the recommended dosage of vitamin A for an underweight person?

(2) What are the health benefits and consequences of taking vitamin A? Could you please list 5 benefits and disadvantages respectively?

(3) What are the consequences of vitamin A deficiency or excess? Could you please list 5 consequences of deficiency or excess, respectively?

(4) Could you please list food items that are considered good sources of vitamin A? Could you please pick one food item and list the exact amount of this food one could take without experiencing an "overdose" on vitamin A?

Task 2: Low blood pressure

A friend has low blood pressure. You are curious about this issue and want to investigate more. Specifically, you want to know:

(1) What are the causes and consequences of low blood pressure? Considering that your friend is on a diet, might this be the cause to your friend's health condition?

(2) What are the differences between low blood pressure and high blood pressure in terms of symptoms and signs? Could you please list 5 entries for differences in symptoms and signs, respectively?

(3) What are the medical treatments for low blood pressure? Which solution would you recommend to your friend if they have a heart condition? Why?

a medical condition of their choosing. The self-generated search tasks will also include sub-questions (Table 2). As with the assigned search tasks, participants will be asked to save webpages which contain answers to sub-questions. Task order will be randomized.

Table 2 Two self-generated search tasks

Task 3: Medication or supplement

You want to find information about a medication or supplement recommended for the health condition you have (or your family member has). Specifically, you want to know:

(1) The names of the medication or supplement

(2) Where to buy them

(3) Side effects

Task 4: Diet

You want to find information about the diet recommended for the health condition you have (or your family member has). Specifically, you want to know:

(1) Food items

(2) Several recipes

(3) Approximate cost

Pre-task, Post-task, and Post-session Questionnaires. Before each task participants will be asked to rate their familiarity with task topic. After each task, participants will be asked to respond to the NASA Task Load Index (NASA-TLX) six-point scale to assess subjective task workload [27]. At the end of the session, we will administer the eHealth Literacy Scale (eHEALS) [28] to assess participants' perceived ability to use technology for health information.

3.4   Data Collection

Computer Interaction: Participants' interactions with the computer will be captured in two ways during the web search tasks. Web page interactions, search queries, and mouse movement on web pages will be recorded using the web browser extension YASBIL [29]. Eye movements (fixations and saccades), pupil diameter, and interactions with the computer will be recorded by iMotions[2] software using Tobii TX-300 remote eye-tracker. Participants will be seated 60–70 cm from the screen and eye-tracker. The screen resolution will be 1920×1080 pixels and the screen size will be 23 inches diagonally. A 9-point calibration will be performed prior to the search tasks. Lab illumination will be kept consistant across all participant sessions. For the subset of HOA participants who complete an additional session, eye movements will be recorded using the Gazepoint GP3 60 Hz eye-tracker. Search task performance and interaction measures include the number of searches/web pages opened, query keyword characteristics (e.g., word usage frequency), time on search engine result pages (SERPs), and time on other web pages. Mouse movement data will include a stream of mouse pointer timestamps and positions on screen. It will be used to calculate total distance traveled by the mouse, time taken to initiate a mouse movement, mouse movement curvature, and time spent idling or pausing between successive mouse movements [7].

Eye-tracking. We plan to use the following eye-tracking measures: (a) fixation measures: count, duration, rate, regressions, and type (reading/scanning) [30], (b) saccade measures: duration, amplitude, velocity, and angle, and (c) pupillometry measures: relative pupil dilation [31, 32]. Data will be analyzed separately for SERPs and content web pages. SERPs will be divided into individual search result areas, while content pages will be divided into "target" areas containing information relevant to search tasks and "non-target" areas. Eye fixations, saccades, and blinks will be detected by the Tobii I-VT algorithm [33, 34] implemented in the iMotions software. We will calculate fixation duration and pupillary measures on "target" areas containing information relevant to search tasks as well as transitions (saccades) between "target" and "non-target" areas. Additional descriptive eye-tracking measures will be calculated (e.g., mean, total, standard deviation, min,

---

[2] https://imotions.com/.

max). The selection of eye-tracking measures is informed by prior work with early-stage Alzheimer's disease participants [5, 6] and studies with healthy participants [35–38].

3.5 Neuropsychological Assessments

We will administer a neuropsychological assessment battery to HOA and CI participants [39] designed to quickly and reliably assess a variety of cognitive domains. Specific tasks in the battery include, but are not limited to the following: the Montreal Cognitive Assessment (screening tool used to detect cognitive impairment that assesses orientation, visuospatial construction, executive functioning, abstraction, short term memory and delayed recall, attention, naming, repetition, and phonemic fluency) [25], copy, recall, and recognition of the Benson complex figure (visuospatial construction, visual memory), forward and backward digit span tasks (working memory, executive functioning) [40, 41] the Stroop task (inhibition, cognitive flexibility) [42–44] a phonemic fluency task [41], a task from the Visual Object and Space Perception battery (spatial perception) [45], an abbreviated 16-item version of the Peabody Picture Vocabulary Test-Revised (receptive vocabulary) [46, 47] the California Verbal Learning Test, Second Edition-Short Form (verbal learning and memory) [47], as well as brief assessments of verbal agility (motor speech), solving mathematical problems (calculation and number processing), sentence repetition (phonological working memory), and sentence comprehension (receptive language). Participants will also complete a subset of tasks from the National Institutes of Health Toolbox Cognition battery including the following: a dimensional change card sort test (set-shifting, executive functioning), a Flanker task (attention, inhibitory control), and a test of pattern comparison processing speed (processing speed) [49]. Lastly, participants will complete the Gray Oral Reading Test-Fourth Edition, a norm-referenced assessment of oral reading skills that provides scores for reading rate, accuracy, comprehension, fluency, and overall reading ability. Neuropsychological assessment data will be used for confirmation of cognitive profiles in the CI group and will allow for exploratory correlation analyses between neuropsychological data, mouse movements, and eye-tracking metrics.

3.6 Planned Analyses

Between-group tests will be used to determine whether significant differences can be found on task performance, mouse movement, and eye-tracking measures in HOA versus YA [38] and HOA versus CI groups. Demographic variables (e.g., age, gender, education level, normal/corrected-to-normal vision status, etc.) will be included as covariates. Inferential statistics (depending on properties of data distributions, analysis of variance, or non-parametric variants) will be performed to test for significant differences on metrics between groups. We also plan to use techniques from machine learning (classification algorithms; e.g., Random Forest, Convolutional Neural Networks) to predict group membership (HOA versus YA; HOA versus CI) and perform ROC analysis (using a battery of metrics: accuracy, precision, recall, false positive, false negative, area under ROC curve and F-measure) to assess classification performance. In this project phase, the group sizes (YA vs. HOA vs. CI) will be unbalanced, and the HOA and CI group size will be small; therefore we will generate additional data samples using two methods: (a) random generation with replacement and (b) synthetic sample generation SMOTE [51]. In the future, with a larger group of HOA and CI participants, we will not need to generate additional data samples.

We will also compare data collected in session 2 for the 10 HOA participants who complete an additional session with the "home-level" eye-tracker. The strength of within-subjects correlational analyses between eye-tracking measures and neuropsychological assessment scores will be examined. Analysis of variance to assess the differences in cognitive scores explained by the two types of eye-tracking measures will also be utilized.

4 Summary

Findings from this study will constitute the first steps toward establishing a web search behavior "phenotype" for older individuals. Once established, this phenotpye may serve as a cognitive metric derived from a naturalistic and highly ecologically-valid context. Our data for individuals with CI will elucidate whether well-documented cognitive changes in MCI correlate with eye-tracking and mouse movement metrics obtained on web search tasks. Once this pattern is established in a larger patient sample, web search behaviors may be used as a scalable proxy for cognitive testing. While we selected web search tasks for this study, we expect that our approach would transfer to other everyday web tasks in business or leisure contexts. As an assessment tool, this naturalistic paradigm has the potential to be administered remotely (via webcam eye-tracking). Efficient, affordable and accessible tele-neuropsychological assessment will enable both early detection and ongoing monitoring of cognitive function in a variety of clinical populations affected by or at risk of cognitive impairment.

Acknowledgements This research has been supported in part by the University of Texas at Austin Associate Professor Experimental program.

References

1. Henderson, L., Maniam, B., & Leavell, H. (2017). The silver tsunami: Evaluating the impact of population aging in the U.S. *Journal of Business and Behavioral Sciences, 29*, 153–169.
2. Rajan, K. B., Weuve, J., Barnes, L. L., McAninch, E. A., Wilson, R. S., & Evans, D. A. (2021). Population estimate of people with clinical Alzheimer's disease and mild cognitive impairment in the United States (2020–2060). *Alzheimer's & Dementia, 17*, 1966–1975. https://doi.org/10.1002/alz.12362
3. Petersen, R. C., Lopez, O., Armstrong, M. J., Getchius, T. S. D., Ganguli, M., Gloss, D., Gronseth, G. S., Marson, D., Pringsheim, T., Day, G. S., Sager, M., Stevens, J., & Rae-Grant, A. (2018). Practice guideline update summary: Mild cognitive impairment: Report of the guideline development, dissemination, and implementation subcommittee of the American academy of neurology. *Neurology, 90*, 126–135. https://doi.org/10.1212/WNL.0000000000004826
4. Ward, A., Tardiff, S., Dye, C., & Arrighi, H. M. (2013). Rate of conversion from prodromal Alzheimer's disease to Alzheimer's dementia: A systematic review of the literature. *Dementia and Geriatric Cognitive Disorders Extra, 3*, 320–332. https://doi.org/10.1159/000354370
5. Seligman, S. C., & Giovannetti, T. (2015). The Potential utility of eye movements in the detection and characterization of everyday functional difficulties in mild cognitive impairment. *Neuropsychology Review, 25*, 199–215. https://doi.org/10.1007/s11065-015-9283-z
6. Oyama, A., Takeda, S., Ito, Y., Nakajima, T., Takami, Y., Takeya, Y., Yamamoto, K., Sugimoto, K., Shimizu, H., Shimamura, M., Katayama, T., Rakugi, H., & Morishita, R. (2019). Novel method for rapid assessment of cognitive impairment using high-performance eye-tracking technology. *Science and Reports, 9*, 1–9. https://doi.org/10.1038/s41598-019-49275-x
7. Seelye, A., Hagler, S., Mattek, N., Howieson, D. B., Wild, K., Dodge, H. H., & Kaye, J. A. (2015). Computer mouse movement patterns: A potential marker of mild cognitive impairment. *Alzheimer's and Dementia: Diagnosis, Assessment and Disease Monitoring, 1*, 472–480. https://doi.org/10.1016/j.dadm.2015.09.006
8. Stringer, G., Couth, S., Brown, L. J. E., Montaldi, D., Gledson, A., Mellor, J., Sutcliffe, A., Sawyer, P., Keane, J., Bull, C., Zeng, X., Rayson, P., Leroi, I. (2018). Can you detect early dementia from an email? A proof of principle study of daily computer use to detect cognitive and functional decline. *International Journal of Geriatric Psychiatry, 33*, 867–874. https://doi.org/10.1002/gps.4863
9. Anderson, M., & Perrin, A. (2017). Technology use among seniors.
10. Austin, J., Hollingshead, K., & Kaye, J. (2017). Internet searches and their relationship to cognitive function in older adults: Cross-sectional analysis. *Journal of Medical Internet Research, 19*, e7671. https://doi.org/10.2196/jmir.7671
11. Crutcher, M. D., Calhoun-Haney, R., Manzanares, C. M., Lah, J. J., Levey, A. I., & Zola, S. M. (2009). Eye tracking during a visual paired comparison task as a predictor of early dementia. *American Journal of Alzheimer's Disease and Other Dementias®, 24*, 258–266. https://doi.org/10.1177/1533317509332093
12. Lagun, D., Manzanares, C., Zola, S. M., Buffalo, E. A., & Agichtein, E. (2011). Detecting cognitive impairment by eye movement analysis using automatic classification algorithms. *Journal of Neuroscience Methods, 201*, 196–203. https://doi.org/10.1016/j.jneumeth.2011.06.027
13. Wilcockson, T. D. W., Mardanbegi, D., Xia, B., Taylor, S., Sawyer, P., Gellersen, H. W., Leroi, I., Killick, R., & Crawford, T. J. (2019). Abnormalities of saccadic eye movements in dementia due to Alzheimer's disease and mild cognitive impairment. *Aging, 11*, 5389–5398. https://doi.org/10.18632/aging.102118
14. Ardila, A., & Rosselli, M. (1989). Neuropsychological characteristics of normal aging. *Developmental Neuropsychology, 5*, 307–320. https://doi.org/10.1080/87565648909540441
15. Ardila, A., Ostrosky-Solis, F., Rosselli, M., & Gómez, C. (2000). Age-related cognitive decline during normal aging: The complex effect of education. *Archives of Clinical Neuropsychology, 15*, 495–315. https://doi.org/10.1093/arclin/15.6.495
16. Ardila, A. (2007). Normal aging increases cognitive heterogeneity: Analysis of dispersion in WAIS-III scores across age. *Archives of Clinical Neuropsychology, 22*, 1003–1011.

17. Salmon, D. P., & Bondi, M. W. (2009). Neuropsychological assessment of dementia. *Annual Review of Psychology, 60*, 257–282. https://doi.org/10.1146/annurev.psych.57.102904.190024
18. Giovannetti, T., Bettcher, B. M., Brennan, L., Libon, D. J., Burke, M., Duey, K., Nieves, C., & Wambach, D. (2008). Characterization of everyday functioning in mild cognitive impairment: A direct assessment approach. *Dementia and Geriatric Cognitive Disorders, 25*, 359–365. https://doi.org/10.1159/000121005
19. Dwolatzky, T., Whitehead, V., Doniger, G. M., Simon, E. S., Schweiger, A., Jaffe, D., & Chertkow, H. (2003). Validity of a novel computerized cognitive battery for mild cognitive impairment. *BMC Geriatrics, 3*, 4. https://doi.org/10.1186/1471-2318-3-4
20. Lu, P. H., Lee, G. J., Raven, E. P., Tingus, K., Khoo, T., Thompson, P. M., & Bartzokis, G. (2011). Age-related slowing in cognitive processing speed is associated with myelin integrity in a very healthy elderly sample. *Journal of Clinical and Experimental Neuropsychology, 33*, 1059–1068. https://doi.org/10.1080/13803395.2011.595397
21. Kramer, J. H., Mungas, D., Reed, B. R., Wetzel, M. E., Burnett, M. M., Miller, B. L., Weiner, M. W., & Chui, H. C. (2007). Longitudinal MRI and cognitive change in healthy elderly. *Neuropsychology, 21*, 412–418. https://doi.org/10.1037/0894-4105.21.4.412
22. Albert, M. S., DeKosky, S. T., Dickson, D., Dubois, B., Feldman, H. H., Fox, N. C., Gamst, A., Holtzman, D. M., Jagust, W. J., Petersen, R. C., Snyder, P. J., Carrillo, M. C., Thies, B., & Phelps, C. H. (2011). The diagnosis of mild cognitive impairment due to Alzheimer's disease: Recommendations from the national institute on aging-Alzheimer's association workgroups on diagnostic guidelines for Alzheimer's disease. *Alzheimer's & Dementia, 7*, 270–279. https://doi.org/10.1016/j.jalz.2011.03.008
23. Dunne, R. A., Aarsland, D., O'Brien, J. T., Ballard, C., Banerjee, S., Fox, N. C., Isaacs, J. D., Underwood, B. R., Perry, R. J., Chan, D., Dening, T., Thomas, A. J., Schryer, J., Jones, A.-M., Evans, A. R., Alessi, C., Coulthard, E. J., Pickett, J., Elton, P., … Burns, A. (2021). Mild cognitive impairment: The Manchester consensus. *Age and Ageing, 50*, 72–80. https://doi.org/10.1093/ageing/afaa228
24. American Psychiatric Association (2013) Neurocognitive disorders. In *Diagnostic and Statistical Manual of Mental Disorders*.
25. Nasreddine, Z. S., Phillips, N. A., Bédirian, V., Charbonneau, S., Whitehead, V., Collin, I., Cummings, J. L., & Chertkow, H. (2005). The montreal cognitive assessment, MoCA: A brief screening tool for mild cognitive impairment. *Journal of the American Geriatrics Society, 53*, 695–699. https://doi.org/10.1111/j.1532-5415.2005.53221.x
26. Borlund, P. (2003). The IIR evaluation model: A framework for evaluation of interactive information retrieval systems. Information Research 8, paper no. 152.
27. Hart, S. G., & Staveland, L. E. (1988). Development of NASA-TLX (Task Load Index): Results of empirical and theoretical research.
28. Norman, C. D., & Skinner, H. A. (2006). eHEALS: The eHealth literacy scale. *Journal of Medical Internet Research, 8*, e27. https://doi.org/10.2196/jmir.8.4.e27
29. Bhattacharya, N., & Gwizdka, J. (2021). YASBIL: Yet another search behaviour (and) interaction logger. In *Proceedings of the 44th International ACM SIGIR Conference on Research and Development in Information Retrieval* (pp. 2585–2589). Association for Computing Machinery. https://doi.org/10.1145/3404835.3462800
30. Cole, M. J., Gwizdka, J., Liu, C., & Belkin, N. J. (2011). Dynamic assessment of information acquisition effort during interactive search. In *Proceedings of the American Society for Information Science and Technology* (pp. 1–10). https://doi.org/10.1002/meet.2011.14504801149
31. Duchowski, A. T., Krejtz, K., Krejtz, I., Biele, C., Niedzielska, A., Kiefer, P., Raubal, M., & Giannopoulos, I. (2018). The index of pupillary activity: Measuring cognitive load Vis-à-vis task difficulty with pupil oscillation. In *Proceedings of the 2018 CHI Conference on Human Factors in Computing Systems* (pp. 282:1–282:13). ACM. https://doi.org/10.1145/3173574.3173856
32. Duchowski, A. T., Krejtz, K., Gehrer, N. A., Bafna, T., & Bækgaard, P. (2020). The low/high index of pupillary activity. In *Proceedings of the 2020 CHI Conference on Human Factors*

in Computing Systems (pp. 1–12). Association for Computing Machinery. https://doi.org/10. 1145/3313831.3376394

33. Salvucci, D. D., & Goldberg, J. H. (2000). Identifying fixations and saccades in eye-tracking protocols. In *Proceedings of the 2000 Symposium on Eye Tracking Research & Applications* (pp. 71–78). ACM. https://doi.org/10.1145/355017.355028

34. Karthik, G., Amudha, J., & Jyotsna, C. (2019). A custom implementation of the velocity threshold algorithm for fixation identification. In *2019 International Conference on Smart Systems and Inventive Technology (ICSSIT)* (pp. 488–492). https://doi.org/10.1109/ICSSIT 46314.2019.8987791.

35. Gwizdka, J. (2014). Characterizing relevance with eye-tracking measures. In *Proceedings of the 5th Information Interaction in Context Symposium* (pp. 58–67). ACM. https://doi.org/10. 1145/2637002.2637011

36. Gwizdka, J., & Dillon, A. (2020). Eye-tracking as a method for enhancing research on information search. In W. T. Fu, H. van Oostendorp (Eds.), *Understanding and Improving Information Search: A Cognitive Approach* (pp. 161–181). Springer International Publishing. https://doi. org/10.1007/978-3-030-38825-6_9

37. Bhattacharya, N., & Gwizdka, J. (2018). Relating eye-tracking measures with changes in knowledge on search tasks. In *Proceedings of the 2018 ACM Symposium on Eye Tracking Research & Applications* (pp. 62:1–62:5). ACM. https://doi.org/10.1145/3204493.3204579

38. Bhattacharya, N., & Gwizdka, J. (2019). Measuring learning during search: Differences in interactions, eye-gaze, and semantic similarity to expert knowledge. In *Proceedings of the 2019 Conference on Human Information Interaction and Retrieval* (pp. 63–71). ACM. https:// doi.org/10.1145/3295750.3298926

39. Staffaroni, A. M., Tsoy, E., Taylor, J., Boxer, A. L., & Possin, K. L. (2020). Digital cognitive assessments for dementia. *Practical Neurology (Fort Washington, Pa.), 2020*, 24–45.

40. Wechsler, D. (1997). *Wechsler adult intelligence scale—III (WAIS-III).* The Psychological Corporation.

41. Kramer, J. H., Jurik, J., Sha, S. J., Rankin, K. P., Rosen, H. J., Johnson, J. K., & Miller, B. L. (2003). Distinctive neuropsychological patterns in frontotemporal dementia, semantic dementia, and alzheimer disease. *Cognitive and Behavioral Neurology, 16*, 211–218.

42. Stroop, J. R. (1935). Studies of interference in serial verbal reactions. *Journal of Experimental Psychology, 18*, 643–662.

43. Delis, D. C., Kaplan, E., & Kramer, J. H. (2001). Delis-Kaplan executive function. *System.* https://doi.org/10.1037/t15082-000

44. Kaplan, E., Goodglass, H., & Weintraub, S. (1983). Boston naming. *TEST.* https://doi.org/10. 1037/t27208-000

45. Warrington, E. K., & James, M. (1991). Visual object and space perception battery. Thames Valley Test Co., Bury St. Edmunds, Suffolk, U.K.

46. Dunn, L. M., & Dunn, L. M. (1981). *Peabody picture vocabulary test-revised.* American Guidance Service Inc.

47. Howard, D., Education, H., & Patterson, K. (1992). Pyramids and Palm Trees Manual. Harcourt Education.

48. Delis, D. (2000). California verbal learning test-second edition. TX.

49. Weintraub, S., Dikmen, S. S., Heaton, R. K., Tulsky, D. S., Zelazo, P. D., Bauer, P. J., Carlozzi, N. E., Slotkin, J., Blitz, D., Wallner-Allen, K., Fox, N. A., Beaumont, J. L., Mungas, D., Nowinski, C. J., Richler, J., Deocampo, J. A., Anderson, J. E., Manly, J. J., Borosh, B., ... Gershon, R. C. (2013). Cognition assessment using the NIH toolbox. *Neurology, 80*, S54–S64. https://doi.org/10.1212/WNL.0b013e3182872ded

50. Pinto, T. C. C., Machado, L., Bulgacov, T. M., Rodrigues-Júnior, A. L., Costa, M. L. G., Ximenes, R. C. C., & Sougey, E. B. (2019). Is the montreal cognitive assessment (MoCA) screening superior to the mini-mental state examination (MMSE) in the detection of mild cognitive impairment (MCI) and Alzheimer's Disease (AD) in the elderly? *International Psychogeriatrics, 31*, 491–504. https://doi.org/10.1017/S1041610218001370
51. Chawla, N. V., Bowyer, K. W., Hall, L. O., & Kegelmeyer, W. P. (2011). SMOTE: Synthetic minority over-sampling technique. https://doi.org/10.1613/jair.953

Mixed Emotions: Evaluating Reactions to Dynamic Technology Feedback with NeuroIS

Sophia Mannina and Shamel Addas

Abstract Technology is increasingly leveraged in the healthcare context to provide instant and personalized feedback that can help users improve their well-being. However, technologies sometimes evoke negative emotions, which can reduce information technology (IT) use. These emotions may be related to the manner in which technologies provide feedback. It is hypothesized that emotional ramifications of feedback can be limited by providing feedback that is empathetic and supportive. An experiment is proposed to evaluate emotional reactions to different types of feedback provided by a chatbot in real-time. The chatbot will generate combinations of outcome, corrective, and personal feedback. Emotions will be continuously monitored throughout the chatbot interaction using emerging NeuroIS tools. The results will indicate how technologies that provide feedback can be designed to optimize emotions and promote continuous IT use.

Keywords Feedback · Emotion · NeuroIS · Facial recognition · Electrodermal activity (EDA) · Feedback intervention theory (FIT) · IT use · Chatbot

1 Introduction

Feedback, or information that indicates how well an individual is meeting their goals [1], is critical for individuals to understand and improve their performance [2, 3]. Feedback is not only important for accomplishing work-related or personal goals, but also for improving one's health and well-being [4]. Digital technologies are valued for their ability to provide personalized health feedback in real-time [5]. For example, chatbots (intelligent agents that communicate through text, voice, and/or animation) are increasingly accepted as effective tools for providing feedback remotely [6].

S. Mannina (✉) · S. Addas
The J.R. Smith School of Business, Queen's University, Kingston, ON, Canada
e-mail: 20scm@queensu.ca

S. Addas
e-mail: shamel.addas@queensu.ca

© The Author(s), under exclusive license to Springer Nature Switzerland AG 2022
F. D. Davis et al. (eds.), *Information Systems and Neuroscience*,
Lecture Notes in Information Systems and Organisation 58,
https://doi.org/10.1007/978-3-031-13064-9_21

Chatbots that provide therapy, counseling, and other mental health services have been shown to help individuals manage stress and depression [7].

Despite the potential for such technologies to help users improve their health, digital technologies that provide feedback are often considered frustrating to use, as opposed to stimulating and encouraging [8, 9]. Prior literature suggests that emotions play a role in health-related technology interactions and identifies a need to evaluate how digital technologies for self-care affect users' emotions [10]. Although different technology characteristics (e.g. usability) can contribute to emotions like frustration [11], feedback itself can evoke emotional reactions. Prior studies indicate that positive feedback correlates with positive emotions, while negative feedback is related to negative emotions [12]. Negative emotions are known to reduce information technology (IT) use [13]. For example, when a mobile app for mental health assessment evokes anxiety, frustration, and boredom, perceived usability (an established antecedent of IT use) decreases [14]. Poor health impacts, such as weakened immunity and cardiovascular functioning, are also observed as outcomes of negative emotions [12].

Since negative feedback (i.e., feedback that conveys a negative discrepancy between one's goals and outcomes [2, 15]) is necessary to help individuals identify actions for improvement, it should be provided in a manner that limits negative emotions. Studies suggest that social support can reduce negative emotions and enhance positive emotions [16, 17]. Ramifications of negative feedback may thus be limited if subsequent feedback that is empathetic and encouraging is provided. Empathetic feedback is also found to increase IT use intentions [18]. A technology's ability to provide empathic feedback is therefore considered to be critical [16]. It can be inferred that the manner in which feedback is provided affects users' emotions, and consequently, their IT use.

A pilot study is proposed to investigate the following research question: How do different types of technology feedback affect users' emotions? Rather than view feedback as a single unabridged construct, this research differentiates between various types of feedback to explore unique effects. We draw upon feedback intervention theory (FIT) to evaluate how different types of feedback may be related to emotions in the context of general health and well-being. This study will respond to the call for a more nuanced understanding of emotional reactions to feedback [19] by continuously assessing users' emotions with NeuroIS tools. NeuroIS is a developing area of research that applies cognitive neuroscience theories, methods, and tools to evaluate IS phenomena [20–22]. Most studies that employ NeuroIS use eye tracking or functional magnetic resonance imaging (fMRI), but other tools can also provide valuable insight [23, 24]. Facial recognition and electrodermal activity (EDA) monitoring will be used in this study to measure immediate changes in emotions as different types of feedback are provided by a chatbot. Since emotions are temporary states, their complex relationships with behaviour should be dynamically assessed to capture changes over time [25, 26].

2   Theoretical Background

2.1   Emotions

Emotions are "a mental state of readiness for action that promote behavioral activation and help prioritize and organize behaviors in ways that optimize individual adjustments to the demands of the environment" [13, p. 690]. According to the dimensional perspective [27], emotions can be evaluated on the basis of valence (the extent to which an emotion is positive or negative) and arousal (the extent to which an individual is activated or excited) [28]. Emotions have important implications for IT use [13]. Interestingly, some studies suggest that a single technology can evoke a range of different emotions in users. In their investigation of a web-based portal for self-management of asthma, Savoli et al. [29] find that some patients experience positive emotions while using the portal, while others feel angry, guilty, or sad. This paper hypothesizes that different types of feedback may be responsible for some of the variance in emotions that users experience when interacting with technologies.

2.2   Feedback

A feedback intervention is defined as an action taken by an external agent to provide information concerning aspect(s) of one's performance of a task [2]. As feedback can be generated and provided in diverse manners, Feedback Intervention Theory (FIT) is an important foundation for understanding the relationship between feedback and a recipient's behaviours [30]. FIT posits five main principles, one of which asserts that feedback interventions change one's locus of attention, thereby affecting behavior [2]. Based on this principle, Brohman et al. [26] identify three types of feedback in the context of digital health that may attract varying levels of attention: outcome (conveys information pertaining to performance results), corrective (conveys information about the process that led to a performance discrepancy and/or means for reducing it), and personal (conveys information in a manner that is caring, appears well-intentioned, is sensitive to emotional and cognitive needs, and offers a pathway to improvement) [30].

From these definitions, it is reasonable to conjecture that different types of feedback provoke various emotions. Corrective feedback is likely to evoke negative emotions because it highlights poor performance. Even so, corrective feedback is critical for performance improvement as it guides individuals to identify actions that can help them to reduce performance discrepancies and achieve their goals [32]. As personal feedback is sensitive to the emotional needs of the recipient and conveys a sense of care, this type of feedback may prompt positive emotions. This is supported by prior studies that find emotional support to be related to reductions in negative affect [16]. Personal feedback may thus be able to limit ramifications of corrective feedback on users' emotions.

3 Hypotheses

A research model is proposed (Fig. 1) to assess how users' emotions evolve over time as different types of feedback are received during a technology interaction. Our hypotheses suggest an overall negative effect of feedback that carries over time. However, if feedback is supported with personal feedback, these negative effects will be mitigated.

Feedback to Emotion at Time 1 (t_1). Literature supports that negative feedback is related to intense negative emotions [33]. Negative outcome feedback may be perceived as a threat, which can prompt negative emotions [34]. Emotions typically evoked by negative outcome feedback, including guilt and embarrassment, are recognized as particularly intense and devastating [35]. Numerous studies also suggest that corrective feedback (which reveals a gap between outcomes and goals) is related to negative emotions [19]. For example, patients experience feelings of frustration when they are made aware of discrepancies in their blood glucose level, which

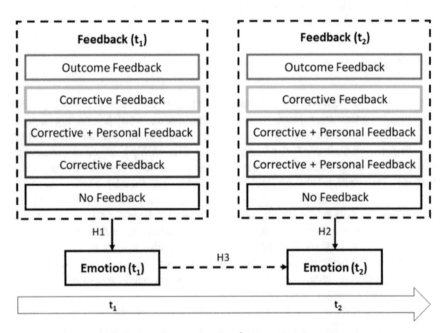

Group 1: Negative Outcome at t_1 → Negative Outcome at t_2
Group 2: Corrective at t_1 → Corrective at t_2
Group 3: Corrective + Personal at t_1 → Corrective + Personal at t_2
Group 4: Corrective at t_1 → Corrective + Personal Feedback at t_2
Group 5: No Feedback at t_1 → No Feedback at t_2

Fig. 1 Research Model

is likely because they are reminded of their failures in managing their health [31]. Polonsky [36] also supports that when technologies indicate unsatisfactory health status, patients tend to feel poorly about themselves. This may trigger negative self-talk and incite feelings of guilt. However, if corrective feedback is accompanied by personal feedback, its negative impacts could be negated. Empathic feedback is found to limit negative affect [16] and may buffer emotions like guilt and stress [37]. Corrective feedback that is encouraging and sympathetic is found to provoke less negative emotions [38]. Negative emotions can be evanescent and superseded by positive emotion [19]. Since supportive feedback delivered with empathic concern is found to be related to positive emotions [34], a combination of corrective and personal feedback can be expected to have a positive effect overall on the recipient's emotions. Thus:

H1: Negative outcome feedback is negatively related to emotion at t_1 (H1a); Corrective feedback is negatively related to emotion at t_1 (H1b); Corrective feedback accompanied by personal feedback is positively related to emotion at t_1 (H1c).

Feedback to Emotion at Time 2 (t_2). Since emotions are dynamic and temporary states [39], new events can prompt individuals to regulate or reappraise their emotions [40]. Prior work suggests that repeated exposure to feedback concerning performance discrepancies or issues is emotionally taxing [3, 30]. This indicates that multiple iterations of negative or corrective feedback may exacerbate negative emotions. Since corrective feedback that encourages and supports recipients is found to reduce negative emotions [38], exposure to corrective feedback accompanied by personal feedback could mitigate negative emotions and prompt positive emotions. Hence:

H2: Repeated exposure to negative outcome feedback is negatively related to emotion at t_2 (H2a); Repeated exposure to corrective feedback is negatively related to emotion at t_2 (H2b); Repeated exposure to corrective feedback accompanied by personal feedback is positively related to emotion at t_2 (H2c); Exposure to corrective feedback at t_1 and corrective feedback accompanied by personal feedback at t_2 is positively related to emotion at t_2 (H2d).

Emotion at Time 1 (t_1) to Emotion at Time 2 (t_2). The duration of an emotion can significantly vary according to a variety of factors [41]. Some emotions may thus be long-lasting and have lagging effects over time [39]. It has been suggested that "carry-over" effects of previous emotions can impact subsequent emotions [42]. Therefore:

H3: Emotion at t_1 is related to emotion at t_2.

4 Methodology

An experimental pilot study will be performed using online chatbots to administer feedback to a convenience sample of 12 graduate students. The chatbots will be programmed to ask questions and provide feedback concerning dietary quality as this is a general area of health that can be easily monitored and understood. All participants will complete the Short Healthy Eating Index (sHEI) survey in Qualtrics to obtain an initial estimate of their dietary quality, or intake of vegetables, dairy, and other food groups. This 22-item tool has been systematically validated and is found to be less burdensome than other instruments used to measure dietary quality [43].

Participants will then be asked to visit a designated lab at Queen's University to interact with a chatbot in a controlled environment. Participants will be randomly assigned to four treatment groups and one control group (Fig. 1). The first treatment group will be exposed to two iterations of negative outcome feedback. The second group will be exposed to two iterations of corrective feedback. The third group will be exposed to two iterations of corrective feedback accompanied by personal feedback. The fourth group will be exposed to corrective feedback, followed by corrective feedback accompanied by personal feedback. The control group will not be exposed to feedback (the chatbot will simply acknowledge their response). Separate chatbots will be programmed to administer each of these treatments. All technology characteristics will be held constant across these chatbots to observe how the feedback provided drives variance in users' emotions. In collaboration with HEC Montréal, the COBALT Research Ecosystem will be used to capture physiological measures of emotion during the chatbot interactions. This includes facial recognition technology to analyze valence and an EDA device to measure arousal through skin conductance [26]. Skin temperature will be continuously monitored during the experiment using the EDA device. A video recording of participants' faces will be captured using a webcam and processed after the experiment using FaceReader software to evaluate emotional valence according to participants' facial expressions. Multilevel analysis will be performed to evaluate differences in emotions between participants in the different treatment groups, and to determine within-subject effects of repeated exposure to feedback at different time points.

5 Conclusion

By evaluating relationships between different types of feedback and emotions in a dynamic experiment, this proposed study will provide insight as to how technologies can be designed to evoke strong positive emotions in users. Well-established relationships between emotion and IT use [13] suggest that this will have important implications for continuous technology use. Future iterations of this study can expand on the present design and evaluate IT use as an outcome of the expected relationship between technology feedback and emotions. Deployment of emerging NeuroIS tools in this study will additionally advance methodological practices as these tools will be tested to obtain physiological measures of emotion in real-time.

Acknowledgements Thank you to the researchers at HEC Montréal for their support and guidance in using the COBALT system.

References

1. Ashford, S. J., & Cummings, L. L. (1983). Feedback as an individual resource: Personal strategies of creating information. *Organizational Behavior and Human Performance, 32*, 370–398.
2. Kluger, A., & DeNisi, A. (1996). The effects of feedback interventions on performance: a historical review, a meta-analysis, and a preliminary feedback intervention theory. *Psychological Bulletin, 119*, 254–284.
3. Ilgen, D. R., Fisher, C. D., & Taylor, M. S. (1979). *Consequences Of Individual Feedback On Behavior In Organizations, 64*, 349.
4. DiClemente, C. C., Marinilli, A. S., Singh, M., & Bellino, L. E. (2001). The role of feedback in the process of health behavior change. *American Journal of Health Behavior, 25*, 217–227.
5. Nisha, N., Iqbal, M., & Rifat, A. (2019). The changing paradigm of health and mobile phones. *Journal of Global Information Management, 27*, 19–46.
6. Sheth, A., Yip, H. Y., & Shekarpour, S. (2019). Extending patient-chatbot experience with internet-of-things and background knowledge: Case studies with healthcare applications. *IEEE Intelligent Systems, 34*, 24–30.
7. Abd-Alrazaq, A. A., Alajlani, M., Ali, N., Denecke, K., Bewick, B. M., & Househ, M. (2021). Perceptions and opinions of patients about mental health chatbots: scoping review. *Journal of Medical Internet Research, 23*, e17828.
8. Wannheden, C., Stenfors, T., Stenling, A., Von Thiele Schwarz, U. (2021) Satisfied or Frustrated? A Qualitative Analysis of Need Satisfying and Need Frustrating Experiences of Engaging With Digital Health Technology in Chronic Care. Frontiers in Public Health 8.
9. Lupton, D. (2014). Critical perspectives on digital health technologies. *Sociology compass, 8*, 1344–1359.
10. Nijland, N., Van Gemert-Pijnen, J., Boer, H., Steehouder, M. F., & Seydel, E. R. (2008). Evaluation of internet-based technology for supporting self-care: problems encountered by patients and caregivers when using self-care applications. *Journal of Medical Internet Research, 10*, e13.
11. Lazar, J., Jones, A., Hackley, M., & Shneiderman, B. (2006). Severity and impact of computer user frustration: A comparison of student and workplace users. *Interacting with Computers, 18*, 187–207.
12. Schwab, S., Markus, S., Hassani, S. (2022). Teachers' feedback in the context of students' social acceptance, students' well-being in school and students' emotions. *Educational Studies* 1–18.
13. Beaudry, P. (2010). The other side of acceptance: studying the direct and indirect effects of emotions on information technology use. *MIS Quarterly, 34*, 689.
14. Liu, Y. S., Hankey, J., Lou, N. M., Chokka, P., & Harley, J. M. (2021). Usability and emotions of mental health assessment tools: comparing mobile app and paper-and-pencil modalities. *Journal of Technology in Human Services, 39*, 193–211.
15. Fishbach, A., Eyal, T., & Finkelstein, S. R. (2010). How positive and negative feedback motivate goal pursuit. *Social and Personality Psychology Compass, 4*, 517–530.
16. Nguyen, H., Masthoff, J. (Year) Designing empathic computers: the effect of multimodal empathic feedback using animated agent. In: *Proceedings Of The 4th International Conference On Persuasive Technology*, (pp. 1–9).
17. De Gennaro, M., Krumhuber, E. G., & Lucas, G. (2020). Effectiveness of an empathic chatbot in combating adverse effects of social exclusion on mood. *Frontiers in psychology, 10*, 3061.
18. Terzis, V., Moridis, C. N., & Economides, A. A. (2012). The effect of emotional feedback on behavioral intention to use computer based assessment. *Computers & Education, 59*, 710–721.
19. Han, Y., & Hyland, F. (2019). Academic emotions in written corrective feedback situations. *Journal of English for Academic Purposes, 38*, 1–13.
20. vom Brocke, J., Hevner, A., Léger, P. M., Walla, P., & Riedl, R. (2020). Advancing a NeuroIS research agenda with four areas of societal contributions. *European Journal of Information Systems, 29*, 9–24.

21. Dimoka, A. (2012). How to conduct a functional magnetic resonance (fMRI) study in social science research. *MIS Quarterly* 811–840.
22. Riedl, R., Banker, R. D., Benbasat, I., Davis, F. D., Dennis, A. R., Dimoka, A., Gefen, D., Gupta, A., Ischebeck, A., & Kenning, P. (2010). On the foundations of NeuroIS: Reflections on the Gmunden Retreat 2009. *Communications of the Association for Information Systems, 27*, 15.
23. Fischer, T., Davis, F. D., Riedl, R. (2019). NeuroIS: A Survey on the Status of the Field. *Information systems and neuroscience,* (pp. 1–10). Springer.
24. Riedl, R., Fischer, T., Léger, P.-M., & Davis, F. D. (2020). A decade of neurois research: progress, challenges, and future directions. *ACM SIGMIS Database: The DATABASE for Advances in Information Systems, 51*, 13–54.
25. Folkman, S., & Lazarus, R. S. (1988). The relationship between coping and emotion: Implications for theory and research. *Social science & medicine, 26*, 309–317.
26. Labonté-LeMoyne, E., Courtemanche, F., Coursaris, C., Hakim, A., Sénécal, S., Léger, P.-M. (Year). Development of a new dynamic personalised emotional baselining protocol for human-computer interaction. In: *NeuroIS Retreat,* (pp. 214–219). Springer.
27. Harmon-Jones, E., Harmon-Jones, C., & Summerell, E. (2017). On the importance of both dimensional and discrete models of emotion. *Behavioral sciences, 7*, 66.
28. Russell, J. A. (1980). A circumplex model of affect. *Journal of Personality and Social Psychology, 39*, 1161–1178.
29. Savoli, A., Barki, H., & Paré, G. (2020). Examining how chronically ill patients' reactions to and effective use of information technology can influence how well they self-manage their illness. *MISQ, 44*, 351–389.
30. Brohman, K., Addas, S., Dixon, J., & Pinsonneault, A. (2020). Cascading feedback: a longitudinal study of a feedback ecosystem for telemonitoring patients with chronic disease. *MIS Quarterly, 44*, 421–450.
31. Toscos, T., Connelly, K., Rogers, Y. (Year). Designing for positive health affect: Decoupling negative emotion and health monitoring technologies. In: *2013 7th International Conference on Pervasive Computing Technologies for Healthcare and Workshops,* (pp. 153–160). IEEE.
32. Zsohar, H., & Smith, J. A. (2009). The power of and and but in constructive feedback on clinical performance. *Nurse Educator, 34*, 241–243.
33. Belschak, F. D., & Den Hartog, D. N. (2009). consequences of positive and negative feedback: the impact on emotions and extra-role behaviors. *Applied Psychology, 58*, 274–303.
34. Young, S. F., Richard, E. M., Moukarzel, R. G., Steelman, L. A., & Gentry, W. A. (2017). How empathic concern helps leaders in providing negative feedback: A two-study examination. *Journal of Occupational and Organizational Psychology, 90*, 535–558.
35. Bidjerano, T. (2010). Self-conscious emotions in response to perceived failure: A structural equation model. *The Journal of Experimental Education, 78*, 318–342.
36. Polonsky, W. (1999). Diabetes burnout: What to do when you can't take it anymore. *American Diabetes Association.*
37. Saintives, C., Lunardo, R. (2016). How guilt affects consumption intention: the role of rumination, emotional support and shame. *Journal of Consumer Marketing.*
38. Baron, R. A. (1988). Negative effects of destructive criticism: Impact on conflict, self-efficacy, and task performance. *Journal of applied psychology, 73*, 199.
39. Lazarus, R. S. (1991). Cognition and motivation in emotion. *American psychologist, 46*, 352.
40. Koval, P., Butler, E. A., Hollenstein, T., Lanteigne, D., & Kuppens, P. (2015). Emotion regulation and the temporal dynamics of emotions: Effects of cognitive reappraisal and expressive suppression on emotional inertia. *Cognition and Emotion, 29*, 831–851.
41. Verduyn, P., & Lavrijsen, S. (2015). Which emotions last longest and why: The role of event importance and rumination. *Motivation and Emotion, 39*, 119–127.

42. Böckenholt, U. (1999). Analyzing multiple emotions over time by autoregressive negative multinomial regression models. *Journal of the American Statistical Association, 94*, 757–765.
43. Colby, S., Zhou, W., Allison, C., Mathews, A. E., Olfert, M. D., Morrell, J. S., Byrd-Bredbenner, C., Greene, G., Brown, O., Kattelmann, K., & Shelnutt, K. (2020). Development and validation of the short healthy eating index survey with a college population to assess dietary quality and intake. *Nutrients, 12*, 2611.

Smart Production and Manufacturing: A Research Field with High Potential for the Application of Neurophysiological Tools

Josef Wolfartsberger and René Riedl

Abstract The concepts of Smart Production and Industry 4.0 refer to the connection of physical production with digital technology and advanced analytics to create a more holistic and flexible ecosystem. The human worker is a central element, who is mentally supported in her daily routine and decision-making processes by data-based assistive systems. Currently, there are few studies that scientifically demonstrate the effectiveness of these systems. While behavioral methods and self-report instruments are commonly used to assess cognitive workload, mental fatigue, and stress, among other variables, neurophysiological tools provide promising complementary insights. This work outlines four important application areas for the use of neurophysiology in smart production and manufacturing. The current work has the goal to instigate further research in the study domain, both theoretical and empirical.

Keywords Cognitive workload · Industry 4.0 · Mental fatigue · Neuroscience · Smart manufacturing · Smart production · Stress

1 Introduction

The concept of Industry 4.0 defines a trend away from mass production towards mass customization. More product variety means significant changes for people working in production planning. Using data-driven production processes, people are supported in their activities. These production assistance systems (as defined by [1]) act on historical data and, combined with the expert knowledge of humans, attempt to generate accurate forecasts of future events. The intention is to relieve the worker

J. Wolfartsberger (✉) · R. Riedl
University of Applied Sciences Upper Austria, Steyr, Austria
e-mail: josef.wolfartsberger@fh-ooe.at

R. Riedl
e-mail: rene.riedl@fh-ooe.at

R. Riedl
Johannes Kepler University Linz, Linz, Austria

physically and, more importantly, mentally from their activities to attenuate or even prevent mental fatigue. Even though we observe a paucity of high-quality scientific studies on the effectiveness of assistive systems in production (however, see for example [2]), various studies deal with the visualization of production data and instructions directly in the operator's field of view using mixed or augmented reality technologies [3, 4]. In such studies, different possibilities of assistance (e.g., during manual assembly) are compared based on cognitive load and effectiveness measures (e.g., the latter is often measured based on errors during the assembly process). The measurement of cognitive load is often based on self-report instruments such as the standardized NASA-TLX questionnaire. However, neurophysiological tools provide promising complementary measurement options. For example, it has been proposed that neurophysiological measurement (e.g., electroencephalography, in short EEG, or functional near-infrared spectroscopy, in short fNIRS) could offer a more accurate measurement of cognitive load during the execution of certain activities [5–8]. Neurophysiological measurements have further been used in several domains which are relevant to production and manufacturing, including user experience design [9] and technostress research [10–14].

Against the presented background, this work outlines four important application areas for the use of neurophysiology in smart production and manufacturing. The current work has the goal to instigate further research in the study domain, both theoretical and empirical. The four application areas were identified based on an existing publication [15], in which the potential of smart production and manufacturing technologies, including data-based assistive systems, predictive maintenance, and mixed reality-based services were investigated. As the present work constitutes research-in-progress, we do not make a claim that our list of application areas is exhaustive. Rather, we consider the four areas as interesting use cases which deserve more systematic investigation in the future. However, as our description also includes first published research in each domain, other researchers' work confirms the appropriateness of the areas (phenomena) as candidates with high potential for the complementary use of physiological tools for their empirical investigation. Ultimately, it is hoped that the current work instigates future research in the study domain.

2 Application Areas

In the following, we describe four manufacturing and production areas in which the use of neurophysiological measurement holds significant value, based on [15]. Moreover, based on a first literature review, we identified empirical studies which applied physiological measurement in these areas. We briefly describe selected studies, thereby substantiating the potential of neurophysiological tools in manufacturing and production.

2.1   Assistive Systems for Decision-Making

Assistive systems in manufacturing aim to support workers (e.g., machine operators) in their decision-making process and thus relieve them physically and mentally. Klocke et al. [1] describe four stages of assistive systems in production engineering, reaching from stage 1 "simple warning lights" to stage 4 "automatically generated recommendations based on historic data and cause-and-effect relationships". To date, there is a paucity of research examining the cognitive effects of assistive systems in manufacturing on workers. Gaining a deep understanding of the neurological effects of assistive systems (along the four stages) might open new ways to organize manufacturing processes in the future. Fruitful exemplary research questions to be studied in the future are:

- Can neurophysiological methods determine the level of stress, or technostress, in an engineering working environment?
- Which level of assistive system in production technology causes the least cognitive load?
- Are assistive systems in manufacturing suitable means to relieve workers mentally in their production environment?

A limited number of studies which could constitute a useful basis for a more systematic future research agenda exist. As an example, Ostberg et al. [16] provide an overview of stress theories and the ways to determine stress in individuals. Also, they outline a methodology to measure stress in controlled experiments that is tailored to software engineering research. Importantly, their methodology explicitly considers neurobiological stress markers. As another example, Weber et al. [17] present a systematic review of brain and autonomic nervous system activity measurement in software engineering. Even though these works have their origin in the development of intangible products (i.e., software programs), these studies also constitute a valuable foundation for development processes of tangible products.

2.2   Assistive Systems for Manual Assembly Work

Products are getting increasingly complex and personalized. A wider variety of products means significant changes for people doing manual assembly tasks and increases the complexity of those tasks. New mechanisms are needed to support the required flexibility, including novel types of vision-based approaches (like in-situ projections and smart glasses with Augmented Reality support). Studies show that especially the usage of smart glasses may cause high cognitive load [3, 4]. The benefits of such systems to reducing mental workload have previously been justified with self-report instruments (e.g., NASA-TLX). Neurophysiological methods have the potential to measure the cognitive load of workers more precisely; for a first example, please see [18]. Thus, the goal is to investigate the potential of neurophysiological methods for measuring the mental workload during manual assembly work. Exemplary research questions are:

- Which neurophysiological tools are appropriate to detect cognitive overload and mental fatigue during manual assembly work?
- Which assistive systems for manual assembly tasks cause the least cognitive load?
- Which impact does the assistive system have on assembly time and error rate and how is this related to the operator's neurophysiological states?

Xiao et al. [19] presented an "approach for mental fatigue detection and estimation of assembly operators in the manual assembly process of complex products, with the purpose of founding the basis for adaptive transfer and demonstration of assembly process information (API), and eventually making the manual assembly process smarter and more human-friendly" (p. 239). Importantly, the proposed approach estimates the mental state of assembly operators based on EEG, specifically the Emotiv EPOC+ headset. Because the Emotiv instrument is not a full research-grade instrument, a recent debate in the NeuroIS community on measurement quality must be considered when interpreting the results [20, 21].

2.3 Virtual Reality-Supported Training of Assembly/Maintenance Tasks

Due to its interactive nature, Virtual Reality has great potential for the training of industrial assembly procedures and maintenance tasks. Training in a virtual environment is potentially easier, safer, and cheaper than real-world training [22]. Finding the ideal mix of challenge and engagement in a virtual learning environment is crucial for keeping the learner in a "state of flow". According to Csíkszentmihályi [23], flow has a documented correlation with high performance in the fields of creativity and learning. In this application domain, the potential of VR is examined with a focus on the extent to which neurophysiological tools can help to increase the learning effect. A first user study on this effect was conducted in [24]. Exemplary research questions are:

- Can neurophysiological tools help to determine the optimal learning state ("state of flow") in a Virtual Reality training situation?
- How can the VR environment react to deviations from the flow state (especially in situations of anxiety and boredom)?
- Which neurophysiological tools are appropriate to detect the "state of flow" in a Virtual Reality training situation?

In this context, an interesting paper was published by Rezazadeh et al. [25]; the authors summarize the contribution of their study as follows: "a virtual crane training system has been developed which can be controlled using control commands extracted from facial gestures and is capable to lift up loads/materials in the virtual construction sites. Then, we integrate affective computing concept into the conventional VR training platform for measuring the cognitive load and level of satisfaction during performance using human's forehead bioelectric-signals. By employing the affective measures and our novel control scheme, the designed interface could be adapted to user's affective status during the performance in real-time. This adaptable user interface approach helps the trainee to cope with the training for long-run performance, leads to gaining more expertise and provides more effective transfer of learning to other operation environments" (p. 289).

2.4 Virtual Reality-Supported Collaborative Design Review

The idea of virtual design review allows users to examine prototypes in a realistic way starting in the earliest design stages. Many companies conduct design reviews to detect errors in their products early on before the physical product is manufactured. The results in [26] indicate that design review can benefit from VR technology, since VR reveals details that may stay unrevealed in a 2D representation on the screen. Nevertheless, design review is a collaborative process, but VR isolates users from their team members sharing the same physical space. We propose to use neurophysiological tools to examine the users' well-being in VR in a collaborative working environment. In a further step, measures could be examined to counteract the feeling of isolation in VR. First user studies are reported in [27, 28]. Exemplary research questions are:

- Can neurophysiological tools determine if VR supports collaborative design review in a shared space?
- Can we measure different effects when team members are geographically separated?
- Can neurophysiological tools detect a feeling of isolation in design review situations, where only one team member uses VR, while others are discussing the design in real space?
- Can we counteract this issue in real-time when a feeling of isolation is detected based on neurophysiological measurement?

We identified relevant research in this application area. For example, Marín-Morales et al. [29] developed an emotion recognition system for affective states evoked through Immersive Virtual Environments. Based on the circumplex model of affects, four virtual rooms were created to elicit four possible arousal-valence combinations. Drawing upon this conceptualization, an experiment using EEG and electro-cardiography (ECG) was carried out. The results indicate that the use of Immersive Virtual Environments may elicit different emotional, which can be automatically detected based on analysis of neural and cardiac measures. In another study, de Freitas et al. [30] summarize the benefits and challenges of VR-based usability testing and design reviews in industry through a patents and articles review.

3 Conclusion

One of the challenges in Industry 4.0 is to put humans properly at the center of smart manufacturing design and to create working environments, which are not characterized by excessive demands and mental stress. Today humans are often not replaced by digitalization, but instead supported in their daily work routine. Based on their experience, humans remain the ones who make decisions, supported by data-based recommendations and automatically generated forecasts. To investigate the effectiveness of these assistive systems in various application areas, more objective measurement tools are needed that allow for more definitive conclusions about outcome variables (e.g., mental workload of the worker). This paper presents four areas from production and manufacturing where we foresee considerable application potential for neurophysiological tools. Research in this domain must not ignore the psychosocial aspects of human–machine collaboration. As stated by Evjemo et al. [31], the synergy of the skills of machines and humans will be even more important in the future to increase productivity and quality, while still maintaining sustainable working conditions and environment, health and safety. Thus, it will be rewarding what future insights research will reveal, and it is hoped that the use of neurophysiological measurement will play a role in this research as the complementary use of such tools—in addition to behavioral and self-report measures—usually comes along with significantly increased insights in the study of human–machine interaction [32–34].

References

1. Klocke, F., Bassett, E., Bönsch, C., Gärtner, R., Holsten, S., Jamal, R., Jurke, B., Kamps, S., Kerzel, U., Mattfeld, P., Shirobokov, A., Stauder, J., Stautner, M., & Trauth, D. (2017). Assistenzsysteme in der Produktionstechnik. Internet of Production für agile Unternehmen. *AWK Aachener Werkzeugmaschinen-Kolloquium*, 287–313.
2. Korn, O., Schmidt, A., & Hörz. T. (2012). Assistive systems in production environments: exploring motion recognition and gamification. In *Proceedings of the 5th International Conference on PErvasive Technologies Related to Assistive Environments (PETRA '12)* (1–5). ACM.
3. Funk, M., Kosch, T., Schmidt, A. (2016). Interactive worker assistance: Comparing the effects of head-mounted displays, in-situ projection, tablet, and paper instructions. In *Proceedings of the 2016 ACM International Joint Conference on Pervasive and Ubiquitous Computing*.
4. Funk, M., Bächler, A., Bächler, L., Kosch, T., Heidenreich, T., & Schmidt, A. (2017). Working with augmented reality? A long-term analysis of in-situ instructions at the assembly workplace. In *Proceedings of the 10th ACM International Conference on Pervasive Technologies Related to Assistive Environments*.
5. Leff, D. R., Orihuela-Espina, F., Elwell, C. E., Athanasiou, T., Delpy, D. T., & Darzi, A. W. (2011). Assessment of the cerebral cortex during motor task behaviours in adults: A systematic review of functional near infrared spectroscopy (fNIRS) studies. *NeuroImage, 54*, 2922–2936.
6. Tinga, A. M., de Back, T. T., & Louwerse, M. M. (2020). Non-invasive neurophysiology in learning and training: Mechanisms and a SWOT analysis. *Frontiers in Neuroscience, 14*, 589.

7. Müller-Putz, G., Riedl, R., & Wriessnegger, S. (2015). Electroencephalography (EEG) as a research tool in the information systems discipline: Foundations, measurement, and applications. *Communications of the Association for Information Systems, 37*, 911–948.
8. Gefen, D., Ayaz, H., & Onaral, B. (2014). Applying functional near infrared (fNIR) spectroscopy to enhance MIS research. *AIS Transactions on Human-Computer Interaction, 6*, 55–73.
9. Angioletti, L., Cassioli, F., & Balconi, M. (2020). Neurophysiological correlates of user experience in Smart Home Systems (SHSs): First evidence from electroencephalography and autonomic measures. *Frontiers in Psychology, 11.*
10. Fischer, T., & Riedl, R. (2020). Technostress measurement in the field: A case report. In F. D. Davis, R. Riedl, J. vom Brocke, P.-M. Léger, A. B. Randolph & T. Fischer (Eds.), *Information Systems and Neuroscience: NeuroIS Retreat 2020* (vol. 43, pp. 71–78). LNISO. Springer, Cham.
11. Riedl, R. (2013). On the biology of technostress: Literature review and research agenda. *ACM SIGMIS Database: The DATA BASE for Advances in Information Systems, 44*, 18–55.
12. Riedl, R., Kindermann, H., Auinger, A., & Javor, A. (2012). Technostress from a neurobiological perspective: System breakdown increases the stress hormone cortisol in computer users. *Business & Information Systems Engineering, 4*(2), 61–69.
13. Riedl, R., Kindermann, H., Auinger, A., & Javor, A. (2013). Computer breakdown as a stress factor during task completion under time pressure: Identifying gender differences based on skin conductance. *Advances in Human-Computer Interaction, 2013*, 1–8.
14. Tams, S., Hill, K., Ortiz de Guinea, A., Thatcher, J., & Grover, V. (2014). NeuroIS—Alternative or complement to existing methods? Illustrating the holistic effects of neuroscience and self-reported data in the context of technostress research. *Journal of the Association for Information Systems, 15*, 723–753.
15. Zenisek, J., Wild, N., & Wolfartsberger, J. (2021). Investigating the potential of smart manufacturing technologies. *Procedia Computer Science, 180*, 507–516.
16. Ostberg, J., Graziotin, D., Wagner, S., & Derntl, B. (2020). A methodology for psychobiological assessment of stress in software engineering. *PeerJ Computer Science, 6*, e286.
17. Weber, B., Fischer, T., & Riedl, R. (2021). Brain and autonomic nervous system activity measurement in software engineering: A systematic literature review. *Journal of Systems & Software, 178*, 110946.
18. Kosch, T., Funk, M., Schmidt, A., & Chuang, L. L. (2018). Identifying cognitive assistance with mobile electroencephalography: A case study with in-situ projections for manual assembly. In *Proceedia ACM Human Computer Interaction, 2*, 1–20.
19. Xiao, H., Duan, Y., Zhang, Z., & Li, M. (2018). Detection and estimation of mental fatigue in manual assembly process of complex products. *Assembly Automation, 38*(2), 239–247.
20. Müller-Putz, G. R., Tunkowitsch, U., Minas, R. K., Dennis, A. R., & Riedl, R. (2021). On electrode layout in EEG studies: A limitation of consumer-grade EEG instruments. In F. D. Davis, R. Riedl, J. vom Brocke, P.-M. Léger, A. B. Randolph & G. Müller-Putz (Eds.), *Information Systems and Neuroscience: NeuroIS Retreat 2021* (Vol. 52, pp. 90–95). LNISO. Springer, Cham.
21. Riedl, R., Minas, R. K., Dennis, A. R., & Müller-Putz, G. (2020). Consumer-grade EEG instruments: Insights on the measurement quality based on a literature review and implications for NeuroIS research. In F. D. Davis, R. Riedl, J. vom Brocke, P.-M. Léger, A. B. Randolph & G. Müller-Putz (Eds.), *Information Systems and Neuroscience: NeuroIS Retreat 2020* (Vol. 43, pp. 350–361). LNISO. Springer, Cham.
22. Wolfartsberger, J., Riedl, R., Jodlbauer, H., Haslinger, N., Hlibchuk, A., Kirisits, A., & Schuh, S. (2022). Virtual Reality als Trainingsmethode: Eine Laborstudie aus dem Industriebereich. *HMD, 59*, 295–308.
23. Csikszentmihalyi, M. (1990). *Flow: The psychology of optimal experience.* Harper & Row, New York.
24. Tremmel, C., Herff, C., Sato, T., Rechowicz, K., Yamani, Y., & Krusienski, D. (2019). Estimating cognitive workload in an interactive virtual reality environment using EEG. *Frontiers in Human Neuroscience, 13*, 401.

25. Rezazadeh, I. M., Wang, X., Firoozabadi, M., & Golpayegani, R. H. (2011). Using affective human–machine interface to increase the operation performance in virtual construction crane training system: A novel approach. *Automation in Construction, 20*, 289–298.

26. Wolfartsberger, J. (2019). Analyzing the potential of virtual reality for engineering design review. *Automation in Construction, 104*, 27–37.

27. Dey, A., Chatburn, A., & Billinghurst, M. (2019). Exploration of an EEG-based cognitively adaptive training system in virtual reality. In *IEEE Conference on Virtual Reality and 3D User Interfaces (VR)* (pp. 220–226), Osaka, Japan.

28. Wolfartsberger, J., Zenisek, J., & Wild, N. (2020). Supporting teamwork in industrial virtual reality applications. *Procedia Manufacturing, 42*, 2–7.

29. Marín-Morales, J., Higuera-Trujillo, J. L., Greco, A., Guixeres, J., Llinares, C., Scilingo, E., Alcañiz, R. M., & Valenza, G. (2018). Affective computing in virtual reality: Emotion recognition from brain and heartbeat dynamics using wearable sensors. *Scientific Reports, 8*, 13657.

30. de Freitas, F. V., Gomes, M. V. M., & Winkler, I. (2022). Benefits and challenges of virtual-reality-based industrial usability testing and design reviews: A patents landscape and literature review. *Applied Sciences, 12*, 1755.

31. Evjemo, L., Gjerstad, T., Grøtli, E., & Sziebig, G. (2020). Trends in smart manufacturing: Role of humans and industrial robots in smart factories. *Current Robotics Reports, 1*, 35–41.

32. Riedl, R., & Léger, P. M. (2016). *Fundamentals of NeuroIS: Information systems and the brain.* Springer, Heidelberg.

33. Dimoka, A., Banker, R. D., Benbasat, I., Davis, F. D., Dennis, A. R., Gefen, D., Gupta, A., Ischebeck, A., Kenning, P., Müller-Putz, G. R., Pavlou, P. A., Riedl, R., vom Brocke, J., & Weber, B. (2012). On the use of neurophysiological tools in IS research: Developing a research agenda for NeuroIS. *MIS Quarterly, 36*, 679–702.

34. Riedl, R. (2009). Zum Erkenntnispotenzial der kognitiven Neurowissenschaften für die Wirtschaftsinformatik: Überlegungen anhand exemplarischer Anwendungen. *NeuroPsychoEconomics, 4*, 32–44.

New Measurement Analysis for Emotion Detection Using ECG Data

Verena Dorner and Cesar Enrique Uribe Ortiz

Abstract Electrocardiography (ECG) offers a lot of information that can be processed to make inferences about levels of arousal, stress, and emotions. One of the most popular measures is the Heart Rate Variability (HRV), a measure of the variation on the heart beats, which is only taken from one heart movement of the cardiac cycle, the R-wave. We explore the other heart movements of the cardiac cycle observed in the ECG with the aim of deriving new proxy measures for stress and arousal to enrich and complement HRV analysis. This article discusses existing approaches, suggests new measurements for stress and arousal detected in an ECG, and examines their potential to contribute new information based on their correlations with two HRV measures.

Keywords ECG · Heart Rate Variability · Algorithm · Experiment

1 Introduction

Heart rate (HR) and heart rate variability (HRV) are often used as proxies for measuring emotions [e.g., 1–4], (techno) stress [e.g., 5–8], and arousal [e.g., 9]. In the NeuroIS field, for instance, HR and HRV have been used to study how technology and interface design affect users' stress levels [8], arousal in electronic auctions [9], or how to provide biofeedback to users [10]. However, not all studies find significant relations between emotions and HRV [11]. As these measures are derived from only a part of the information provided by the ECG, there is scope to look at the complexity of the heart cycle and examine other movements that can be meaningful in an arousal,

V. Dorner · C. E. Uribe Ortiz (✉)
Vienna University of Economics and Business, Vienna, Austria
e-mail: cesar.enrique.uribe.ortiz@wu.ac.at

V. Dorner
e-mail: verena.dorner@wu.ac.at

C. E. Uribe Ortiz
Vienna University of Technology, Vienna, Austria

© The Author(s), under exclusive license to Springer Nature Switzerland AG 2022   219
F. D. Davis et al. (eds.), *Information Systems and Neuroscience*,
Lecture Notes in Information Systems and Organisation 58,
https://doi.org/10.1007/978-3-031-13064-9_23

stress, and emotion detection to enrich and complement HRV analysis [12]. With the improvements in processing speed and refinement of computations tools, analyzing all the waves shown in the ECG becomes feasible even for real-time monitoring and analysis [13, 14]. Also, the market for wearable devices is consolidating and increasing worldwide, which makes ECG-based measures for field studies and real-time applications much more accessible, but also poses some new challenges for data analysis [15].

In terms of the NeuroIS research agenda, we seek to contribute to the areas of emotion research and neuro-adaptive systems [17]. We expect that expanding HRV analysis to other movements of the heart will help further develop robust diagnostic indices for users' mental (or emotional) state (e.g., stress) states, and contribute to refining approaches to deriving these states from bio-signals in real-time [17]. To this end, this paper explores the possibilities of using other heart movements (or "waves" as represented in the ECG) in the cardiac cycle as proxies in the evaluation of emotion, arousal, and stress.

2   Background

2.1   *HRV and Other Waves in the Cardiac Cycle*

Studies suggest that emotions and heart movements are related, specifically that the duration of the heart's movements is related to different levels of arousal, stress, and valence [e.g., 18–22]. The HRV is measured as the variability of the distance, in milliseconds (ms), between two consecutive R peaks in a predetermined lapse of time [16]. Analyzing variability in one part of the cardiac cycle only might lead to erroneous conclusion that the heart movements have not changed, which would indicate that levels of emotions or stress have not changed either[1] (see [15] for a detailed explanation of the physiology of the heart rate). Figure 1 shows a situation where we observe the same values for R-to-R distance, but the other waves presented on the cardiac cycle (the QRS complex, the PQ segment and the QT segment) are very different.

With HRV analysis alone, we would not be able to detect variations, let alone differentiate between reasons for them, which might be explained by the changes in the length of any of the waves or a combination of them. Several authors have used other waves and segments to describe relations between emotional arousal and heart movements better, and to improve predictions of emotions. Hansen et al. [24] found that adrenaline caused a prolongation of QT duration and a flattening of the T-wave amplitude. Dweck et al. [25] showed that noxious arousal causes changes in the T-wave. Yavuzkir et al. [13] found a positive relation between prolonged anxiety

---

[1] Prior to computing indices of interest, ECG signals must be checked and pre-processed to identify non-normal beats (e.g., arrhythmic events or ectopic beats), which would generate noise at the moment of using the HR and HRV values [21].

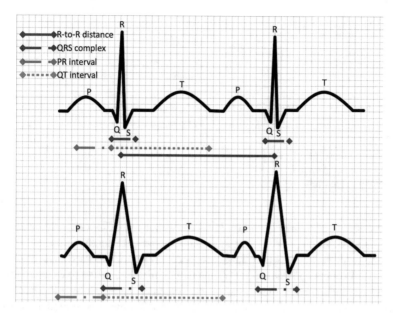

Fig. 1 Same HRV but different waves and segment's duration (*Source* Own elaboration based on Weinhaus & Roberts [23])

and PWD and no significant difference between groups for HR. Andrássy et al. [14] found a correlation between stress and the length of the QT segment. Dousty et al. [26] found evidence that R-wave amplitude increases when people listen to music, in comparison to silence. Lampert et al. [27] found evidence that negative emotions and stress trigger a decrease or absence of P-waves and happiness protects the heart from such condition.

2.2 Short Review of Software Libraries for ECG Analysis

First, analyzing the ECG to acquire the P-, Q- and T-peaks and S and R troughs is necessary. Table 1 contains the results of an ongoing review of currently available software libraries and programs that offer the required functionality. The review procedure started with an internet search for available libraries/tools that would permit coding new or adapting existing algorithms and have reasonably fast processing times[2]; this search is ongoing. The next step will continue with an analysis of NeuroIS articles like [8] to find out which tools have been generally most used in academic studies.

---

[2] There are providers of proprietary software (e.g. Kubios) that does not allow to freely modify their analysis, which is why we exclude them from the review.

Table 1 Short review on libraries for ECG analysis

Library	Language	Detects	Author(s)
NeuroKit2 toolbox	Python	All peaks and troughs Possible beginning and end of waves	[28]
Online adaptive QRS detector	Matlab	Q, R, S and T waves	[29]
Complete implementation of Pan Tompkins	Matlab	QRS complexes	[30]
Python online and offline ECG QRS detector based on the Pan-Tomkins algorithm	Python	QRS complexes in a real-time and pre-recorded ECG signal dataset	[31]

In particular, NeuroKit2 toolbox is a handy tool for ECG analysis because can identify all the waves presented on an ECG. However, the toolbox does not make it easy to adapt or switch out algorithms, or to access the variables used as identifiers for the waves. As our purpose is to identify and work with those values so we can compare them with standard HRV measures, we implemented an algorithm[3] to identify the peaks and troughs of the 5 waves (P-, Q-, R-, S- and T-waves) presented in the ECG.

3 Method

3.1 Procedure

Next, defining and computing proxies for arousal follows. Prior research suggests that the heart, as a muscle, reacts in different ways according to the situation. It is reasonable to conclude that, apart from the HRV, measures based on the other waves might also generate good proxies for emotion recognition [20]. We decided to use three segments of the heart's movements to define three potential proxies for arousal—the QRS complex variation, the PR interval, and the QT interval. Due to the complication of finding exactly when the waves start and finish, we used the distance between peaks and troughs as a proxy for those segments (e.g., QT segment goes from the first moment of the Q-wave until the latest moment of the T-wave, we measured it instead from the Q-trough until the T-peak).

---

[3] In summary, the algorithm aims to find the R values by segmenting the data frame in 10 s intervals and looking at the maximum of each interval. Once the maximum value is obtained, it uses a 15% amplitude window to finds other peaks. Finally, considering the data as a whole and using max values of each wave's duration mentioned in Weinhaus & Roberts [23], it generates a duration window going from R to P and from R to T. For further questions about the algorithm, please contact the authors. As of the day of the submission, the algorithm has not been formally tested for accuracy. In a preliminary test, we visually analyzed the results and concluded that the fit appears to be very good. Formal testing is planned next.

The QRS complex variation is exactly the moment when both ventricles of the heart contract [23]. We use the standard deviation of QRS complex length as a proxy[4]:

$$\text{SDQRS} = \sqrt{\frac{1}{T} \sum_{t=1}^{T} \left(d(Q_t, S_t) - \overline{d} \right)^2} \tag{1}$$

The PR interval contains the starting of the heart cycle and finishes at the beginning of the QRS complex. We use the standard deviation of PR interval length as a proxy:

$$\text{SDPR} = \sqrt{\frac{1}{T} \sum_{t=1}^{T} \left(d(P_t, Q_t) - \overline{d} \right)^2} \tag{2}$$

The QT interval represents the length between the start of the QRS complex until the heart is back in its resting membrane potential [23]. We use the standard deviation of QT interval length as a proxy:

$$\text{SDQT} = \sqrt{\frac{1}{T} \sum_{t=1}^{T} \left(d(Q_t, T_t) - \overline{d} \right)^2} \tag{3}$$

3.2 Evaluation

As Fig. 1 illustrates, we might observe the same values for the HRV measures while the other waves presented on the cardiac cycle are very different. This might indicate differences in the level of emotional arousal that we would not detect with HRV measures alone. The first step in the evaluation is to determine whether the information contained in our proposed proxies could feasibly complement HRV analysis. To this end, we compute correlation coefficients of our proxies with two standard HRV measures. One standard HRV measures, according to a review by Shaffer & Ginsberg [32], is the standard deviation of normal heartbeats:

$$\text{SDNN} = \sqrt{\frac{1}{T} \sum_{t=1}^{T} \left(d(R_{t+1}, R_t) - \overline{d} \right)^2} \tag{4}$$

Another standard measure is the root mean square of successive RR interval differences (RMSSD; [32]):

$$\text{RMSSD} = \sqrt{\frac{1}{T-1} \sum_{t=1}^{T} \left(d(R_{t+2}, R_{t+1}) - d(R_{t+1}, R_t) \right)^2} \tag{5}$$

---

[4] d is the Euclidean distance function in $\mathbb{R}$ for period t; T is the last period analyzed in each segment; $\overline{d}$ is the average of the distance between the segments.

While there are other measures that could be used, we chose these two due to their suitability for short-term (~5 min) and ultra-short-term (<5 min) HRV measurement [32]. The dataset we use in the next section for evaluating the proxies does not contain long-term ECG. The second step in the evaluation is to determine how well changes in our proxies correlate to external stimuli (i.e., changes in the situation we would expect to incur changes in the level of emotional arousal) and subjective perceptions of emotional arousal levels.

4 Results

We used a dataset from an experiment conducted by Dorner et al. [33],[5] in which participants were exposed to arousal-inducing stimuli. ECG was acquired with bioPLUX sensors. The dataset also contained self-reported levels of arousal and valence with the self-assessment manikin scores [34] before and after each stimulus.

The experiment investigates how interface elements like live bio-feedback affect financial risk-taking in online robo-advisory [33]. The experiment was conducted in 2019 in Germany with a student sample. The data was checked for quality and completeness, leaving us with a full dataset for 125 participants and 1,862 observations for HRV measures and our proxies. 49.73% of participants were male. The average age of the participants was 23.5 years (SD = 5.38, min = 18, max = 59).

Our aim in this preliminary analysis is to determine how strongly our proxies correlate with the HRV measures. If our supposition—that information contained in our proxies can complement the information contained in the HRV measures—is true, we would expect to see correlations well below 1 (above −1). Otherwise, if our proxies cannot bring any additional information for arousal and stress detection, we would expect to see correlations close to 1 (−1). Table 2 clearly shows that the HRV measures do not correlate highly with our proxies. The two HRV measures RMSSD and SDNN are positively correlated. Both are positively but substantially less strongly correlated with our three proposed proxies SDQRS, SDPR, and SDQT. RMSSD shows near-zero correlations with SDQRS and SDPR (Table 2). The three proxies are positively correlated with each other. As expected, SDQRS and SDQT show one of the highest correlations (0.4963), since the QRS complex is contained in the QT segment. SDQT is moderately correlated with SDPR (0.5636).

---

[5] Unpublished study; submission in preparation. For further details on the experiment, please contact the authors.

Table 2 Pairwise Pearson correlations

		HRV		New proxies		
		SDNN	RMSSD	SDQRS	SDPR	SDQT
HRV	SDNN	1				
	RMSSD	0.6960***	1			
New proxies	SDQRS	0.2062***	0.0615**	1		
	SDPR	0.1475***	0.0796***	0.3679***	1	
	SDQT	0.3059***	0.1895***	0.4963***	0.5636***	1

Note Significance level of pairwise correlations in parenthesis. *** $p < 0.01$, ** $p < 0.05$, * $p < 0.1$

5 Discussion and Summary

The heart's reactions can reflect arousal and stress in more ways than in changing R-to-R variability. Combining HRV measures with measures that shed light on the other movements of the heart offers the potential for more robust and in-depth analyses. Preliminary evidence suggests that three proxies based on other movements of the heart (QRS complex, PR interval, and QT interval) could serve to enrich and complement the information contained in the HRV measures.

Next, we will evaluate the correlations between our suggested proxies and self-reported perceived levels of arousal. One limitation regarding the dataset is that it does not contain long-term ECG measurements and does not attempt to differentiate between emotions beyond levels of arousal and valence; also, with the given data set it is not possible to control for heart diseases or prescription drugs that could generate a bias in our analysis. Additionally, it is just one data set. Further research will be required to evaluate the applicability and robustness of our proxies for other datasets.

We believe that our research will help further develop neuro-adaptive systems based on ECG real-time analysis, for instance live bio-feedback applications [10, 17]. Unimodal neuro-adaptive systems have been criticized due to the difficulties of identifying a user's mental or emotional state from only one bio-signal [17]. Integrating multi-modal signal measurements into real-life applications has, however, been rarely feasible so far. Exceptions are health and sports applications like the Fitbit smartwatch, which combine electrodermal activity (EDA) and ECG measures [35]. Our research is intended to help extract as much information as possible from the ECG signal, in effect producing several measures from one bio-signal, with the hope that this will make it easier to differentiate between user states and define more robust metrics.

In research settings, adding sensors to acquire other bio-signals like EDA might not always be an option, especially in field studies. In addition, concerns have been raised regarding the reliability and validity of heart rate measurements by wearable devices [15]. Future research could investigate whether, in combination with advanced smoothing and imputation algorithms, our proposed approach could contribute to improving information acquisition from low-quality signals.

References

1. Lane, R. D., McRae, K., Reiman, E. M., Chen, K., Ahern, G. L., & Thayer, J. F. (2009). Neural correlates of heart rate variability during emotion. *NeuroImage, 44*(1), 213–222.
2. Appelhans, B. M., & Luecken, L. J. (2006). Heart rate variability as an index of regulated emotional responding. *Review of General Psychology, 10*(3), 229–240.
3. Agrafioti, F., Hatzinakos, D., & Anderson, A. K. (2011). ECG pattern analysis for emotion detection. *IEEE Transactions on Affective Computing, 3*(1), 102–115.
4. Mather, M., & Thayer, J. F. (2018). How heart rate variability affects emotion regulation brain networks. *Current Opinion in Behavioral Sciences, 19*, 98–104.
5. Kim, H. G., Cheon, E. J., Bai, D. S., Lee, Y. H., & Koo, B. H. (2018). Stress and heart rate variability: A meta-analysis and review of the literature. *Psychiatry Investigation, 15*(3), 235.
6. Thayer, J. F., Åhs, F., Fredrikson, M., Sollers, J. J., III., & Wager, T. D. (2012). A meta-analysis of heart rate variability and neuroimaging studies: Implications for heart rate variability as a marker of stress and health. *Neuroscience & Biobehavioral Reviews, 36*(2), 747–756.
7. Schubert, C., Lambertz, M., Nelesen, R. A., Bardwell, W., Choi, J. B., & Dimsdale, J. E. (2009). Effects of stress on heart rate complexity—a comparison between short-term and chronic stress. *Biological Psychology, 80*(3), 325–332.
8. Baumgartner, D., Fischer, T., Riedl, R., & Dreiseitl, S. (2019). Analysis of heart rate variability (HRV) feature robustness for measuring technostress. In F. D. Davis, R. Riedl, J. vom Brocke, P.-M. Léger, & A. B. Randolph (Eds.), *Information systems and neuroscience: NeuroIS retreat 2018* (pp. 221–228). Springer.
9. Teubner, T., Adam, M. T. P., & Riordan, R. (2015). The impact of computerized agents on immediate emotions, overall arousal and bidding behavior in electronic auctions. *Journal of the Association for Information Systems, 16*, 838–879.
10. Lux, E., Adam, M. T., Dorner, V., Helming, S., Knierim, M. T., & Weinhardt, C. (2018). Live biofeedback as a user interface design element: A review of the literature. *Communications of the Association for Information Systems, 43*(1), 18.
11. Silvia, P. J., Jackson, B. A., & Sopko, R. S. (2014). Does baseline heart rate variability reflect stable positive emotionality? *Personality and Individual Differences, 70*, 183–187.
12. King, M. G., Burrows, G. D., & Stanley, G. V. (1983). Measurement of stress and arousal: Validation of the stress/arousal adjective checklist. *British Journal of Psychology, 74*(4), 473–479.
13. Gradl, S., Kugler, P., Lohmüller, C., & Eskofier, B. (2012). Real-time ECG monitoring and arrhythmia detection using Android-based mobile devices. In *2012 annual international conference of the IEEE engineering in medicine and biology society* (pp. 2452–2455). IEEE
14. Xia, H., Asif, I., & Zhao, X. (2013). Cloud-ECG for real time ECG monitoring and analysis. *Computer methods and programs in biomedicine, 110*(3), 253–259.
15. Stangl, F. J., & Riedl, R. (2022). Measurement of heart rate and heart rate variability with wearable devices: A systematic review. In *Internationale Tagung Wirtschaftsinformatik*.
16. Camm, A. J., Malik, M., Bigger, J. T., Breithardt, G., Cerutti, S., Cohen, R. J., Coumel, P., Fallen, E. L., Kennedy, H. L., Kleiger, R. E., & Lombardi, F. (1996). Heart rate variability: standards of measurement, physiological interpretation and clinical use. Task Force of the European Society of Cardiology and the North American Society of Pacing and Electrophysiology. *Circulation, 93*, 1043–1095
17. vom Brocke, J., Hevner, A., Léger, P. M., Walla, P., & Riedl, R. (2020). Advancing a NeuroIS research agenda with four areas of societal contributions. *European Journal of Information Systems, 29*(1), 9–24.

18. Yavuzkir, M., Atmaca, M., Dagli, N., Balin, M., Karaca, I., Mermi, O., Tezcan, E., & Aslan, I. N. (2007). P-wave dispersion in panic disorder. *Psychosomatic Medicine, 69*(4), 344–347.
19. Andrássy, G., Szabo, A., Ferencz, G., Trummer, Z., Simon, E., & Tahy, Á. (2007). Mental stress may induce QT-interval prolongation and T-wave notching. *Annals of Noninvasive Electrocardiology, 12*(3), 251–259.
20. Jing, C., Liu, G., & Hao, M. (2009). The research on emotion recognition from ECG signal. In *2009 international conference on information technology and computer science* (vol. 1, pp. 497–500). IEEE.
21. Xu, Y., & Liu, G. Y. (2009). A method of emotion recognition based on ECG signal. In *2009 international conference on computational intelligence and natural computing* (vol. 1, pp. 202–205). IEEE.
22. Jerritta, S., Murugappan, M., Wan, K., & Yaacob, S. (2013). Emotion detection from QRS complex of ECG signals using hurst exponent for different age groups. In *2013 Humaine association conference on affective computing and intelligent interaction* (pp. 849–854). IEEE.
23. Weinhaus, A. J., & Roberts K. P. (2005). *Anatomy of the human heart. Handbook of cardiac anatomy, physiology, and devices.* Humana Press.
24. Hansen, O., Johansson, B. W., & Gullberg, B. (1991). Metabolic, hemodynamic, and electro-cardiographic responses to increased circulating adrenaline: Effects of pretreatment with class 1 antiarrhythmics. *Angiology, 42*(12), 990–1001.
25. Dweck, M. R., Lang, C. C., Neilson, J. M., & Flapan, A. D. (2006). Noxious arousal induces T-wave changes in healthy subjects. *Journal of Electrocardiology, 39*(3), 324–330.
26. Dousty, M., Daneshvar, S., & Haghjoo, M. (2011). The effects of sedative music, arousal music, and silence on electrocardiography signals. *Journal of electrocardiology, 44*(3), 396-e1.
27. Lampert, R., Jamner, L., Burg, M., Dziura, J., Brandt, C., Liu, H., Li, F., Donovan, T., & Soufer, R. (2014). Triggering of symptomatic atrial fibrillation by negative emotion. *Journal of the American College of Cardiology, 64*(14), 1533–1534.
28. Makowski, D., Pham, T., Lau, Z. J., Brammer, J. C., Lespinasse, F., Pham, H., Schölzel, C., & Chen, S. H. (2021). NeuroKit2: A Python toolbox for neurophysiological signal processing. *Behavior research methods, 53*(4), 1689–1696.
29. Sedghamiz, H. (2022). An online algorithm for R, S and T wave detection. (https://www.mathworks.com/matlabcentral/fileexchange/45404-an-online-algorithm-for-r-s-and-t-wave-detection). MATLAB Central File Exchange. Retrieved 15 Mar 2022.
30. Sedghamiz, H. (2022). Complete Pan Tompkins implementation ECG QRS detector. (https://www.mathworks.com/matlabcentral/fileexchange/45840-complete-pan-tompkins-implement ation-ecg-qrs-detector. MATLAB Central File Exchange. Retrieved 15 Mar 2022.
31. Sznajder, M., & Łukowska, M. (2017). Python online and offline ECG QRS detector based on the Pan-Tomkins algorithm (v1.1.0). Zenodo. https://doi.org/10.5281/zenodo.826614
32. Shaffer, F., & Ginsberg, J. P. (2017). An overview of heart rate variability metrics and norms. *Frontiers in Public health, 258.*
33. Dorner V., Iliewa, Z. Weber, M., & Weinhardt, C. (2022). Can a robo-advisor be a money doctor?. Unpublished study.
34. Bradley, M. M., & Lang, P. J. (1994). Measuring emotion: The self-assessment manikin and the semantic differential. *Journal of behavior therapy and experimental psychiatry, 25*(1), 49–59.
35. Advanced Health Smartwatch | Fitbit Sense. Accessed 23rd April 2021. https://www.fitbit.com/global/us/products/smartwatches/sense

Caption and Observation Based on the Algorithm for Triangulation (COBALT): Preliminary Results from a Beta Trial

Pierre-Majorique Léger, Alexander J. Karran, Francois Courtemanche,
Marc Fredette, Salima Tazi, Mariko Dupuis, Zeyad Hamza,
Juan Fernández-Shaw, Myriam Côté, Laurène Del Aguila,
Chantel Chandler, Pascal Snow, Domenico Vilone, and Sylvain Sénécal

Abstract We report on the development progress of a closed beta trial of the COBALT ecosystem, a suite of instruments and software developed to make psychophysiological research accessible to the NeuroIS community and beyond. We provide an overview of the ecosystem and early feedback derived from interview data. The current synthesis of these data indicates that the COBALT ecosystem is

P.-M. Léger · A. J. Karran (✉) · F. Courtemanche · M. Fredette · S. Tazi · M. Dupuis · Z. Hamza ·
J. Fernández-Shaw · M. Côté · L. Del Aguila · C. Chandler · P. Snow · D. Vilone · S. Sénécal
Tech3Lab, HEC Montréal, Montreal, Canada
e-mail: alexander-john.karran@hec.ca

P.-M. Léger
e-mail: pierre-majorique.leger@hec.ca

F. Courtemanche
e-mail: francois.courtemanche@hec.ca

M. Fredette
e-mail: marc.fredette@hec.ca

S. Tazi
e-mail: salima.tazi@hec.ca

M. Dupuis
e-mail: mariko.dupuis@hec.ca

Z. Hamza
e-mail: zeyad.hamza@hec.ca

J. Fernández-Shaw
e-mail: juan-antonio.fernandez-shaw-morales@hec.ca

M. Côté
e-mail: myriam.2.cote@hec.ca

L. Del Aguila
e-mail: laurene.del-aguila@hec.ca

C. Chandler
e-mail: chantel.chandler@hec.ca

© The Author(s), under exclusive license to Springer Nature Switzerland AG 2022   229
F. D. Davis et al. (eds.), *Information Systems and Neuroscience*,
Lecture Notes in Information Systems and Organisation 58,
https://doi.org/10.1007/978-3-031-13064-9_24

utile, easy to use and understand, and can be implemented quickly within a range of experimental contexts to provide vastly improved time-to-insight through workflow efficiency gains.

Keywords NeuroIS · Psychophysiology · Beta test · Technology · Experimental design

1 Introduction

Over the past few years, researchers from the NSERC-Prompt Industrial Research Chair in User Experience (UX Chair) and Tech3Lab have developed the COBALT research ecosystem (Caption and Observation Based on the Algorithm for Triangulation). COBALT research ecosystem is a set of scientific tools and software that enables the triangulated analysis of a user's psychophysiological states during ecologically valid human–machine interactions aimed directly at neuroIS researchers to allow for in-field ecologically valid studies. The need for ecological validity and independence from a laboratory environment was identified by [1]. The platform we present in this manuscript aims to provide an easy to configure hardware sensing toolkit for psychophysiological data collection, inference, and user experience research in the field, which costs orders of magnitude less than laboratory grade sensors, supports user interface research at a distance such as in the home or workplace, records and outputs signal that is of sufficient fidelity for valid psychophysiological inference to answer research questions, contributes an easy-to-use software toolkit for designing experiments and analyzing experimental data and decreases time to insight through easy to apply data processing tools.

The platform is based around a mobile psychophysiological sensor package and associated cloud-based experimental design and data analysis to decrease time to insight for user experience testing in multiple domains ranging from entertainment to industry. COBALT development was funded by the Natural Sciences and Engineering Research Council of Canada (NSERC), Prompt Innov, Fonds de Recherche du Québec (FRQ), and Fondation HEC Montréal.

A beta trial that aims to evaluate the Tech3Lab COBALT research ecosystem is currently being conducted with the help of NeuroIS researchers worldwide to assess the usefulness of this new instrument and the feasibility of making it available to the research community. Twelve researchers are currently participating in ongoing testing of the COBALT ecosystem, which began in Fall 2021. They have agreed to be

P. Snow
e-mail: pascal.snow@hec.ca

D. Vilone
e-mail: domenico.vilone@hec.ca

S. Sénécal
e-mail: sylvain.senecal@hec.ca

interviewed at different project stages. This article reports preliminary results from this beta test.

2   COBALT Ecosystem

The COBALT ecosystem is a set of psychophysiological instruments and analysis software designed and developed by HEC Montréal Tech3Lab and UX Chair. The COBALT ecosystem has been developed to support research related to human–computer interactions and NeuroIS. Compared to other commercial solutions available in the field, its main advantage is that it was developed to specifically support the study of psychophysiological reactions during dynamic, non-linear interactions with technology. The COBALT ecosystem described below is comprised of three components: COBALT BLUEBOX, COBALT CAPTURE and COBALT PHOTOBOOTH.

COBALT BLUEBOX. The COBALT BLUEBOX (see Fig. 1) [2] is a custom low-cost psychophysiological sensing device based on the BITalino hardware reference, which was tested against an established medical grade gold standard [3]. The device allows the recording of electrodermal activity (EDA), electrocardiogram (EKG), respiration (RSP), electromyography (EMG), and acceleration (ACC). Future developments will include other modalities such as GPS location and user inputs. The main characteristics of BLUEBOX are (i) Costs orders of magnitude less than laboratory grade sensors; (ii) Supports user interface research at a distance such as in the home or workplace (iii) Records and outputs signals that are of sufficient fidelity for valid psychophysiological inference and research.

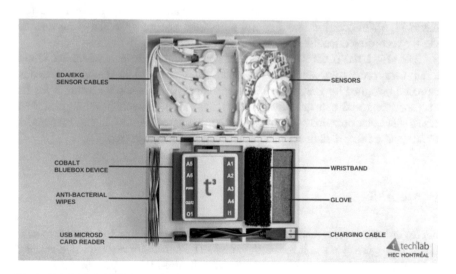

Fig. 1   COBALT BLUEBOX package

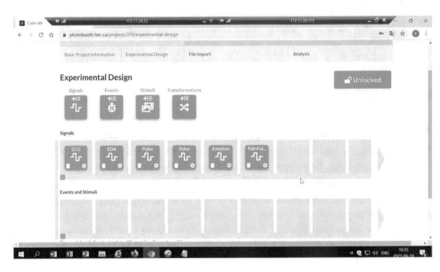

Fig. 2 COBALT PHOTOBOOTH

COBALT CAPTURE. Is a web-based software application that can be used to conduct remote experiments with the BLUEBOX and allows recording a participants' computer screen and webcam in parallel. Both video feeds are synchronized automatically with the BLUEBOX. Additionally, COBALT CAPTURE [2] allows the presentation of video/image stimuli and customizable questionnaires. The web-based software platform is used to conduct an experimental session; the CAPTURE platform records a participant's computer screen(s) and webcam and synchronizes these data with BLUEBOX and is used to annotate relevant events after an experimental session. The COBALT ecosystem can be used in both laboratory and remote experiments [4, 5]. In addition, the platform allows the researcher to add event markers into the recording once the session is completed.

COBALT PHOTOBOOTH. The final component of the ecosystem is a cloud-based data processing and analysis platform (see Fig. 2). Using a patented approach designed by HEC Montréal researchers [6], this software allows for the postprocessing and synchronization of behavioral and psychophysiological data. Researchers can download the synchronized processed data locally in preparation for analyzing these data in their preferred statistical package (see [7]).

3   Methods

Twelve university researchers in the field of NeuroIS were selected to participate in this beta program via a call for participation. The university researchers consented to use the COBALT ecosystem in their research projects and provide feedback to

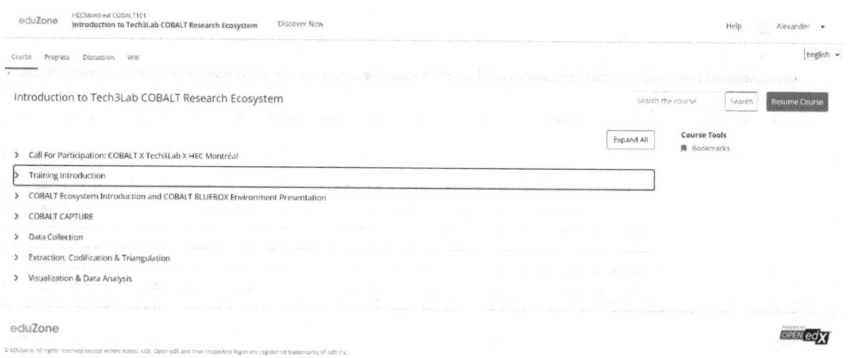

Fig. 3 COBALT training environment

the HEC Montréal Tech3Lab development team. This project was approved by the ethics committee of our institution (Certificate number 2022-4554).

Our team provided detailed online training and remote support to the researchers using the COBALT research ecosystem. The training included the following:

- A complete module about the COBALT device and accessories (unboxing, connection, technical parts overview, installation, troubleshooting, etc.);
- A module on research project preparation, pretesting, and launch;
- A module focusing on data collection;
- A module about extraction, codification, and triangulation; and finally,
- A module detailing data analysis and visualization procedures using specific tools.

The training is illustrated fully through real experimental examples performed by the scientific and research professionals of Tech3Lab (see Fig. 3). The training is available at the following address: https://cours.eduzone.ca/courses/course-v1:HEC Montreal+COBALT101+2021/course/.

4 Preliminary Results

After collating and synthesizing the interview data, it was found that the interviewed researchers placed a high value on the possibilities offered by the combination of tools provided by the COBALT ecosystem package. Six respondents went so far as to indicate their preference for the COBALT ecosystem as an end-to-end tool for advanced psychophysiological research, and one described a preference for the Bluebox sensor package alone. Data from the remaining five researchers is currently undergoing analysis and synthesis.

The current sample of participants contains researchers from diverse academic backgrounds, from business management and marketing to data science and information technology to design thinking and media communication. Despite this diversity, in response to the question: "How does the COBALT ecosystem fit your needs?", all of the participants responded that they perceived the COBALT ecosystem to be useful when applied to their respective research studies and were able to use the tools to collect psychophysiological data in different contexts. For some researchers, this was their first opportunity to use such tools to collect data that is not typically evidenced in their fields. This is particularly the case for researchers in fields that rely on subjective self-reported measurements, such as surveys and questionnaires. Accordingly, in answer to the question: "Could you tell me about your research project?" One researcher responded that *"when it comes to understanding the effect of persuasion and intent in communication studies, we typically rely on self-report measures of what someone might do in a scenario, but this biodata gives us their actual response. It's not what they think they might do based on a hypothetical scenario; it's how they actually feel whether they know it or not and whether they want to tell us or not. It reduces bias"* (P06).

Concerning the training provided to all researchers using the COBALT ecosystem. Researchers reported that the material was comprehensive, straightforward, practical and easily applicable. Moreover, the synthesis of the interviews showed that having the learning materials available online proved to be a valuable reference resource during testing and implementation of COBALT within their research protocols, contributing to a sense of self-sufficiency and confidence in the platform. The interview data showed a clear trend in which the COBALT ecosystem was easy to implement into existing or new experimental designs and that the learning material was developed and delivered in a way that helped clarify and simplify this technology.

In response to the question: "Have you completed the training?" And "how was the process?" one researcher responded: *"There's many tools and moving pieces and things to do, but I think that given the complexity, it's delivered in a way that greatly simplifies it, versus us having to learn each of these tools by ourselves"*—(P05).

During testing the COBALT ecosystem has been used to evaluate participants' psychophysiological responses when interacting with diverse artifacts such as chatbots, haptic chairs and social media news sites.

5   Conclusion and Next Steps

When asked about future needs, the interviewed researchers indicated that certain elements of the COBALT ecosystem could use some refinement. A common theme that manifested was the need for additional guidance for when things do not go as expected and that a compilation of encountered errors and associated solutions into a troubleshooting guide would prove most beneficial, reducing the time required to find solutions. Another theme that arose was electronic data interchange format for statistical analysis, mismatches in the ability to quickly process data files for analysis within different applications was identified as reducing the benefits of the integrated COBALT ecosystem moderately. There are research and development activities currently underway to address this issue.

References

1. Greif-Winzrieth, A., et al. (2021). *Exploring the potential of NeuroIS in the wild: Opportunities and challenges of home environments.* Springer International Publishing.
2. Courtemanche, F., Sénécal, S., Fredette, M., & Léger, P.-M. (2022). In H. Montréal (Ed.), *COBALT-Bluebox: Multimodal user data wireless synchronization and acquisition system.*
3. Batista, D., et al. (2019). Benchmarking of the BITalino biomedical toolkit against an established gold standard. *Healthcare Technology Letters, 6*(2), 32–36.
4. Giroux, F., et al. (2021). Guidelines for collecting automatic facial expression detection data synchronized with a dynamic stimulus in remote moderated user tests. In *International Conference on Human-Computer Interaction.* Springer.
5. Vasseur, A., Léger, P.-M., & Sénécal, S. (2019). Eyetracking for IS research: A literature review. In *Proceedings of SIGHCI.*
6. Courtemanche, F., Fredette, M., Senecal, S., Leger, P. M., Dufresne, A., Georges, V., & Labonté-Lemoyne, E. (2019). *Method of and system for processing signals sensed from a user.*
7. Léger, P.-M., et al. (2019). A cloud-based lab management and analytics software for triangulated human-centered research. *Information systems and neuroscience* (pp. 93–99). Springer.

Leveraging Affective Friction to Improve Online Creative Collaboration: An Experimental Design

Maylis Saigot

Abstract Emotional contagion is a pillar of social interaction. As such, it has immense potential to facilitate communication and improve collaboration. In the context of remote collaboration, it is especially important that working partners can build trust and a sense of cohesiveness. While digital media may complicate socio-affective communication, we argue that some media capabilities are better able to support processes of affective alignment. We define affective friction as an affective misalignment between the members of a workgroup that may result in diverging affective responses to shared experiences. We propose that affective friction is a central element of affective alignment and a driving force of creative collaboration. As a result, the capacity of a medium to make affective friction perceptible to working partners is essential for successful remote collaboration. We suggest a two-stage experimental design to test our hypotheses.

Keywords Online collaboration · Emotional contagion · Affective friction · Media richness · Social presence · Media synchronicity · Media naturalness

1 Introduction

Advances in technology have enabled the rise of communication technology that has transformed the way people work together. More recently, the COVID-19 pandemic precipitated this transition. As we come out of the emergency response from the corporate world, many organizations are shifting towards hybrid work arrangements with a substantial part of work tasks being conducted online. Academic research needs to support this transition and offer frameworks to optimize and sustain it. Importantly, employees find it challenging to feel a sense of cohesiveness and community while sitting in different spaces than their colleagues [1–3]. Essential attributes of collaboration like cooperativeness [4] or trustworthiness [5] rely heavily on cues that are naturally present in face-to-face settings. Rich media are those that support

M. Saigot (✉)
Department of Digitalization, Copenhagen Business School, Copenhagen, Denmark
e-mail: msa.digi@cbs.dk

© The Author(s), under exclusive license to Springer Nature Switzerland AG 2022
F. D. Davis et al. (eds.), *Information Systems and Neuroscience*,
Lecture Notes in Information Systems and Organisation 58,
https://doi.org/10.1007/978-3-031-13064-9_25

the exchange of cues that are most similar to face-to-face (e.g., video-conference), including characteristics such as immediate feedback, personalness, etc. [6–8]. They enable parties to communicate great amounts of affective cues [9] and have traditionally been associated with the effective exchange of equivocal information [6, 7]. Leaner media, on the other hand, are less personal (e.g., e-mail) and provide less support for affective cues [8]. They are usually associated with the effective communication of non-equivocal information [6, 7]. As lean media struggle to render some of these affective cues, they can inhibit social connection [10] and make social interaction and relationship-building more difficult. Because trust and shared meaning are fundamental to successful collaboration [11, 12], it is important to overcome these challenges and support social connection even during online interaction. Emotional contagion is a primitive mechanism that facilitates effective communication, helping people experience closeness and social connection [13, 14]. This suggests that hindered emotional contagion could be a contributing factor to some of the negative interpersonal experiences of remote collaboration. In this study, we propose an experimental design to understand how affective alignment between remote work partners impacts their collaborative outcomes and how different media affect the relationship between affective alignment and collaborative outcomes.

2   Theoretical Development

As working partners engage in a collaborative task remotely, they are likely to be in different moods from foregoing events and experiences [15, 16]. These assorted affective states create *affective friction* between the partners (i.e., an affective misalignment between workgroup members that may result in diverging affective responses to shared experiences), leading to miscommunication, hindered collective strategies, and interpretation discrepancies [17]. The digital nature of collaboration media can make it more challenging for partners to notice the friction [18]. Thus, the ability to recognize affective friction relies on a medium's capability to render assorted affective states. We expect that *affective media*, those traditionally considered to make mediated interaction feel "unmediated" [19] and often associated with high levels of social presence [19], richness [6], synchronicity [20], and naturalness [21], are particularly suited to affective communication and thus more likely to make affective friction perceptible. On the other hand, *non-affective media* tend to muffle them, because they offer limited capability to render users' core affect.

P1. Media affectivity increases perceived affective friction.

Affective friction is likely to result in repeated instances of unpleasant experiences (e.g., disagreements, misunderstandings, misjudgments, etc.) [17]. A well-documented phenomenon known as the "sensitization hypothesis" posits that repeated exposure to interparental conflict have a strong negative impact on children's long-term conflict appraisal, and results in increased negative emotionality, exacerbated perception of conflicts, and overall heightened sensitization to conflicts [22–26]. While empirical evidence of the phenomenon has been constrained to parent-children relationships and child-adolescent age ranges, we argue that repeated exposure to symptomatic affective friction is likely to have similar effects for adult work relationships. As tension resulting from affective friction accumulates, colleagues may start noticing the friction and its effects, and become more sensitive to it. In doing so, they start paying increased attention to their diverging perceptions and ideas—we call this process *affective transitioning* (i.e., the formulation and communication of non-congruent core affect among team members that increase perceived affective friction).

P2. Affective transitioning increases perceived affective friction, and perceived affective friction feeds affective transitioning.

As working partners experience *affective transitioning*, they are likely to open themselves to new and conflicting perspectives, resulting in divergent thinking, which "allows one to explore in different directions from the initial problem state, in order to discover many possible ideas and idea combinations that may serve as solutions" [27]. This process is generative, as new possibilities and suggested compromises act as "probes" that force individuals to confront their differences and reconsider what is essential [28]. It encourages team members to reconsider past decisions and new possibilities in light of new information and uncertainties [29].

P3. Affective transitioning improves performance in divergent-thinking tasks.

A fundamental next step for success is that working partners engage in *affective harmonizing*. Noticing the efforts made during the *affective transitioning* phase and their potential successful outcomes creates a positive feedback loop that fuels *convergent thinking* [30–32]. At this stage, working partners are likely to be motivated to resolve their affective friction and may start engaging in socio-affective interaction to create a sense of collectivity (e.g., help-seeking/giving, reinforcing, etc. [33]). In doing so, working partners may start experiencing matching affective reactions. Importantly, having gone through *transitioning* before *harmonizing* prevents the group from suffering the pitfalls of groupthink (i.e., a mode of thinking where a team's desire to reach unanimity overpowers their ability and motivation to seek the best possible outcome instead of the most consensual one) [32]. Groupthink is likely to occur when a team has elevated perceived collective efficacy. The lack of friction within a group is thus likely to result in elevated perceived collective efficacy, which is a risk factor for groupthink [32]. When a team engages in *harmonizing*, the propensity to think alike and the motivation to reach a consensus may jeopardize the quality of the outcome. However, if the team has gone through *transitioning* first, it is predisposed to better assess the situation and not let its decisions be impacted by in-group pressures.

P4. Affective harmonizing improves performance in convergent-thinking tasks when it is preceded by affective transitioning.

3 Material and Method

We suggest a two-phase experimental approach: Study I aims at testing whether affective friction explains creative and collaborative performance through a combination of affective processes, and Study II aims at testing whether the affectivity of the medium moderates this effect.

The task. Participants take part in an online collaborative creative task [34] based on the following scenario: *after two years of remote teaching due to the COVID-19 pandemic, the colleges of a university located on the West Coast of the United States decide to join forces to welcome students and staff back on campus. The university organizes a large-scale competition to find the best solution to the following challenge: how might we strengthen our university community, as campus reopens, with solutions that reconnect people and enhance collective wellbeing, teaching, and learning?* This scenario is based on a real challenge that was posted on openIDEO.org, an online platform that leverages crowdsourcing to foster open innovation. Using a real-life challenge strengthens its relevance, timeliness, and concreteness. Placing the challenge in a local university further increases personal relevance and self-efficacy for the participants [35].

The task is divided into two sub-activities. For the *first activity*, participants are given 15 min to brainstorm and come up with as many creative ideas [34] as possible to address the challenge. They are informed they will be evaluated both on the number of ideas and the creativity of each idea. For the *second activity*, participants are given 10 min to select the idea they agree is the best overall. Similar studies typically provide 15 min for brainstorming and 10–15 min for idea selection [34–37].

Data Collection. All interactions will be recorded (video, audio, and text) for both studies using COBALT CAPTURE [38] to obtain facial expression, tone, speech and log data, and additional physiological data will be recorded for Study II (COBALT BLUEBOX for EDA and EKG [39] and Tobii Pro X3 120 eye tracker).

Measures and Instruments. *Affective Friction* is measured through a Perceived Affective Friction questionnaire designed by the author for the purpose of this study (PAF). *Affective Processes* are measured through physiological synchrony. *Divergent thinking* is operationalized as the first activity of the task (brainstorm) and its *Performance* is measured based on common creativity evaluation techniques (quantity, originality, and paradigm relatedness of the ideas generated [34, 40, 41]) and self-reports of perceived group creativity (PGC) [42, 43]. *Convergent thinking* is operationalized through the second activity of the task (idea selection) [35] and its *Performance* is determined based on a blind assessment of the selected ideas conducted by two external evaluators selected for their subject-matter expertise (i.e., pedagogical professionals who are likely to make this kind of decision in a real-life situation), and a cognitive group consensus (CGC) questionnaire [43, 44]. In addition, overall *Performance* includes related measures of team processes using self-reporting of collaboration quality [43, 45], team performance [46], personal success [46], satisfaction [47] (TP). See Table 1 for a full overview.

Table 1 Overview of measures and instruments

Type	Description
Physiological	Cardiac activity (ECG)
Physiological	Electrodermal activity (EDA)
Behavioral	Eye gaze
Behavioral	Facial expression
Self-report	Affective state: PANAS [48]
Self-report	Cognitive group consensus (CGC) [43, 44]
Self-report	Team processes (TP): Collaboration quality [43, 45], Personal success [46], Team performance [46], Satisfaction [47], Perceived group creativity [42, 43], Team performance [46]
Self-report	Perceived affective friction (developed by author)
Evaluation	Idea quality: external assessment by recruited panel of experts
Evaluation	Idea creativity: quantity, originality, and paradigm relatedness [34, 40, 41]

3.1 Study I

Participants. The experiment uses a one-factor, between-subjects design (see Table 2), with the independent variable being *Affective Processes*. We will recruit 80 participants who will each collaborate with a confederate actor, making for 20 dyads per condition. To prevent fatigue and reduce the risk of carryover effects, 8 different actors will be recruited and randomly assigned across conditions.

Confederate. In natural settings, dyads may need multiple interactions over longer periods of time to start noticing dysfunctions. In addition, some people might be uncomfortable sharing affective displays in the presence of strangers and/or an experimenter. This type of confounds can be controlled for but not manipulated, and are likely to introduce variance when it comes to perceived affective friction. In this study, we suggest inducing artificial and perceptible affective friction with the help

Table 2 Conditions of Study I

Condition	First activity (brainstorm)	Second activity (evaluation and selection)
TH	Affective transitioning	Affective harmonizing
TT	Affective transitioning	Affective transitioning
HH	Affective harmonizing	Affective harmonizing
HT	Affective harmonizing	Affective transitioning

Table 3 Hypotheses of study I

Hypotheses	Measures
H1. Affective transitioning increases perceived affective friction	PAF questionnaire
H2. Affective harmonizing decreases perceived affective frictions	PAF questionnaire
H3. TH will yield the best performance	PGC, CGC, TP, experts

of a confederate actor. While controlled collaboration commonly uses conversational agents [49–53], the use of a human agent is preferred for this study given the complexity and dynamism of socio-affective interaction. Although uncommon, using an actor as a confederate is a promising method that has yielded unique insights in prior experimental research on collaborative behavior [54, 55].

Each participant will be paired with an actor, thinking they are being paired with another participant. The confederate is a trained actor with good improvisation skills and elevated emotional intelligence [56]. Additional training (including partial scripts, instructions on verbal and non-verbal affective cues to display, and a sample of pre-determined solutions to the challenge) will be provided to ensure the actor can stick to their role under various circumstances. These instructions will be carefully crafted to ensure symmetrical cognitive stimulation across dyads.

Conditions. The *Affective Processes* vary on four levels and rely on *Affective Transitioning* and *Affective Harmonizing*. In *Affective Transitioning*, the confederate actor manifests evident divergences from their partner both in terms of affective cues and idea generation. In *Affective Harmonizing*, the actor manifests convergence with their partner. The conditions vary in the alternating of each of these processes for each of the two activities in the task through various combinations (e.g., *Affective Transitioning* during brainstorming followed by *Affective Harmonizing* during idea selection). All interactions occur through video conferencing on Zoom (with the possibility to use embedded emoji reactions), and the partners can always see and hear each other, as well as see a shared Google Document where both partners record their ideas. Table 3 describes the hypotheses and measures for Study I.

3.2 Study II

Participants. The experiment uses a 2 (media affectivity: high vs. low) × 2 (affective friction: high vs. low) between-subjects design. We will recruit 200 participants, resulting in 100 collaborative dyads in total or 25 per condition (see Table 4).

Conditions. Dyads will either communicate using a *High Affectivity Medium* (Zoom video conference) or a *Low Affectivity Medium* (Zoom chat with camera and microphones turned off). Their screen will be split between their interaction medium and a Google Document to which they can both contribute. Prior to starting the collaborative task, participants will take part in a mood induction pre-task. Participants will either play an intentionally buggy Pac-man game (f-pacman) [57, 58] or

Table 4 Conditions of study II

Condition	Affective friction	Media affectivity
HH	High (f-pacman, r-pacman)	High (Zoom)
HL	High (f-pacman, r-pacman)	Low (Chat)
LH	Low (r-pacman, r-pacman OR f-pacman, f-pacman)	High (Zoom)
LL	Low (r-pacman, r-pacman OR f-pacman, f-pacman)	Low (Chat)

a regular Pac-man game (r-pacman) for 5 min. They will be told that we are interested in player performance while wearing non-intrusive sensors. After the game, they are asked to write down their thoughts about the game in a few sentences. 50 of the dyads will experience the *High Affective Friction* condition (within each dyad, one participant plays the control game and the other plays the frustrating game, by random assignment). The remaining 50 experience the *Low Affective Friction* condition (within each dyad, both participants play the same game—either the control or the frustrating version, by random assignment).

Mood Manipulation. Each version of the game (f-pacman and r-pacman) was pre-tested by subjects recruited through Amazon Mechanical Turk, a web service that coordinates demand and supply for a variety of tasks that require human intelligence. Once the MTurk workers accepted the task, they were taken to a Qualtrics survey. The participants' baseline mood was measured through self-report using the PANAS scale prior to starting the task [48]. They were then redirected to the Pac-man game[1] which they played for 5 min. Once the 5 min were up, the participants were asked to return to the main survey, where they answered an open-ended question ("Please write down any thoughts you had about the game you just played") and the PANAS scale again, as a manipulation check. We controlled for game completion using an embedded timer as well as a completion code disclosed after 5 min of playtime and included an attention check to further maximize the quality of the responses.

Physiological Analysis Strategy. To test Study II hypotheses (See Table 5), we propose *Affective Processes* as a measure of affective instability (*transitioning*) and stabilization (*harmonizing*) over a time period. This concept is closely related to the concepts of physiological synchrony, often referred to as physiological linkage or compliance [e.g., 59–61]. The physiological data of each participant is recorded using the COBALT Bluebox, which uses a system of LED signals to synchronize the recordings of the screen, camera, audio, cardiac and electrodermal activity within each participant. Cardiac activity will be recorded through 2 electrodes positioned on the upper chest of the participants and 1 electrode on the left lower rib. Electrodermal activity will be recorded through 2 electrodes placed on the non-dominant hand of the participant and secured with a hand-glove and wristband. For each dyad, we plan to time-synchronize the data by sending a parallel trigger to both devices at the beginning of each collaborative task. Within each dyad, we will use a set of analytical techniques to measure physiological synchrony, including cross-correlation [59–63],

---

[1] GitHub repositories: https://github.com/maxkonrad/frustrating-pacman, https://github.com/max konrad/regular-pacman.

Table 5 Hypotheses of study II

Hypotheses	Measures
H1. Perceived affective friction will be highest in HH	PAF questionnaire
H2. HH will yield the best performance	PGC, CGC, TP, experts
H3. Affective processes moderate this relationship	Cardiac activity, EDA, gaze

coherence, cross-recurrence and delayed coincidence count [61, 64]. We will further augment our findings by analyzing gaze data, which have been linked to affective expressions and intimacy [65, 66]. Prior to the experiment, areas of interest (AOIs) will be predefined as the interlocutor's (Zoom), the conversation (Chat), and the shared document. Using gaze overlap signals [67, 68], we will measure shared and mutual gaze (when both participants look at the same area of the screen or when they look at each other through the video). Finally, we will analyze speech and voice tone data to look for evidence of vocal mimicry [69].

4 Preliminary Findings

Mood Induction Pre-tasks. First, we test the mood induction pre-task. 83 subjects (46 for f-pacman and 37 for r-pacman) were recruited for the pre-test. The selection criteria were based on MTurk quality criteria (master's qualification and at least 100 completed tasks with an approval rate above 95%) and having at least obtained a US High School diploma.

Based on our attention and completion checks, we dropped 6 responses for the regular version (n = 31), and 2 responses for the frustrating version (n = 44).

F-pacman. We operationalized "frustration" using 3 items of the PANAS questionnaire ("hostile", "irritable", and "upset"), and the value for Cronbach's Alpha for the frustration construct was $\alpha = 0.87$. D'Agostino and Pearson's test and the Shapiro–Wilk test (using SciPy [70]) both rejected the null hypothesis that our data comes from a normal distribution. Consequently, we tested our hypothesis (*Ha: frustration scores are greater post-treatment*) using a Wilcoxon signed-rank test, which is considered to be the non-parametric equivalent of the paired T-test [71] using the Python package Pinguin [72]. The difference between the pre and post-treatment levels of frustration was significant (W = 46.5, p < 0.0001, with a right-tailed alternative). We report a matched pairs rank-biserial correlation of r = -0.834 [73] and a common language effect size of CL = 0.697 [74, 75], indicating that most of our participants reported feeling more frustrated after playing the game. The difference in median frustration pre/post-treatment is shown in Fig. 1.

R-pacman. To verify the absence of a frustrating effect in our regular Pac-man, we applied the same statistical tests as the frustrating condition, albeit we used a

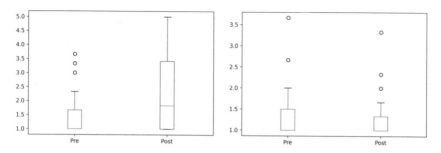

Fig. 1 Box-plot of self-reported frustration before and after playing the game (5-point Likert scale, frustration being the mean of the scores for "hostile", "irritable", and "upset"). The left box-plot shows the results for f-pacman (p < 0.0001), the right box-plot shows the results for the r-pacman (p = 0.68)

two-sided hypothesis (*H0: the median of differences between pre and post-treatment is 0*). The difference between the pre and post-treatment levels of frustration was non-significant (W = 23, p = 0.68, with a two-sided alternative). We thus accept the null hypothesis that there is no significant difference between the pre and post-treatment frustration levels of the participants for the r-pacman.

5 Expected Contributions and Conclusion

The suggested experimental design compares different combinations of affective processes to understand their impact on a creative task's outcome. Importantly, this work addresses questions that are fundamental in the post-pandemic corporate world. It explores how medium affectivity impacts the perception of affective friction among team members and subsequently results in differentiated affective processes that improve or worsen collaborative outcomes. This is an important avenue for improving team processes and management in the context of increased hybrid and remote work arrangements. Moreover, psychophysiological data have been under-represented as a method to capture emotional alignment during online creative collaboration. Capturing and measuring affective friction and processes would constitute an important methodological contribution to the field of NeuroIS. Moreover, the use of an online Pac-man game is compatible with online experiments, allowing other researchers to design experiments that can be conducted remotely. Finally, while using a confederate actor is common in experimental research, it is rare to use an actor for a collaborative task—especially in studies about affective processes. The success of this study would thus democratize the use of confederate actors in research on team processes and online collaboration, potentially opening a wealth of much-needed research in the context of a complete re-invention of work relationships.

Acknowledgements This research is partially funded by a Carlsberg Foundation Young Researcher Fellowship awarded to Rob Gleasure, with the title "Mood synchronicity, collaboration media, and task outcomes". I also thank Rob for his clever ideas and valuable feedback, the COBALT team at the HEC Montreal Tech3Lab for providing the hardware and software used in this study, Brian N. Huh for coordinating the practical aspects of the study and facilitating access to the Behavioral Research Lab at the Marshall School of Business (USC), and Furkan Selek for developing f-pacman and r-pacman.

References

1. Cummings, J. N., Espinosa, J. A., & Pickering, C. K. (2009). Crossing spatial and temporal boundaries in globally distributed projects: A relational model of coordination delay. *Information Systems Research, 20*, 420–439. https://doi.org/10.1287/isre.1090.0239
2. O'Leary, M. B., & Cummings, J. N. (2007). The spatial, temporal, and configurational characteristics of geographic dispersion in teams. *MIS Quarterly, 31*, 433–452. https://doi.org/10.2307/25148802
3. Sarker, S., & Sahay, S. (2004). Implications of space and time for distributed work: An interpretive study of US–Norwegian systems development teams. *European Journal of Information Systems, 13*, 3–20. https://doi.org/10.1057/palgrave.ejis.3000485
4. Swaab, R. I., Galinsky, A. D., Medvec, V., & Diermeier, D. A. (2012). The communication orientation model: Explaining the diverse effects of sight, sound, and synchronicity on negotiation and group decision-making outcomes. *Personality and Social Psychology Review, 16*, 25–53. https://doi.org/10.1177/1088868311417186
5. Riedl, R., Mohr, P. N. C., Kenning, P. H., Davis, F. D., & Heekeren, H. R. (2014). Trusting humans and avatars: A brain imaging study based on evolution theory. *Journal of Management Information Systems, 30*, 83–114. https://doi.org/10.2753/MIS0742-1222300404
6. Daft, R. L., & Lengel, R. H. (1986). Organizational information requirements, media richness and structural design. *Management Science, 32*, 554–571. https://doi.org/10.1287/mnsc.32.5.554
7. Daft, R. L., Lengel, R. H., & Trevino, L. K. (1987). Message equivocality, media selection, and manager performance: Implications for information systems. *MIS Quarterly, 11*, 355–366. https://doi.org/10.2307/248682
8. Ferry, D. L., Kydd, C. T., & Sawyer, J. (2001). Measuring facts of media richness. *Journal of Computer Information Systems, 41*, 69–78. https://doi.org/10.1080/08874417.2001.11647026
9. Kahai, S., & Cooper, R. (2003). Exploring the core concepts of media richness theory: The impact of cue multiplicity and feedback immediacy on decision quality. *Journal of Management Information Systems, 20*, 263–299. https://doi.org/10.1080/07421222.2003.11045754
10. Ijsselsteijn, W., van Baren, J., & van Lanen, F. (2003). Staying in touch: Social presence and connectedness through synchronous and asynchronous communication media. *Human-Computer Interaction: Theory and Practice (Part II), 2*, e928.
11. Bjørn, P., & Ngwenyama, O. (2009). Virtual team collaboration: Building shared meaning, resolving breakdowns and creating translucence. *Information Systems Journal, 19*, 227–253. https://doi.org/10.1111/j.1365-2575.2007.00281.x
12. Kahai, S. S., Carroll, E., & Jestice, R. (2007). Team collaboration in virtual worlds. *SIGMIS Database, 38*, 61–68. https://doi.org/10.1145/1314234.1314246
13. Hatfield, E., Cacioppo, J., & Rapson, R. (1992). Primitive emotional contagion. *Current Directions in Psychological Science*, 151–177
14. Chartrand, T. L., & Dalton, A. N. (2008). Mimicry: Its ubiquity, importance, and functionality. In: *Oxford handbook of human action* (pp. 458–483). Oxford University Press.
15. Frijda, N. H. (1993). Moods, emotion episodes, and emotions. *Handbook of emotions* (pp. 381–403). The Guilford Press.

16. Weiss, H. M., & Cropanzano, R. (1996). Affective events theory. *Research in Organizational Behavior, 18*, 1–74.
17. van den Berg, W., Curseu, P. L., & Meeus, M. T. H. (2014). Emotion regulation and conflict transformation in multi-team systems. *International Journal of Conflict Management, 25*, 171–188. https://doi.org/10.1108/IJCMA-05-2012-0038
18. Bartel, C. A., & Saavedra, R. (2000). The collective construction of work group moods. *Administrative Science Quarterly, 45*, 197–231. https://doi.org/10.2307/2667070
19. Lombard, M., & Ditton, T. (1997) At the heart of it all: The concept of presence. *Journal of Computer-Mediated Communication, 3*, 321. https://doi.org/10.1111/j.1083-6101.1997.tb00072.x
20. Dennis, A. R., Fuller, R. M., & Valacich, J. S. (2008). Media, tasks, and communication processes: A theory of media synchronicity. *MIS Quarterly, 32*, 575–600. https://doi.org/10.2307/25148857
21. Kock, N. (2005). Media richness or media naturalness? The evolution of our biological communication apparatus and its influence on our behavior toward E-communication tools. *IEEE Transactions on Professional Communication, 48*, 117–130. https://doi.org/10.1109/TPC.2005.849649
22. David, K. M., & Murphy, B. C. (2004). Interparental conflict and late adolescents' sensitization to conflict: The moderating effects of emotional functioning and gender. *Journal of Youth and Adolescence, 33*, 187–200. https://doi.org/10.1023/B:JOYO.0000025318.26238.40
23. Davies, P., Myers, R., Cummings, E., & Heindel, S. (1999). Adult conflict history and children's subsequent responses to conflict: An experimental test. *Journal of Family Psychology, 13*, 610–628. https://doi.org/10.1037/0893-3200.13.4.610
24. El-Sheikh, M. (1997). Children's responses to adult–adult and mother–child arguments: The role of parental marital conflict and distress. *Journal of Family Psychology, 11*, 165–175. https://doi.org/10.1037/0893-3200.11.2.165
25. Goeke-Morey, M. C., Papp, L. M., & Cummings, E. M. (2013). Changes in marital conflict and youths' responses across childhood and adolescence: A test of sensitization. *Development and Psychopathology, 25*, 241–251. https://doi.org/10.1017/S0954579412000995
26. Grych, J. H., & Fincham, F. D. (1990). Marital conflict and children's adjustment: A cognitive-contextual framework. *Psychological Bulletin, 108*, 267.
27. Finke, R. A., Ward, T. B., & Smith, S. M. (1992). *Creative cognition: Theory, research, and applications.* The MIT Press.
28. Jarvenpaa, S., Standaert, W., & Vlerick Business School. (2018). Digital probes as opening possibilities of generativity. *JAIS*, 982–1000. https://doi.org/10.17705/1jais.00516
29. Rerup, C., & Feldman, M. S. (2011). Routines as a source of change in organizational schemata: The role of trial-and-error learning. *AMJ, 54*, 577–610. https://doi.org/10.5465/amj.2011.61968107
30. Hannan, M. T., & Freeman, J. (1984). Structural inertia and organizational change. *American Sociological Review, 49*, 149–164.
31. Tushman, M. L., & Romanelli, E. (1985). Organizational evolution: A metamorphosis model of convergence and reorientation. *Research in Organizational Behavior, 7*, 171–222.
32. Whyte, G. (1998). Recasting Janis's groupthink model: The key role of collective efficacy in decision fiascoes. *Organizational Behavior and Human Decision Processes, 73*, 185–209. https://doi.org/10.1006/obhd.1998.2761
33. Hargadon, A. B., & Bechky, B. A. (2006). When collections of creatives become creative collectives: A field study of problem solving at work. *Organization Science, 17*, 484–500. https://doi.org/10.1287/orsc.1060.0200
34. Goncalo, J. A., & Staw, B. M. (2006). Individualism–collectivism and group creativity. *Organizational Behavior and Human Decision Processes, 100*, 96–109. https://doi.org/10.1016/j.obhdp.2005.11.003
35. Cheng, X., Fu, S., de Vreede, T., de Vreede, G.-J., Seeber, I., Maier, R., & Weber, B. (2020). Idea convergence quality in open innovation crowdsourcing: A cognitive load perspective. *Journal of Management Information Systems, 37*, 349–376. https://doi.org/10.1080/07421222.2020.1759344

36. Dennis, A. R., Valacich, J. S., Carte, T. A., Garfield, M. J., Haley, B. J., & Aronson, J. E. (1997). Research report: The effectiveness of multiple dialogues in electronic brainstorming. *Information Systems Research, 8*, 203–211. https://doi.org/10.1287/isre.8.2.203

37. Pearsall, M. J., Ellis, A. P. J., & Evans, J. M. (2008). Unlocking the effects of gender faultlines on team creativity: Is activation the key? *Journal of Applied Psychology, 93*, 225–234. https://doi.org/10.1037/0021-9010.93.1.225

38. Courtemanche, F., Sénécal, S., Léger, P.-M., & Fredette, M. (2022). COBALT—Capture: Real-time bilateral exchange system for multimodal user data via the internet

39. Courtemanche, F., Sénécal, S., Fredette, M., & Léger, P.-M. (2022). COBALT—Bluebox: Multimodal user data wireless synchronization and acquisition system

40. Hender, J. M., Dean, D. L., Rodgers, T. L., & Nunamaker, J. F., Jr. (2002). An examination of the impact of stimuli type and GSS structure on creativity: Brainstorming versus non-brainstorming techniques in a GSS environment. *Journal of Management Information Systems, 18*, 59–85. https://doi.org/10.1080/07421222.2002.11045705

41. Hender, J. M., Rodgers, T. L., Dean, D. L., & Nunamaker, J. F. (2001). Improving group creativity: Brainstorming versus non-brainstorming techniques in a GSS environment. In *Proceedings of the 34th Annual Hawaii International Conference on System Sciences* (10 pp)

42. Nunamaker, J. F., Applegate, L. M., & Konsynski, B. R. (1987). Facilitating group creativity: Experience with a group decision support system. *Journal of Management Information Systems, 3*, 5–19. https://doi.org/10.1080/07421222.1987.11517775

43. O'Leary, K., Gleasure, R., O'Reilly, P., & Feller, J. (2022). Introducing the concept of creative ancestry as a means of increasing perceived fairness and satisfaction in online collaboration: An experimental study. *Technovation, 110*, 102369. https://doi.org/10.1016/j.technovation.2021.102369

44. Mohammed, S., & Ringseis, E. (2001). Cognitive diversity and consensus in group decision making: The role of inputs, processes, and outcomes. *Organizational Behavior and Human Decision Processes, 85*, 310–335. https://doi.org/10.1006/obhd.2000.2943

45. Wu, I.-L., & Chiu, M.-L. (2018). Examining supply chain collaboration with determinants and performance impact: Social capital, justice, and technology use perspectives. *International Journal of Information Management, 39*, 5–19. https://doi.org/10.1016/j.ijinfomgt.2017.11.004

46. Hoegl, M., & Gemuenden, H. G. (2001). Teamwork quality and the success of innovative projects: A theoretical concept and empirical evidence. *Organization Science, 12*, 435–449. https://doi.org/10.1287/orsc.12.4.435.10635

47. Valacich, J. S., Dennis, A. R., & Nunamaker, J. F. (1992). Group size and anonymity effects on computer-mediated idea generation. *Small Group Research, 23*, 49–73. https://doi.org/10.1177/1046496492231004

48. Watson, D., Clark, L. A., & Tellegen, A. (1988). Development and validation of brief measures of positive and negative affect: The PANAS scales. *Journal of Personality and Social Psychology, 54*, 1063–1070. https://doi.org/10.1037/0022-3514.54.6.1063

49. Bittner, E., & Shoury, O. (2019). Designing automated facilitation for design thinking: A chatbot for supporting teams in the empathy map method. In *Proceedings of the 52nd Hawaii International Conference on System Sciences*

50. Elshan, E., & Ebel, P. (2020). Let's team up: Designing conversational agents as teammates. In *International Conference on Information Systems (ICIS)*. Online/Hyderabad, India.

51. Hayashi, Y., & Ono, K. (2013). Embodied conversational agents as peer collaborators: Effects of multiplicity and modality. In: *2013 IEEE RO-MAN* (pp. 120–125).

52. Hwang, A.H.-C., & Won, A. S. (2021). IdeaBot: Investigating social facilitation in human-machine team creativity. In *Proceedings of the 2021 CHI Conference on Human Factors in Computing Systems* (pp. 1–16.) ACM, Yokohama, Japan.

53. Rich, C., & Sidner, C. L. (1998). COLLAGEN: A collaboration manager for software interface agents. In S. Haller, A. Kobsa, & S. McRoy (Eds.), *Computational models of mixed-initiative interaction* (pp. 149–184). Springer.

54. Barsade, S. G. (2002). The Ripple effect: Emotional contagion and its influence on group behavior. *Administrative Science Quarterly, 47*, 644–675. https://doi.org/10.2307/3094912
55. Baron, R. A. (1984). Reducing organizational conflict: An incompatible response approach. *Journal of Applied Psychology, 69*, 272–279. https://doi.org/10.1037/0021-9010.69.2.272
56. Mayer, J. D., Roberts, R. D., & Barsade, S. G. (2008). Human abilities: Emotional intelligence. *Annual Review of Psychology, 59*, 507–536. https://doi.org/10.1146/annurev.psych.59.103006. 093646
57. Reuderink, B., Nijholt, A., & Poel, M. (2009). Affective Pacman: A frustrating game for brain-computer interface experiments (pp. 221–227)
58. Reuderink, B., Mühl, C., & Poel, M. (2013). Valence, arousal and dominance in the EEG during game play. *International Journal of Autonomous and Adaptive Communications Systems, 6*, 45–62. https://doi.org/10.1504/IJAACS.2013.050691
59. Chanel, G., Kivikangas, J. M., & Ravaja, N. (2012). Physiological compliance for social gaming analysis: Cooperative versus competitive play☆. *Interacting with Computers, 24*, 306–316. https://doi.org/10.1016/j.intcom.2012.04.012
60. Järvelä, S., Kivikangas, J. M., Kätsyri, J., & Ravaja, N. (2014). Physiological linkage of dyadic gaming experience. *Simulation & Gaming, 45*, 24–40. https://doi.org/10.1177/104687811351 3080
61. Verdiere, K. J., Albert, M., Dehais, F., & Roy, R. N. (2020). Physiological synchrony revealed by delayed coincidence count: Application to a cooperative complex environment. *IEEE Transactions on Human-Machine Systems, 50*, 395–404. https://doi.org/10.1109/THMS.2020.298 6417
62. Dehais, F., Vergotte, G., Drougard, N., Ferraro, G., Somon, B., Ponzoni Carvalho Chanel, C., & Roy, R. (2021). AI can fool us humans, but not at the psycho-physiological level: A hyperscanning and physiological synchrony study.
63. Elkins, A. N., Muth, E. R., Hoover, A. W., Walker, A. D., Carpenter, T. L., & Switzer, F. S. (2009). Physiological compliance and team performance. *Applied Ergonomics, 40*, 997–1003. https://doi.org/10.1016/j.apergo.2009.02.002
64. Mønster, D., Håkonsson, D. D., Eskildsen, J. K., & Wallot, S. (2016). Physiological evidence of interpersonal dynamics in a cooperative production task. *Physiology & Behavior, 156*, 24–34. https://doi.org/10.1016/j.physbeh.2016.01.004
65. Kret, M. E. (2015). Emotional expressions beyond facial muscle actions. A call for studying autonomic signals and their impact on social perception. *Frontiers in Psychology, 6*. https://doi.org/10.3389/fpsyg.2015.00711
66. Janssen, J. H., Bailenson, J. N., IJsselsteijn, W. A., Westerink, J. H. D. M. (2010). Intimate heartbeats: Opportunities for affective communication technology. *IEEE Transactions on Affective Computing, 1*, 72–80. https://doi.org/10.1109/T-AFFC.2010.13
67. Schneider, B., & Pea, R. (2017). Real-time mutual Gaze perception enhances collaborative learning and collaboration quality. In M. Orey & R. M. Branch (Eds.), *Educational media and technology yearbook* (pp. 99–125). Springer International Publishing.
68. D'Angelo, S., & Begel, A. (2017). Improving communication between pair programmers using shared Gaze awareness. In *Proceedings of the 2017 CHI Conference on Human Factors in Computing Systems* (pp. 6245–6290). ACM, Denver Colorado USA.
69. Chartrand, T. L., & van Baaren, R. (2009). Human mimicry. *Advances in experimental social psychology* (pp. 219–274). Academic Press/Elsevier.
70. Virtanen, P., Gommers, R., Oliphant, T. E., Haberland, M., Reddy, T., Cournapeau, D., Burovski, E., Peterson, P., Weckesser, W., Bright, J., van der Walt, S. J., Brett, M., Wilson, J., Millman, K. J., Mayorov, N., Nelson, A. R. J., Jones, E., Kern, R., Larson, E., … van Mulbregt, P. (2020). SciPy 1.0: Fundamental algorithms for scientific computing in Python. *Nature Methods, 17*, 261–272. https://doi.org/10.1038/s41592-019-0686-2
71. Wilcoxon, F. (1992). Individual comparisons by ranking methods. In *Breakthroughs in statistics* (pp. 196–202). Springer.
72. Vallat, R. (2018). Pingouin: Statistics in Python. *Journal of Open Source Software, 3*, 1026. https://doi.org/10.21105/joss.01026

73. Kerby, D. S. (2014). The simple difference formula: An approach to teaching nonparametric correlation. *Comprehensive Psychology, 3*, 11.IT.3.1. https://doi.org/10.2466/11.IT.3.1
74. McGraw, K. O., & Wong, S. P. (1992). A common language effect size statistic. *Psychological Bulletin, 111*, 361–365. https://doi.org/10.1037/0033-2909.111.2.361
75. Vargha, A., & Delaney, H. D. (2000). A critique and improvement of the CL common language effect size statistics of McGraw and Wong. *Journal of Educational and Behavioral Statistics, 25*, 101–132. https://doi.org/10.3102/10769986025002101

A Look Behind the Curtain: Exploring the Limits of Gaze Typing

Marius Schenkluhn, Christian Peukert, and Christof Weinhardt

Abstract Text entry in Augmented and Virtual Reality applications remains a major challenge for a comprehensive and ubiquitous usability of this technology. Therefore, researchers have proposed several solutions for text entry in Augmented Reality. Hands-free gaze typing is a popular approach to ensure mobility and both fast and accurate typing. However, prior studies struggle with the missing expertise of participants. We propose a study design that eliminates the learning process and gives an outlook to future gaze typing performance. In particular, we focus on the time-dependent effects of different dwell times on typing performance. By focusing on user-specific limitations, the results of this study are a prerequisite to user-adaptive gaze typing without fatigue.

Keywords Gaze typing · Augmented reality · Dwell time · Eye tracking · NeuroIS

1 Introduction

Eye tracking is becoming increasingly relevant in research[1] and business as sensors are becoming more powerful and cost-effective [1]. Primarily, eye tracking technology is used to study human attention allocation behavior to improve user experiences, marketing campaigns, or detect fatigue while driving [2]. However, eye

---

[1] Number of yearly results on Scopus for "Eye Tracking" is steadily increasing since 2004 from 294 to 3,074 in 2021.

M. Schenkluhn (✉) · C. Peukert · C. Weinhardt
Karlsruhe Institute of Technology, Karlsruhe, Germany
e-mail: marius.schenkluhn@kit.edu

C. Peukert
e-mail: christian.peukert@kit.edu

C. Weinhardt
e-mail: christof.weinhardt@kit.edu

M. Schenkluhn
Robert Bosch GmbH, Plochingen, Germany

© The Author(s), under exclusive license to Springer Nature Switzerland AG 2022
F. D. Davis et al. (eds.), *Information Systems and Neuroscience*,
Lecture Notes in Information Systems and Organisation 58,
https://doi.org/10.1007/978-3-031-13064-9_26

251

tracking can also be leveraged as active element of user interaction. Gaze typing allows users to type solely by using their eyes, i.e., by fixating the respective letters on a virtual keyboard for a certain time [3]. This hands-free interaction mode is particularly useful for future Augmented Reality (AR) and Virtual Reality (VR) applications as traditional input devices are usually not available [3]. With built-in eye tracking sensors, no additional controllers, keyboards, or gloves are required. Therefore, gaze typing enables mobility and concurrent task execution.

Various studies explore the parameters of gaze typing and its achievable speeds to improve usability and performance [3–5]. Moreover, longitudinal studies show the importance of the learning effect on typing speeds [6, 7]. However, the performance development while writing longer texts has not yet received much attention. Exploring this effect with users, who are unfamiliar with the system, is confounded as they need to simultaneously put effort into learning the system. Still, a hypothetical expert gaze typist will likely experience fatigue over extended periods of time at their peak typing speed thus limiting the long-term performance.

Hence, we will approach the issue from a different perspective. By excluding the learning effect and measuring time-dependent performance at varying dwell times during typing in a realistic AR context, we explore human limitations in gaze typing. The overall goal is to create gaze typing that is proactively adapting instead of retrospectively reacting to user fatigue. This would enable users to type short texts at their peak performance and economically utilizing cognitive resources for long texts. With the proposed study design in this article, we want to make a first step towards reaching this overarching research goal and seek to answer the following research question: *What is the influence of dwell time and text length on gaze typing performance?*

In this research-in-progress paper, we propose a design for a laboratory study to answer the research question and demonstrate its implementation in an AR system.

2   Theoretical Background

Eye Tracking. Eye movement can be largely characterized by fixations and saccades [8]. According to Pannasch et al. [8], saccades are defined as "fast sequential movements, [that] are necessary to bring the fovea from one point to another" [p. 1] and fixations as "periods in between of saccades, when the eyes are relatively stable" [p. 1]. Saccades can be triggered by different events and are further categorized into visually guided and memory-guided saccades among others [9]. Visually guided saccades are either reflexive to a sudden visual event or scanning unknown areas. Memory-guided saccades move the eyes to gaze towards a memorized location without external stimulus [9].

Today, eye tracking is usually performed by capturing the corneal reflection with video cameras [10]. Eye trackers are used for interactive and diagnostic purposes in application domains such as neuroscience, aviation, automobile driving, print advertising, and user experience research [2]. The interactive use of eye trackers, called gaze-based interaction, has also gained relevance in VR and AR applications when traditional means of input are not available or when user mobility would be limited [3]. Gaze-based interaction can be the sole mode of input or used for target selection in combination with a controller or gesture to confirm a gaze-selected action.

Gaze Typing. New device factors often ask for new types of text entry, e.g., numpads on mobile phones, touch keyboards on smart phones, and voice recognition on smart speakers. There are several approaches to text entry in AR and VR ranging from physical keyboards to wrist- or gesture-based, and novel 3D techniques [3]. Each technique has benefits and drawbacks in different contexts. New layouts require more training than layouts a user is already familiar with [11]. In general, the goal of new layouts is to maximize the text entry rate and minimizing the error rate at the same time [3].

Gaze typing relies solely on eye tracking to enter text. A common form is the display of a virtual keyboard on a screen. By gazing at a key for more than a given threshold, the eyes "press" the key and type the respective character [3]. The main advantage of gaze typing is its hands-free character, which has been leveraged in several accessibility studies [12, 13]. Additionally, the eyes can move quickly and accurately. Studies therefore indicate a potential of high text entry speeds and low error rates [6, 14, 15].

The time between the beginning of the initial fixation and the activation of the key press is called *dwell time* and is an important variable when designing gaze typing interaction. The dwell time is necessary to differentiate meaningful from unmeaningful gaze events, i.e., intended typing from a character search. As soon as the first fixation is registered within the Area of Interest (AOI) of one key, the system measures the visit time on this particular key. If one of the successive fixations targets an area outside the AOI before the dwell time threshold has been reached, the key press is aborted. Otherwise, if all successive fixations stay within the AOI longer than the dwell time, the key is pressed once.

Short dwell times can cause the Midas touch effect, that is, the unintended selection of every key the user is passing over [4, 16]. Thus, longer dwell times reduce the number of errors. However, overly long dwell times lead to adverse effects as the user is not able to hold the gaze for long periods ("gaze-hold problem") [4]. Additionally, long dwell times slow down text entry speeds as the user must stay fixated to one character until the dwell time has passed. Longitudinal studies show that trained users are able to deal with shorter dwell times around 200 ms while still maintaining low error rates [6, 7]. Hence, this tradeoff between entry speed, error rate, and training determines the success of the application of eye gazing in VR and AR. Feasible dwell time depends on different factors. Following the human performance model of Kristensson and Vertanen [15], the gaze typing interaction consists of overhead time and dwell time. The overhead time includes saccades to transition between keys and error correction. While the dwell time is internal to the application, the overhead time depends on the user.

3   Design Considerations for an Experimental Design to Study Time-Dependent Gaze Typing Performance

To explore the limitations of gaze typing of future experienced typists, we plan to conduct a laboratory experiment. In general, the typing performance is highly influenced by the dwell time [15, 17]. Accordingly, it should be set as low as possible to support fast text entry. In contrast, the ability of users to concentrate is limited. Therefore, we presume that proficient gaze typists experience fatigue while typing a text section and cannot maintain the speed associated with a low dwell time.

Task. MacKenzie [18] discusses different tasks for text entry evaluation. While a text creation task is closer to typical usage, it has several issues for performance measurement. Text creation includes aspects unrelated to the keyboard interaction such as thinking about content, phrasing, and grammar [18]. Additionally, error detection is more complex as the user intention cannot be inferred during text entry [18]. Finally, text creation complicates comparability between subjects because of differences in vocabulary and, therefore, the distribution of letters and words [18].

For these reasons, scientists typically rely on text copying tasks that try to mimic text creation [17–20]. A typical task is as follows: The study participant is presented with one sentence for example from the phrase set of MacKenzie and Soukoreff [21]. After memorizing the sentence, the participant is asked to write the text as quickly as possible with the given keyboard [18].

Gaze typing experiments usually study the performance of prototypes with respect to typing speed and accuracy [3]. Due to typically novel keyboard designs and the unfamiliarity of participants with them, the performance of a potential expert typists cannot be trialed in one session. Longitudinal studies show that the average typing speed increases over multiple sessions, although participants cannot be considered experts afterwards [6, 7]. Novice typists have to search for characters on the keyboard during typing and are not accustomed to the dwell time. During this process, the fast eye movements can be considered as scanning saccades in combination with longer fixations required for pattern identification. In contrast, potential expert typists can use memory-guided saccades to quickly and accurately jump between keys after sufficient training. Thus, this study eliminates the training effect regarding interaction type, keyboard layout, and typing task, by approximating memory-guided saccades with reflexive saccades. Instead of displaying a phrase that the participant must type, the keyboard visually highlights the next key that shall be "pressed." This highlighting cues a reflexive saccade. In essence, participants follow the highlighted keys to gaze type sentences from the phrase sets of MacKenzie and Soukoreff [21] at varying dwell times without relying on their gaze typing skill.

Text copying tasks introduce additional subtasks such as comparing the specified text with the typed text, which reduces text entry performance [18]. However, this study design does not require participants to memorize and compare texts when following the highlighted keys. Thus, we expect a higher external validity regarding real text creation.

Evaluation Procedure. The experiment begins with a calibration of the eye tracker. The participant is introduced to the task and performs a trial round. The task is to follow and focus the highlighted character as quickly and accurately as possible. After the dwell time threshold has been reached the next character of the sentence is highlighted on the keyboard. Afterwards, the task is performed at decreasing dwell time levels. The levels are 600, 450, 300, 150, and 0 ms [17]. This could be adjusted after our pretest. There are two consecutive trials at each dwell time level to account for inter-treatment fatigue. As we want to measure the time-dependent effects on intra-treatment performance, there is a relaxation phase of 3 min between each treatment to reduce effects on inter-treatment performance [14].

Each treatment ends regularly after 5 min. Participants can abort the treatment if they are not able to maintain the increasingly fast speeds. If the error rate exceeds 15% the participant is unlikely to keep up and the treatment ends as well. This threshold requires finetuning during the pretest, too.

Measurements. The *performance* is measured as a function of time to derive a relation between dwell time, text length, and performance. Measures include the over-head time between dwells, the number of lost focuses for one keypress, the minimal distance between the gaze-intersection with the keyboard and the key border, and the error rate. After each round, the participants answer the NASA-TLX questionnaire to supplement self-reported task loads [22].

Participants and Compensation. Participants will be recruited from the participant panel of a large European university. For the 100-min experiment, we plan to compensate each participant with 20 €. Further, the three participants with the highest performance will be rewarded with an additional 5 € to motivate concentration. We decided against an entirely performance-based compensation to avoid pushing participants over their limits in an unrealistic manner.

Application. We chose to implement the experiment in an Augmented Reality context. Most research in the domain of virtual keyboards is currently conducted using Virtual Reality technology [3]. However, Augmented Reality systems are more dependent on mobile means of text entry as VR systems are limited in their mobility in order not to interfere or collide with real objects. In a professional VR application, a hardware keyboard can most likely be used with a mapped virtual representation within the VR environment. The application is implemented in Unity 2020.3 for an off-the-shelf Microsoft HoloLens 2 Augmented Reality head-mounted display using the Mixed Reality Toolkit 2.7.3. The HoloLens 2 is a state-of-the-art standalone AR device with build-in eye tracking capabilities. The eye tracker updates at 30 Hz, is specified with a spatial accuracy of 1.5°, and has been evaluated previously for accuracy and precision [23]. Although it is not as capable as external sensors, it is sufficiently accurate for this use case. The API only grants access to the combined fixation point. The video stream of the IR cameras, independent data per eye, and eye blinking data is not available. While this is a drawback, future eye trackers in consumer devices will likely comprise of the same limitations for privacy reasons. We implemented the application based on Microsoft's usability recommendations[2]

---

[2] https://docs.microsoft.com/en-us/windows/mixed-reality/design/ (Accessed: 23.04.2022).

Fig. 1 Augmented reality application with highlighted character

for colors, contrast, object positioning and size, and audio. Figure 1 depicts the gaze typing keyboard in a simulated environment. A video[3] of the application shows different dwell time levels combined with an overhead time that simulates the human reaction time. The measurement display and the overhead time are included for demonstration purposes and will be deactivated for the experiment. Furthermore, Fig. 2 lists all relevant measurements of the keyboard and its components in an isometric projection. The keyboard fits within the field of view of the HoloLens 2 without the need for users to turn their heads.

4 Discussion

This paper proposes a novel experiment design to explore human limits to inform future gaze typing implementations. Depending on the dwell time level, we expect that participants will start to struggle concentrating on the task after some time. The performance measures will likely represent this effect. For lower dwell time levels, the effect is expected to appear earlier during the task.

By conducting two consecutive trials at each dwell time level, we expect an approximated step function. The second round on the same level might show a slight decrease in performance. If this effect is too prominent in the pretest, the relaxation phase will be extended. As the overhead time can vary between participants, we do not expect that there will be one cutoff point where all participants experience concentration loss. However, understanding these differences between individuals will be the prerequisite for designing proactive user-adaptive gaze typing. A proactive

---

[3] https://www.youtube.com/watch?v=GQayxnlKVqU.

Fig. 2 Dimensions of the gaze typing keyboard

system would be able to increase dwell time to decrease task load before typists experience fatigue.

There are some limitations to this experiment design. The limited spatial and temporal resolution of the eye tracker was already mentioned. Moreover, the overhead time in this experiment does not contain the tasks of character processing and finding on the keyboard layout and error correction. Thus, this factor must be added when comparing the results with other gaze typing studies even if expert typists are able to minimize it. Furthermore, there is a possible secondary training effect on concentration to type longer texts that this study cannot eliminate.

5 Concluding Notes and Future Research

The results of this study will enable the development of proactive user-adaptive eye gazing systems by complementing previous studies with a different perspective. Additionally, future systems do not have to rely on the exact fixation of singular characters [14, 15]. Intelligent dwell-free gaze typing similar to swipe keyboards on smartphones could even improve gaze typing performance [14]. The particularization of the human performance model [15] similar to the keystroke-level model [24] with a focus on the cognitive processes could also help to unveil cognitive limitations. By better understanding human limits, the usability and comfort of these systems can be improved with the results of this study leveraging gaze typing as attractive and competitive means of text entry in AR and VR environments.

References

1. Market research future: Eye tracking market, by type (mobile, remote), by applications (human-computer interaction, virtual reality), by verticals (aerospace, retail, automotive, government & defense)—Forecast 2027 (2021). Accessed 23 April 2022.
2. Duchowski, A. (2002). A breadth-first survey of eye-tracking applications. *Behavior Research Methods, Instruments, & Computers, 34*, 455–470.
3. Dube, T. J., & Arif, A. S. (2019). Text entry in virtual reality: A comprehensive review of the literature. In *International Conference on Human-Computer Interaction* (pp. 419–437). Springer.
4. Penkar, A. M., Lutteroth, C., & Weber, G. (2012). Designing for the eye: Design parameters for dwell in gaze interaction. In *Proceedings of the 24th Australian Computer-Human Interaction Conference* (pp. 479–488).
5. Majaranta, P., Aula, A., & Räihä, K.-J. (2004). Effects of feedback on eye typing with a short dwell time. In *Proceedings of the 2004 symposium on Eye Tracking Research & Applications* (pp. 139–146).
6. Majaranta, P., Ahola, U.-K., & Špakov, O. (2009). Fast gaze typing with an adjustable dwell time. In *Proceedings of the SIGCHI Conference on Human Factors in Computing Systems* (pp. 357–360).
7. Tuisku, O., Majaranta, P., Isokoski, P., & Räihä, K.-J. (2008). Now Dasher! Dash away! Longitudinal study of fast text entry by eye gaze. In *Proceedings of the 2008 Symposium on Eye Tracking Research & Applications* (pp. 19–26).
8. Pannasch, S., Helmert, J. R., Roth, K., Herbold, A.-K., & Walter, H. (2008). Visual fixation durations and saccade amplitudes: Shifting relationship in a variety of conditions. *Journal of Eye Movement Research, 2*.
9. Rommelse, N. N. J., van der Stigchel, S., & Sergeant, J. A. (2008). A review on eye movement studies in childhood and adolescent psychiatry. *Brain and Cognition, 68*, 391–414.
10. Duchowski, A. (2007). *Eye tracking techniques. Eye tracking methodology* (pp. 51–59). Springer.
11. Schenkluhn, M., Peukert, C., & Weinhardt, C. (2022). Typing the future: Designing multimodal AR keyboards. In *Wirtschaftsinformatik 2022 Proceedings*.
12. Hansen, J. P., Tørning, K., Johansen, A. S., Itoh, K., & Aoki, H. (2004). Gaze typing compared with input by head and hand. In *Proceedings of the 2004 Symposium on Eye Tracking Research & Applications* (pp. 131–138).
13. Zhang, X., Kulkarni, H., & Morris, M. R. (2017). Smartphone-based gaze gesture communication for people with motor disabilities. In *Proceedings of the 2017 CHI Conference on Human Factors in Computing Systems* (pp. 2878–2889).
14. Kurauchi, A., Feng, W., Joshi, A., Morimoto, C., & Betke, M. (2016). EyeSwipe: Dwell-free text entry using gaze paths. In *Proceedings of the 2016 CHI Conference on Human Factors in Computing Systems* (pp. 1952–1956).
15. Kristensson, P. O., & Vertanen, K. (2012). The potential of dwell-free eye-typing for fast assistive gaze communication. In *Proceedings of the Symposium on Eye Tracking Research and Applications* (pp. 241–244).
16. Jacob, R. J. K. (1991). The use of eye movements in human-computer interaction techniques: What you look at is what you get. *ACM Transactions on Information Systems (TOIS), 9*, 152–169.
17. Rajanna, V., & Hansen, J. P. (2018). Gaze typing in virtual reality: Impact of keyboard design, selection method, and motion. In *Proceedings of the 2018 ACM Symposium on Eye Tracking Research & Applications* (pp. 1–10).
18. MacKenzie, I. S. (2010). Evaluation of text entry techniques. In I. S. MacKenzie & K. Tanaka-Ishii (Eds.), *Text entry systems: Mobility, accessibility, universality* (pp. 75–101). Elsevier.
19. Yu, D., Fan, K., Zhang, H., Monteiro, D., Xu, W., & Liang, H.-N. (2018). PizzaText: Text entry for virtual reality systems using dual thumbsticks. *IEEE Transactions on Visualization and Computer Graphics, 24*, 2927–2935.

20. Kuester, F., Chen, M., Phair, M. E., & Mehring, C. (2005). Towards keyboard independent touch typing in VR. In *Proceedings of the ACM Symposium on Virtual Reality Software and Technology* (pp. 86–95).
21. MacKenzie, I. S., & Soukoreff, R. W. (2003). Phrase sets for evaluating text entry techniques. In *CHI'03 extended abstracts on human factors in computing systems* (pp. 754–755).
22. Hart, S. G., & Staveland, L. E. (1988). *Development of NASA-TLX (task load index): Results of empirical and theoretical research. Advances in psychology* (Vol. 52, pp. 139–183). Elsevier
23. Aziz, S. D., & Komogortsev, O. V. (2021). *An assessment of the eye tracking signal quality captured in the HoloLens 2.* arXiv:2111.07209
24. Card, S. K., Moran, T. P., & Newell, A. (1980). The keystroke-level model for user performance time with interactive systems. *Communications of the ACM, 23,* 396–410.

Towards Mind Wandering Adaptive Online Learning and Virtual Work Experiences

Colin Conrad and Aaron J. Newman

Abstract NeuroIS researchers have become increasingly interested in the design of new types of information systems that leverage neurophysiological data. In this paper we describe the results of machine learning analysis which validates a method for the passive detection of mind wandering. Following the presentation of the results, we describe ways that this technique could be applied to create a neuroadaptive online learning and virtual meeting tool which may improve users' retention of information by providing auditory feedback.

Keywords Electroencephalography (EEG) · Machine learning applications · Neuro-adaptive systems · Mind wandering

1 Introduction

Since its inception as a field that leverages neurotechnology to give insights into information systems (IS) phenomena, NeuroIS has developed a range of specific interests. Researchers have flagged emotion, attention, and decision making as promising areas of inquiry [1]. These areas promise to contribute to making more human-centered systems, or even designing systems that can adapt to a user's cognitive states [2]. Attention-adaptive systems, in particular, have the potential to contribute to radically new information technology (IT) use experiences, and have begun to gain traction in the community [3].

Mind wandering is an attention-related phenomenon which describes when conscious experience becomes detached from an external environment toward one's internal thoughts or feelings [4]. It is known to have various effects on creativity, attention, and other cognitive processes [5, 6]. Moreover, it has been identified as a topic of interest by IS researchers [7–10] as well as in learning systems [5, 11]. A system adaptive to mind-wandering episodes could contribute to IT use experiences

C. Conrad (✉) · A. J. Newman
Dalhousie University, Halifax, Canada
e-mail: colin.conrad@dal.ca

© The Author(s), under exclusive license to Springer Nature Switzerland AG 2022
F. D. Davis et al. (eds.), *Information Systems and Neuroscience*,
Lecture Notes in Information Systems and Organisation 58,
https://doi.org/10.1007/978-3-031-13064-9_27

and might be particularly important for systems related to improved online learning, or remote meeting technologies, as well as other applications.

In this paper, we take the first steps towards such a system by identifying machine learning algorithms which can reliably detect the mind wandering based on its EEG signal correlates. We describe an offline machine learning experiment that leverages previously published data [11]. The techniques used in this experiment apply those used by brain-computer interfaces, with an aim towards improved systems design [12, 13]. The goal of the experiment is to apply classifiers to unlabeled data to demonstrate the technique's feasibility. Following the description of our methods and results, we discuss and provide details about how these findings can be applied to develop an adaptive mind wandering system, which could be applied to create novel online classrooms, personal performance tools or cognitive wellness systems.

2 Methods

2.1 Data Acquisition and Description

The data analyzed in this experiment were previously disseminated in the literature and we encourage readers to refer to the paper for more details on the methods and justifications for design choices [11]. Participants attended a long lecture as two tones were played in the background, one at 500 Hz played 80% of the time and one at 1000 Hz played 20% of the time. Participants were not given a task to complete related to the tones. They were prompted 10 times at pre-programmed intervals about their degree of experienced mind wandering, based on a Likert scale from 1 ("completely on task") to 5 ("completely mind wandering"). The data of interest comprised 1.2 s epochs extending from 200 ms prior, to 1000 ms after, the onset of each auditory tone. Epochs which occurred in the 10 s preceding an experience sample prompt were labelled according to the degree of reported mind wandering. The remaining data represented nearly 73 min of the total experiment and were unlabeled. Of the 52 participants who were described in the original study, we included in the present analyses only the 11 who used the full range of the Likert scale (i.e. participants whose minds both wandered and remained on task at various times in the experiment). Each participant's data was divided into two subsets for the machine learning analysis. The classification dataset consisted of the epochs that were labelled with the extremes of the Likert scale; epochs that preceded a response of 1 and those that preceded a response of 5. The unlabeled dataset consisted of the approximately 3500 epochs that were unlabeled throughout the experiment, to which we applied the classifier.

2.2 Data Preparation and Classification

Data from both datasets were converted to power spectral density using the multitaper method [14]. The tapers generated consisted of frequencies between 1 and 30 Hz over 32 electrodes, which were then normalized and transformed into a two-dimensional array. To ensure a balanced dataset for machine learning, we conducted up-sampling of the minority classes, which generated synthetic data based on the distribution of the data [15]. Each participant's data was analyzed individually. Machine learning classifiers were created for each participant's data from the classification dataset. All classifiers were trained to discriminate between "completely on task" and "completely mind wandering" trials, using Scikit-Learn's linear discriminant analysis classifier with 2/3 of the labelled data. The classifiers were tested for accuracy on 1/3 of the labelled data. The classifiers were then applied to the unlabeled data from each participant.

2.3 Assessment and Visualization

Each classifier was assessed individually for accuracy. We created visualizations using the classifications of the unlabeled data were retrieved and averaged and smoothed across 100 s segments to assist with visualization and interpretability. Visualizations were created on the smoothed data and compared to the original Likert scale responses made throughout the experiment [16].

3 Results

The results of the up-sampling and classification tasks are provided in Table 1. The classifiers trained on 6 of the 11 participants performed with over 80% accuracy. All visualizations and analysis are provided as a Jupyter notebook appendix, which is available online.[1]

4 Discussion

The results of the machine learning classifiers suggest that they can accurately classify mind wandering based on the limited data that they were given. The application of the classifiers to the unlabeled data demonstrate that the classifiers might predict mind wandering, though we did not test these relationships statistically. We are nonetheless led to believe that we have some evidence that the classifiers (1)

---

[1] https://github.com/cdconrad/2022-towards-adaptive.

Table 1 Summary of labels, total support (after up-sampling) and classification accuracy for predicting labelled data

Participant	Likert 1 labels[†]	Likert 5 labels[†]	Total support[‡]	Accuracy
1	56	14	112	0.973
2	13	15	30	1.000
3	13	13	26	0.444
4	9	29	58	0.900
5	29	14	58	0.700
6	26	11	52	0.833
7	28	26	56	0.789
8	22	13	44	0.800
9	14	35	70	0.750
10	36	38	76	0.846
11	13	28	56	0.789

[†] Denotes actual trials
[‡] Denotes data investigated with up-sampling; a balanced number for each class

performed classification with a degree of accuracy and that (2) they were applied in a way that demonstrates their use for creating an adaptive mind wandering system. It is not surprising that the linear discriminant analysis technique worked well, as it is commonly used in the development of brain-computer interfaces to achieve a mechanically similar task [10]. Nevertheless, the findings would benefit from future research, ideally on a larger sample. Future researchers can apply our proposed technique and make improvements to it to create adaptive interfaces that can improve users' experience.

A potential limitation of our findings is that there is no way to truly validate the results of the machine learning classifiers on the unlabeled data that we retrieved from the original experiment. However, this is likely true of many applications of a live adaptive system, and the samples that we had labels for suggested that the classifiers, on average, performed similarly to many brain-computer interface paradigms. Future experiments can be conducted to validate their accuracy over time with various labeling methods. A second limitation of our findings is that the data is that we had to rely on synthetic data generated by up-sampling to gather sufficient data for classification. Though up-sampling is leveraged by many machine learning researchers to conduct similar classification tasks, this criticism warrants consideration; it is possible that the classification results were the product of overfitting.

# 5	Towards a Mind Wandering Adaptive Experience

The findings described by the experiment suggest that it is possible to create computer programs that are adaptive to a users' mind wandering state. In this final section, we will outline some of the characteristics of such a system for future work.

## 5.1	Proposed System Design

Like brain-computer interfaces, a useful neuroadaptive system would consist of two phases: a training phase, and a test phase [3, 12]. In the case of a mind wandering system, a training phase should involve the creation of the classifiers. The experiment described in the previous sections outline a feasible design for creating such classifiers by leveraging the experience sampling technique over an extended period of time [5, 13]. A major limitation of this approach is the amount of time that would be required to generate sufficient labels for the task. Furthermore, the Likert scale did not guarantee that any given user would generate sufficient labels for an adaptive system.

An alternative approach is to ask participants to conduct a very boring task that is likely to trigger mind wandering over a period of 20–30 min. Such an approach could similarly use mind wandering prompts, though ideally adapted from the Likert scale to give a binary classification (i.e., yes/no), or could use a behavioural measure such as missed cues. The labels should be sufficient to conduct machine learning, though the up-sampling technique described in this paper can be used to overcome imbalanced datasets.

Once the classifiers have been created, they can be applied to determine whether a computer should administer a stimulus. Given that applications of adaptive systems to virtual workplaces or classrooms should encourage productivity, it is important that the stimulus is minimally disruptive, and ideally does not require the modification of the workplace software. One approach could be to create an auditory stimulus, such as a beep, that is administered by the neuro-adaptive program. Such stimuli could help remind participants to attend to the task, potentially encouraging them to return to a meta conscious state where they are aware of their surroundings. A study can then be conducted to determine whether the adaptive system helped participants perform better at information retention. Given that we have demonstrated the classifiers can be developed in the Python programming language, such an application could be made using common Python-based interface development tools. Figure 1 summarizes the design of such a system.

Fig. 1 Overview of a simple
auditory feedback
mechanism for a virtual
meeting

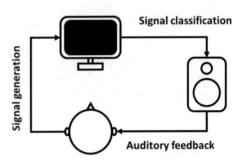

5.2 Other System Applications

If successful, such mind wandering adaptive information systems could have additional applications beyond reminders. For example, wearable technologies are increasingly employing data visualization to encourage desirable behaviour. The system could be similarly applied to provide feedback to participants about their ability to attend to long videos. The system could alternatively be used to passively measure mind wandering as various information system designs are prototyped; it could be that by keeping Zoom videos on, harmful mind wandering is limited.

Regardless of application, the generic system will leverage the training routine, which will continue to limit the system's applicability. Future work on this topic would benefit by identifying techniques for generating labels of mind wandering events in as little time as possible. It would also benefit by identifying labels that can distinguish varieties of mind wandering, some of which may not be harmful to a user's productive experience with a technology.

References

1. Riedl, R., Fischer, T., Léger, P. M., & Davis, F. D. (2020). A decade of NeuroIS research: Progress, challenges, and future directions. *ACM SIGMIS Database: The DATABASE for Advances in Information Systems, 51*(3), 13–54.
2. vom Brocke, J., Hevner, A., Léger, P. M., Walla, P., & Riedl, R. (2020). Advancing a NeuroIS research agenda with four areas of societal contributions. *European Journal of Information Systems, 29*(1), 9–24.
3. Demazure, T., Karran, A., Léger, P. M., Labonté-LeMoyne, É., Sénécal, S., Fredette, M., & Babin, G. (2021). Enhancing sustained attention. *Business & Information Systems Engineering, 63*(6), 653–668.
4. Fox, K. C., & Christoff, K. (Eds.). (2018). *The Oxford handbook of spontaneous thought: Mind-wandering, creativity, and dreaming.* Oxford University Press.
5. Wammes, J. D., Boucher, P. O., Seli, P., Cheyne, J. A., & Smilek, D. (2016). Mind wandering during lectures I: Changes in rates across an entire semester. *Scholarship of Teaching and Learning in Psychology, 2*(1), 13.
6. Yanko, M. R., & Spalek, T. M. (2014). Driving with the wandering mind: The effect that mind-wandering has on driving performance. *Human Factors, 56*(2), 260–269.

7. Klesel, M., Oschinsky, F. M., Conrad, C., & Niehaves, B. (2021). Does the type of mind wandering matter? Extending the inquiry about the role of mind wandering in the IT use experience. *Internet Research, 31*(3), 1018–1039.
8. Sullivan, Y., & Davis, F. (2020). Self-regulation, mind wandering, and cognitive absorption during technology use. In *Proceedings of the 53rd Hawaii International Conference on System Sciences.*
9. Baumgart, T. L., Klesel, M., Oschinsky, F. M., & Niehaves, B. (2020). Creativity loading– please wait! Investigating the relationship between interruption, mind wandering and creativity. In *Proceedings of the 53rd Hawaii International Conference on System Sciences.*
10. Weinert, T., Thiel de Gafenco, M., & Börner, N. (2021). Fostering interaction in higher education with deliberate design of interactive learning videos. In *International Conference on Information Systems.*
11. Conrad, C., & Newman, A. (2021). Measuring mind wandering during online lectures assessed with EEG. *Frontiers in Human Neuroscience, 455.*
12. Krusienski, D. J., Sellers, E. W., McFarland, D. J., Vaughan, T. M., & Wolpaw, J. R.: Toward enhanced P300 speller performance. *Journal of Neuroscience Methods, 167*(1), 15–21.
13. Müller-Putz, G. R., Riedl, R., & Wriessnegger, S. (2015). Electroencephalography (EEG) as a research tool in the information systems discipline: Foundations, measurement, and applications. *Communications of the Association for Information Systems, 37*(1), 46.
14. Gramfort, A., Luessi, M., Larson, E., Engemann, D. A., Strohmeier, D., Brodbeck, C., Parkkonen, L., & Hämäläinen, M. S. (2014). MNE software for processing MEG and EEG data. *Neuroimage, 86,* 446–460.
15. Pedregosa, F., Varoquaux, G., Gramfort, A., Michel, V., Thirion, B., Grisel, O., Blondel, M., Prettenhofer, P., Weiss, R., Dubourg, V., Vanderplas, J., Passos, A., Cournapeau, D., Brucher, M., Perrot, M., & Duchesnay, E. (2011). Scikit-learn: Machine learning in python. *Journal of Machine Learning Research, 12,* 2825–2830.
16. Hunter, J. D. (2007). Matplotlib: A 2D graphics environment. *Computing in Science & Engineering, 9*(3), 90–95.

Measurement of Heart Rate and Heart Rate Variability: A Review of NeuroIS Research with a Focus on Applied Methods

Fabian J. Stangl and René Riedl

Abstract Neuro-Information-Systems (NeuroIS) research contributes to a better understanding of cognitive and affective processes related to the development, adoption, and use of digital technologies. Among others, heart rate (HR) and heart rate variability (HRV) can be used to measure physiological states—more specifically, autonomic nervous system (ANS) activity. Based on a previous systematic literature review in which we surveyed the existing NeuroIS literature on HR and HRV (Stangl and Riedl, 2022 [1]), in the current paper we review completed empirical studies with a focus on the papers' methodological aspects. Thus, this review provides methodological insights to advance the research on HR and HRV with a focus on NeuroIS research.

Keywords Heart rate (HR) · Heart rate variability (HRV) · NeuroIS · Methodological practices · Methodological review

1 Introduction

Neuro-Information-Systems (NeuroIS) research contributes to a better understanding of users' cognitive and affective processes that explain why and how certain effects occur in the use of information and communication technologies [2, 3]. Heart rate (HR) and heart rate variability (HRV) are two important physiological indicators which are relevant to IS research (e.g., [2–5]). To advance the field of HR and

---

The original version of this chapter was revised. Belated corrections have been incorporated. The correction to this chapter can be found at https://doi.org/10.1007/978-3-031-13064-9_36

F. J. Stangl (✉) · R. Riedl
University of Applied Sciences Upper Austria, Steyr, Austria
e-mail: fabian.stangl@fh-steyr.at

R. Riedl
e-mail: rene.riedl@fh-steyr.at

R. Riedl
Johannes Kepler University Linz, Linz, Austria

F. D. Davis et al. (eds.), *Information Systems and Neuroscience*,
Lecture Notes in Information Systems and Organisation 58,
https://doi.org/10.1007/978-3-031-13064-9_28

HRV in NeuroIS research from a methods perspective, this paper provides insights from a methodological perspective, which refers according to Chong and Reinders [6] to "*a type of systematic secondary research (i.e., research synthesis) which focuses on summarising the state-of-the-art methodological practices of research in a substantive field or topic*" (p. 2).

The foundation of this paper is a recently published systematic literature review which surveyed the existing NeuroIS literature on HR and HRV with a focus on measurement based on wearable devices [1]. However, this current paper goes beyond this original article by analyzing methodological aspects of existing empirical literature in a more substantive way. Thus, the main objective of the current paper is to support the mainstream IS researcher by analyzing the previous research in detail from a methods perspective. Based on the results of our systematic analysis of the NeuroIS literature published in peer-reviewed academic journals and conference proceedings in our previous work [1], we address the following research question: *How were the existing empirical NeuroIS studies that used HR and HRV measurements methodologically designed?* More specifically, our analyses reveal aspects related to measurement focus, measurement task, measurement method, measurement metrics, research setting, and study participants.

The remainder of this paper is structured as follows: The following Sect. 2 outlines methodological fundamentals of HR and HRV, focusing on methods commonly used to determine HRV. The knowledge presented in this section summarizes major HR and HRV insights, with the main goal to provide brief guidance for mainstream IS researchers in choosing suitable HRV measurement methods for their own research. Section 3 describes the research methodology consisting of a brief overview of our literature search and the paper selection process. Results are presented in Sect. 4. Finally, in Sect. 5, we make concluding remarks and address implications for future research.

2 Methodological Fundamentals of HR and HRV

The human heart is a remarkable organ with four unique morphological and functional chambers [7]. However, the heart of a healthy individuum beats irregularly [8] due to variations in cardiovascular variables such as heart rate, arterial blood pressure, and stroke volume [9], which may affect brain activity and subsequently sensory and cognitive performance [10]. Also, it is possible that information processing in the brain affects HR and HRV, signifying the relationship between the central nervous system (here with a focus on the brain) and the autonomic nervous system (ANS) [11]. Moreover, this variability of heartbeats may be affected by the temporal variation and magnitude variation of the pulse wave [12]. This means that even if the pulse is constant, the heartbeats do not follow each other according to a constant length of time, which consequently leads to variable intervals between the heartbeats of healthy subjects [13]. HRV is the oscillation in the interval between consecutive heartbeats as well as the oscillations between consecutive instantaneous HRs [14].

The variation in time intervals between successive heartbeats is referred to as sinus RR intervals over time, hereafter *RR interval* [15], and can reflect changes of the ANS (e.g., [16, 17]). In this context, *NN intervals* refer to RR intervals adjusted for ectopic beats (i.e., heartbeat is too fast or too slow due to depolarizations of the sinoatrial node) [14]. However, a higher variation of the intervals between the heartbeats indicates a greater the ability of the ANS to regulate itself [18]. For further details on HR fundamentals and measurement, such as information on how ANS regulates bodily functions, we refer the reader to Stangl and Riedl [1].

Various methods can be used to measure and consequently analyze the variations between heartbeats, such as time domain, frequency domain, time–frequency domain, and nonlinear methods [19]. Although the methods share some common characteristics, they are not considered directly equivalent [20]. We briefly summarize the main characteristics of the methods commonly used to determine HRV in the remainder of this section.

Time Domain Method. This relatively simple method to assess autonomic function [21] can be applied to measure consecutive HR samples or sequences of RR interval durations [22]. Measures recommended for this method include SDNN and HRV triangle index (both estimate total HRV), RMSSD (estimates short-term components of HRV), and SDANN (estimates long-term components of HRV) [14]. As an example, Umetani et al. [23] investigated the effects of age and gender effects on the normal range of time domain HRV over nine decades in healthy subjects. For further methodological information to perform time domain methods, we refer the reader to [14

Frequency Domain Method. This method, which is also known as power spectrum analysis, decomposes the HR signal into its frequency components and quantifies them in terms of their relative intensity [24], providing insight on how power (variance) distributes as a function of frequency [14]. Among the measures recommended for this method is the oscillatory pattern, which characterizes the spectral profile of short-term variability in heart rate and arterial pressure associated with vasomotor and respiratory activity [25]. The oscillatory pattern can be classified as ultra low frequency (ULF, ≤ 0.003 Hz), very low frequency (VLF, from 0.003 to 0.04 Hz), low frequency (LF, from 0.04 to 0.15 Hz), and high frequency (HF, >0.15–0.4 Hz) oscillations [14]. This method provides insights into the neural control of the cardiovascular system embedded in the time series of biological signals [25]. For further methodological information to perform frequency domain methods, we refer the reader to [14, pp. 1045–1048].

Time–Frequency Method. This method combines elements of time domain and the frequency domain methods. Specifically, it consists of the frequency components of the frequency domain method and considers the sequences of RR intervals between the consecutive heartbeats of the time domain method [26]. For example, Konok et al. [27] used specific time domain (i.e., SDNN) and frequency domain metrics (i.e., VLF, LF, HF, LF/HF) in their research to measure sympathetic and parasympathetic activity of the ANS. Another method is the calculation of the continuous wavelet transform, which can provide insight into ANS activity and its control mechanism via the calculation of time–frequency maps of time-varying coherence [28].

Nonlinear Method. Due to the spontaneity and adaptability of the heart beat regulation, methods in the time and frequency domain often cannot adequately capture the complex dynamics of the heart [29]. Voss et al. [30] argue that nonlinear methods can contribute to HRV analysis because they can describe beat-to-beat dynamics and renormalized entropy, which compares the complexity of power spectra at a normalized energy level. Also, a combination of the different methods is possible. As an example, Gao et al. used a combination of time domain, frequency domain, and nonlinear metrics to analyze the fatigue state of agricultural workers [31]. However, methods recommended for nonlinear methods include correlation measure, divergent fluctuation analysis, approximate entropy, and sample entropy, even though some of these methods are not strictly nonlinear [19]. For further methodological information to perform nonlinear methods, we refer the reader to [14, p. 1050].

3 Review Methodology

The starting point for our review was a recently published review by Riedl et al. [32], which investigated the development of the NeuroIS research field during the period 2008–2017. For the subsequent literature search, we considered peer-reviewed publications included in the Senior Scholars' Basket of the Association for Information Systems (AIS), existing NeuroIS Retreat conference proceedings as well as further academic journals and AIS conferences. For our literature search, we used generic terms that represent the NeuroIS field and terms representing the various methods of measuring HR and HRV highlighted in the publications. The review process was based on existing recommendations for conducting literature searches [33–35], and it comprised a literature base of 23 completed empirical studies which were published in the period January 2011 to 2021. For further details on the review methodology, please see [1].

The following exclusion criteria (EC) were applied to select papers for further analyses.

EC1: We excluded papers if, based on current knowledge, the used device is not recommended for HR and HRV measurement for consumer, clinical, or research purposes according to validation studies (i.e., we therefore excluded [36–39]).

EC2: We excluded papers if the validity of the measurement method used for HR and HRV measurement could not be ensured continuously (i.e., we therefore excluded [40]).

Finally, we merged overlapping papers that used the same research methodology (i.e., identical research setting) with same research objective and overlapping research outcome in each completed empirical study. In particular, we merged Adam et al. [41, 42] as well as Lutz et al. [43, 44], considering the more comprehensive paper version for further analysis (i.e., [42, 44]). As a result of the application of the exclusion criteria and the merging of papers, we ended-up with 16 publications that were included in our review, including 12 peer-reviewed journal papers and 4 peer-reviewed conference proceedings papers.

4 Review Results

In this section, we present the main findings of our literature review, guided by our research question. As indicated in Table 1, we analyzed six metrics: measurement focus, measurement task, measurement method, measurement metrics, research setting, and study participants.

Measurement Focus. To get an overview of the measurement focus, Table 2 shows details with the corresponding references. Specifically, we found the following results: Out of the 16 papers, the measurement focus was on HR in 9 papers (56%), on HRV in 3 papers (19%), and on both in 4 papers (25%).

Measurement Task. To get an overview of the various tasks used to measure HR and HRV in NeuroIS research, Table 3 classifies the tasks with the corresponding measurement purpose of the $N = 16$ HR and HRV publications along the identified measurement focus.

Measurement Method. Based on the analysis of $N = 16$ publications, we identified three different measurement methods to perform HRV measurements in NeuroIS research (see Table 4). Specifically, we found the following results: Out of the 7 papers that used methods to calculate HRV, time domain methods were used in 4 papers (57%), time–frequency methods were used in 2 papers (29%), and frequency domain method was used in 1 paper (14%). Notably, the 2 papers that used a time–frequency method (i.e., [27, 55]) each used a combination of specific time domain and frequency domain metrics in their research.

Measurement Metrics. The analysis of $N = 16$ publications revealed that eight different metrics related to the HRV calculation have been used in NeuroIS research (see Table 5). Specifically, we found the following results: Out of the 7 papers that used methods to calculate HRV, 4 different metrics were used to perform a time domain method (i.e., Bavesky Stress Index, NN50 Index, RMSSD, SDNN), and 4 different metrics were used to perform a frequency domain method (i.e., VLF, LF,

Table 1 Overview of research question and metrics

Research question	Metrics
How were the existing empirical NeuroIS studies that used HR and HRV measurements methodologically designed?	• Measurement focus • Measurement task • Measurement method • Measurement metrics • Research setting • Study participants

Table 2 Measurement focus in HR and HRV papers in reviewed publications

Measurement focus	References	Sum	Percentage (%)
HR	[42, 45–52]	9	56
HRV	[44, 53, 54]	3	19
Both	[27, 55–57]	4	25

Table 3 Task with measurement purpose in HR and HRV papers in reviewed publications

Measurement focus	Task with measurement purpose	References
HR	Bidders' arousal during auction	[42]
	Bidders' arousal during auction to increase performance	[45]
	Participants' humor appraisal of content	[46]
	Participants' perceived anxiety when being unable to answer mobile phone calls while performing cognitive tasks	[47]
	Participants' arousal while playing a pattern recognition game	[48]
	Participants' arousal when various IT events happen while writing an essay	[49]
	Participants' arousal during computer mediated and face to face communication	[50]
	Bidders' levels of arousal during bid submission in auctions	[51]
	Participants' arousal and valence when reading news from a mobile device, listening to pre-recorded audio track of someone reading the news, and listening to an actual person presenting the news, i.e., who reads out loud	[52]
HRV	Participants' affective information processing when reading real and fake news	[44]
	Participants' perceived stress during daily life	[53]
	Participants' arousal while doing computer-simulated tasks with different demand level manipulations	[54]
Both (HR + HRV)	Participants' perceived anxiety when being separated from the mobile phone while doing cognitive tasks	[27]
	Participants' effect of respiration when performing tasks on a PC and on a mobile device	[55]
	Participants' perceived stress when doing a public speaking task	[56]
	Participants' perceived cognitive absorption while doing an Enterprise Resource Planning (ERP) system training session	[57]

Table 4 Measurement methods in HR and HRV papers in reviewed publications

Measurement method	References	Sum	Percentage (%)
Time domain method	[44, 53, 56, 57]	4	57
Time–frequency method	[27, 55]	2	29
Frequency domain method	[54]	1	14

Table 5 Overview and description of HRV metrics in reviewed publications

HRV metric	Description	References
Bavesky stress index	A HRV metric reflecting cardiovascular system stress, whereat high SI values indicate lowered HR variability and increased sympathetic activity	[66, 67]
HF	Power of the signal in the high-frequency range (0.15–0.4 Hz)	[14]
LF	Power of the signal in the low-frequency range (0.04–0.15 Hz)	[14]
LF/HF	Ratio LF [ms^2]/HF [ms^2]	[14]
NN50 index	Number of consecutive RR intervals that differ more than 50 ms	[62]
RMSSD	Square root of the mean of the sum of the squares of differences between adjacent NN intervals	[14]
SDNN	Standard deviation of all NN intervals	[14]
VLF	Power of the signal in the very low-frequency range (0.003–0.04 Hz)	[14]

HF, LF/HF). Interestingly, there are more than 70 published metrics for calculating HRV in the scientific literature [19]. In fact, Smith et al. [58] surveyed the existing literature on published methods of HRV analysis for very short-term analysis and identified 136 methods, 70 of which were unique and appropriate for very short-term HRV analysis. Bravi et al. [59] identified more than 70 HRV analysis techniques currently used in the biomedical field to analyze physiological signals in the clinical setting. In this context, it is worth mentioning that Konok et al. [27] and Tozman et al. [54] used the analysis software from Kubios Oy[1] to calculate HRV. This software is considered as an advanced and easy to use HRV analysis software [60] and has also been used in other scientific publications (e.g., [61–65]).

Research Setting. Based on the analysis of N = 16 HR and HRV publications, we identified two different research settings to perform heart-related measurements (see Table 6). With the exception of Buettner et al. [53], all studies were conducted in the laboratory or laboratory-like settings. With this research setting, *"the researcher can control to a great extent, the laboratory can block or minimize certain stimuli hat might interfere with observing the effects of the variables of interest"* [68, p. 435]. For an overview of the tools used in NeuroIS research with a discussion of the strengths and weaknesses of each method per research setting, we refer IS researchers to Riedl and Léger [5, pp. 47–72]; for a more detailed discussion of methods used in cognitive neuroscience, we refer the reader to Senior et al. [69].

Study Participants. The sample sizes of the completed empirical studies (N = 16) ranged from 17 to 851 participants with a median of 66. Female participation ranged from 19% to 73% with a median of 49% (note that information on female participation was not provided in 3 out of 16 studies). Participants' age range was

---

[1] Kubios Oy, https://www.kubios.com (accessed on March 13, 2022).

Table 6 Research setting in HR and HRV papers in reviewed publications

Research setting	References	Sum	Percentage (%)
Laboratory setting	[27, 42, 44–52, 54–57]	15	94
Natural setting	[53]	1	6

from 18 to 55 with a median age of 23 years (note that information on age was not provided in 10 out of 16 studies). When analyzing sample sizes per measurement focus, we found the following sample sizes: HR (73, based on 9 studies; please note that one study (i.e., [45]) reports two laboratory experiments with different sample sizes), HRV (42, based on 3 studies), and HR *and* HRV (51, based on 4 studies). Most of the studies (12 out of 17 studies) relied on student subjects. One study recruited university staff in addition to students (i.e., [46]). One study used professionals from a hospital (i.e., [55]). Finally, two studies (i.e., [49, 53]) did not provide further information on study population. Table 7 summarized details on study participants.

Table 7 Study participant information in reviewed publications (N = 16)

ID	Participants	Male	Female	Age	Mean age	Study population
[27]	93	42	51	18–26	21	Students
[42]	91	79	17	–	22.64	Students
[44]	42	–	–	–	–	Students
[45]	36	24	12	20–28	23.39	Students
	68	52	16	18–27	22.06	Students
[46]	25	13	12	20–35	26.3	Students and staff
[47]	40	27%	73%	–	21.1	Students
[48]	156	76	80	–	–	Students
[49]	103	–	–	–	–	–
[50]	73	21	52	–	–	Students
[51]	103	80	23	–	–	Students
[52]	17	5	12	20–30	22.47	Students
[53]	851	490	361	–	43.7	–
[54]	18	12	6	–	19.23	Students
[55]	28	12	16	19–55	29	Professionals
[56]	66	33	33	20–33	24.17	Students
[57]	36	–	–	–	–	Students

5   Implications and Concluding Remarks

From a methodological perspective, HR and HRV can be measured with different measurement methods. We surmise that technological progress will create even more opportunities for researchers working in this field. For example, there are studies showing that HR and HRV can be measured with novel methods, such as smart clothing [70, 71], smart speaker [72], or through the ear canal [73]. To the best of our knowledge, such measurement methods are not yet used in NeuroIS research to collect HR and HRV data. However, in addition to the capabilities of measurement methods for data acquisition, it is essential for NeuroIS researchers to evaluate the data collection method against the general quality criteria for measurement methods in psychometrics and psychophysiology. In this context, Riedl et al. [74] proposed six quality criteria to be considered in NeuroIS research, which are outlined in Table 8.

Based on the analysis of the state-of-the-art methodological practices, we can draw two major implications for NeuroIS research on HR and HRV. *First*, there is no scientific standardized state-of-the-art methodological practice for calculating HRV. As outlined, Smith et al. [58] identified 70 unique HRV methods for very short-term (30 beat) analysis. Additionally, Bravi et al. [59] identified more than 70 HRV analyses techniques suitable for clinical applications. Although the number of completed empirical studies using HRV as a metric is small and hence definitive conclusions are hard to draw, standardized reporting of the methods is fundamental, especially to conduct potential replication studies; in this context, consider the following statement: "*[replications provide] an even better test of the validity of a phenomenon*" [75, p. 521]. However, regardless of the numerous methods used to calculate HRV and given the possible consequences from the replication crisis point of view, this finding

Table 8 Quality criteria for measurement methods in NeuroIS research

Criterion	Description
Reliability	The extent to which a measurement instrument is free of measurement error, and therefore yields the same results on repeated measurement of the same construct
Validity	The extent to which a measurement instrument measures the construct that it purports to measure
Sensitivity	A property of a measure that describes how well it differentiates values along the continuum inherent in a construct
Diagnosticity	A property of a measure that describes how precisely it captures a target construct as opposed to other constructs
Objectivity	The extent to which research results are independent from the investigator and reported in a way so that replication is possible
Intrusiveness	The extent to which a measurement instrument interferes with an ongoing task, thereby distorting the investigated construct

*Source*Riedl et al. [74, p. xxix]

regarding the calculation of HRV is remarkable. Indeed, such a scientifically reliable and valid metric for HRV assessment would allow research studies to be better compared with each other and, together with large prospective longitudinal studies, would further contribute "*to determine the sensitivity, specificity, and predictive value of HRV in the identification of individuals at risk for subsequent morbid and mortal events*" [14, p. 1060]. NeuroIS researchers should therefore provide methodological details of the empirical work (e.g., online appendices on the journal's website or at least during the review process), as the variety of research design decisions, including possible effects of methodological aspects such as the metric used to calculate HRV, may affect the confirmation and/or rejection of the research hypothesis [74].

Second, the range of study participants in completed empirical studies of HR and HRV in NeuroIS research is extensive. Indeed, the study with the smallest number of study participants was 17 [52] and the study with the largest number of study participants was 851 [53]. To have a benchmark, the average sample size in NeuroIS studies between 2008 and 2017 is 45 study participants, ranging from a minimum of 5 to a maximum of 166 with a median of 30 study participants [32]. To estimate the effect size and the required sample size, a power analysis could be performed prior to data collection, for instance with G*Power [76]. Notably, we found that 3 papers (i.e., [27, 47, 55]) have performed such an analysis in advance. From an ethical perspective, such an analysis would subsequently also prevent studies from having either too little or too much statistical power [77].

This paper reviewed completed HR and HRV empirical studies with a focus on the papers' methodological aspects, centered on the following research question: *How were the existing empirical NeuroIS studies that used HR and HRV measurements methodologically designed?* Specifically, we analyzed measurement focus (see Table 2), measurement task (see Table 3), measurement method (see Table 4), measurement metrics (see Table 5), research setting (see Table 6), and study participants (see Table 7). Together with our recently published systematic literature review on a related topic [1], the aim of the current paper is to contribute to HR and HRV research in NeuroIS by advancing the field from a methodological perspective.

Acknowledgements This research was funded by the Austrian Science Fund (FWF) as part of the project "Technostress in Organizations" (project number: P 30865) at the University of Applied Sciences Upper Austria.

References

1. Stangl, F. J., & Riedl, R. (2022). Measurement of heart rate and heart rate variability with wearable devices: A systematic review. In *Proceedings of the 17th International Conference on Wirtschaftsinformatik*.
2. Riedl, R., Davis, F. D., Banker, R. D., & Kenning, P. H. (2017). *Neuroscience in information systems research: Applying knowledge of brain functionality without neuroscience tools.* Springer, Cham. https://doi.org/10.1007/978-3-319-48755-7
3. Riedl, R., Banker, R. D., Benbasat, I., Davis, F. D., Dennis, A. R., Dimoka, A., Gefen, D., Gupta, A., Ischebeck, A., Kenning, P. H., Müller-Putz, G. R., Pavlou, P. A., Straub, D. W., vom Brocke, J., & Weber, B. (2010). On the foundations of NeuroIS: Reflections on the Gmunden Retreat 2009. *Communications of the Association for Information Systems, 27*, 243–264. https://doi.org/10.17705/1CAIS.02715
4. Dimoka, A., Davis, F. D., Gupta, A., Pavlou, P. A., Banker, R. D., Dennis, A. R., Ischebeck, A., Müller-Putz, G. R., Benbasat, I., Gefen, D., Kenning, P. H., Riedl, R., vom Brocke, J., & Weber, B. (2012). On the use of neurophysiological tools in IS research: Developing a research agenda for NeuroIS. *MIS Quarterly, 36*, 679–702. https://doi.org/10.2307/41703475
5. Riedl, R., & Léger, P.-M. (2016). *Fundamentals of NeuroIS: Information systems and the brain.* Springer, Heidelberg. https://doi.org/10.1007/978-3-662-45091-8
6. Chong, S. W., & Reinders, H. (2021). A methodological review of qualitative research syntheses in CALL: The state-of-the-art. *System, 103*, 102646. https://doi.org/10.1016/j.system.2021.102646
7. Litviňuková, M., Talavera-López, C., Maatz, H., Reichart, D., Worth, C. L., Lindberg, E. L., Kanda, M., Polanski, K., Heinig, M., Lee, M., Nadelmann, E. R., Roberts, K., Tuck, L., Fasouli, E. S., DeLaughter, D. M., McDonough, B., Wakimoto, H., Gorham, J. M., Samari, S., ... Teichmann, S. A. (2020). Cells of the adult human heart. *Nature, 588*, 466–472. https://doi.org/10.1038/s41586-020-2797-4
8. Chialvo, D. R. (2002). Unhealthy surprises. *Nature, 419*, 263–263. https://doi.org/10.1038/419263a
9. Appel, M. L., Berger, R. D., Saul, J. P., Smith, J. M., & Cohen, R. J. (1989). Beat to beat variability in cardiovascular variables: Noise or music? *Journal of the American College of Cardiology, 14*, 1139–1148. https://doi.org/10.1016/0735-1097(89)90408-7
10. Park, H.-D., Correia, S., Ducorps, A., & Tallon-Baudry, C. (2014). Spontaneous fluctuations in neural responses to heartbeats predict visual detection. *Nature Neuroscience, 17*, 612–618. https://doi.org/10.1038/nn.3671
11. Riedl, R. (2013). On the biology of technostress: Literature review and research agenda. *ACM SIGMIS Database: The DATA BASE for Advances in Information Systems, 44*, 18–55. https://doi.org/10.1145/2436239.2436242
12. Watanabe, T., Hoshide, S., & Kario, K. (2022). Noninvasive method to validate the variability of blood pressure during arrhythmias. *Hypertension Research, 45*, 530–532. https://doi.org/10.1038/s41440-021-00835-7
13. Lim, G. B. (2022). Pacing with respiratory sinus arrhythmia improves outcomes in heart failure. *Nature Reviews Cardiology, 19*, 209–209. https://doi.org/10.1038/s41569-022-00681-1
14. Task Force of the European Society of Cardiology the North American Society of Pacing Electrophysiology (1996). Heart rate variability: Standards of measurement, physiological interpretation, and clinical use. *Circulation, 93*, 1043–1065. https://doi.org/10.1161/01.CIR.93.5.1043
15. Cowan, M. J. (1995). Measurement of heart rate variability. *Western Journal of Nursing Research, 17*, 32–48. https://doi.org/10.1177/019394599501700104
16. Alabdulgader, A., McCraty, R., Atkinson, M., Dobyns, Y., Vainoras, A., Ragulskis, M., & Stolc, V. (2018). Long-term study of heart rate variability responses to changes in the solar and geomagnetic environment. *Science and Reports, 8*, 2663. https://doi.org/10.1038/s41598-018-20932-x

17. Shi, K., Steigleder, T., Schellenberger, S., Michler, F., Malessa, A., Lurz, F., Rohleder, N., Ostgathe, C., Weigel, R., & Koelpin, A. (2021). Contactless analysis of heart rate variability during cold pressor test using radar interferometry and bidirectional LSTM networks. *Science and Reports, 11*, 3025. https://doi.org/10.1038/s41598-021-81101-1

18. Pizzoli, S. F. M., Marzorati, C., Gatti, D., Monzani, D., Mazzocco, K., & Pravettoni, G. (2021). A meta-analysis on heart rate variability biofeedback and depressive symptoms. *Science and Reports, 11*, 6650. https://doi.org/10.1038/s41598-021-86149-7

19. Quintana, D. S., Alvares, G. A., & Heathers, J. A. J. (2016). Guidelines for reporting articles on psychiatry and heart rate variability (GRAPH): Recommendations to advance research communication. *Translational Psychiatry, 6*, e803–e803. https://doi.org/10.1038/tp.2016.73

20. Chemla, D., Young, J., Badilini, F., Maison-Blanche, P., Affres, H., Lecarpentier, Y., & Chanson, P. (2005). Comparison of fast Fourier transform and autoregressive spectral analysis for the study of heart rate variability in diabetic patients. *International Journal of Cardiology, 104*, 307–313. https://doi.org/10.1016/j.ijcard.2004.12.018

21. Kleiger, R. E., Stein, P. K., Bosner, M. S., & Rottman, J. N. (1992). Time domain measurements of heart rate variability. *Cardiology Clinics, 10*, 487–498. https://doi.org/10.1016/S0733-865 1(18)30230-3

22. Malik, M. (1997). Time-domain measurement of heart rate variability. *Cardiac Electrophysiology Review, 1*, 329–334.

23. Umetani, K., Singer, D. H., McCraty, R., & Atkinson, M. (1998). Twenty-four hour time domain heart rate variability and heart rate: Relations to age and gender over nine decades. *Journal of the American College of Cardiology, 31*, 593–601. https://doi.org/10.1016/S0735-1097(97)005 54-8

24. Őri, Z., Monir, G., Weiss, J., Sayhouni, X., & Singer, D. H. (1992). Heart rate variability: Frequency domain analysis. *Cardiology Clinics, 10*, 499–533. https://doi.org/10.1016/S0733-8651(18)30231-5

25. Montano, N., Porta, A., Cogliati, C., Costantino, G., Tobaldini, E., Casali, K. R., & Iellamo, F. (2009). Heart rate variability explored in the frequency domain: A tool to investigate the link between heart and behavior. *Neuroscience and Biobehavioral Reviews, 33*, 71–80. https://doi.org/10.1016/j.neubiorev.2008.07.006

26. Bozhokin, S. V., & Suslova, I. B. (2014). Analysis of non-stationary HRV as a frequency modulated signal by double continuous wavelet transformation method. *Biomedical Signal Processing and Control, 10*, 34–40. https://doi.org/10.1016/j.bspc.2013.12.006

27. Konok, V., Pogány, Á., & Miklósi, Á. (2017). Mobile attachment: Separation from the mobile phone induces physiological and behavioural stress and attentional bias to separation-related stimuli. *Computers in Human Behavior, 71*, 228–239. https://doi.org/10.1016/j.chb.2017.02.002

28. Keissar, K., Davrath, L. R., Akselrod, S. (2009). Coherence analysis between respiration and heart rate variability using continuous wavelet transform. *Philosophical Transactions of the Royal Society A: Mathematical, Physical and Engineering Sciences, 367*, 1393–1406. https://doi.org/10.1098/rsta.2008.0273

29. Shi, B., Wang, L., Yan, C., Chen, D., Liu, M., & Li, P. (2019). Nonlinear heart rate variability biomarkers for gastric cancer severity: A pilot study. *Science and Reports, 9*, 13833. https://doi.org/10.1038/s41598-019-50358-y

30. Voss, A., Kurths, J., Kleiner, H. J., Witt, A., & Wessel, N. (1995). Improved analysis of heart rate variability by methods of nonlinear dynamics. *Journal of Electrocardiology, 28*, 81–88. https://doi.org/10.1016/S0022-0736(95)80021-2

31. Gao, R., Yan, H., Duan, J., Gao, Y., Cao, C., Li, L., & Guo, L. (2022). Study on the nonfatigue and fatigue states of orchard workers based on electrocardiogram signal analysis. *Science and Reports, 12*, 4858. https://doi.org/10.1038/s41598-022-08705-z

32. Riedl, R., Fischer, T., Léger, P.-M., Davis, F. D. (2020). A decade of NeuroIS research: Progress, challenges, and future directions. *ACM SIGMIS Database: The DATA BASE for Advances in Information Systems, 51*, 13–54. https://doi.org/10.1145/3410977.3410980

33. Webster, J., & Watson, R.T. (2002). Analyzing the past to prepare for the future: Writing a literature review. *MIS Quarterly, 26*, xiii–xxiii.

34. Kitchenham, B., & Charters, S. (2007). *Guidelines for performing systematic literature reviews in software engineering version 2.3* (EBSE Technical Report EBSE-2007-01). Keele University and University of Durham.

35. vom Brocke, J., Simons, A., Niehaves, B., Riemer, K., Plattfaut, R., & Cleven, A. (2009). Reconstructing the giant: On the importance of rigour in documenting the literature search process. In S. Newell, E. A. Whitley, N. Pouloudi, J. Wareham & L. Mathiassen (Eds.), *Proceedings of the 17th European Conference on Information Systems* (pp. 2206–2217).

36. Gaskin, J., Jenkins, J., Meservy, T., Steffen, J., & Payne, K. (2017). Using wearable devices for non-invasive, inexpensive physiological data collection. In *Proceedings of the 50th Hawaii International Conference on System Sciences* (pp. 597–605). https://doi.org/10.24251/HICSS.2017.072

37. Jensen, M., Piercy, C., Elzondo, J., Twyman, N., Valacich, J., Miller, C., Lee, Y.-H., Dunbar, N., Bessarabova, E., Burgoon, J., Adame, B., & Wilson, S. (2016). Exploring failure and engagement in a complex digital training game: A multi-method examination. *AIS Transactions on Human-Computer Interaction, 8*, 1–20. https://doi.org/10.17705/1thci.08102

38. Öksüz, N., Biswas, R., Shcherbatyi, I., & Maass, W. (2018). Measuring biosignals of overweight and obese children for real-time feedback and predicting performance. In F. D. Davis, R. Riedl, J. vom Brocke, P.-M. Léger & A. B. Randolph (Eds.), *Information Systems and Neuroscience: Gmunden Retreat on NeuroIS 2017* (Vol. 25, pp. 185–193). LNISO. Springer, Cham. https://doi.org/10.1007/978-3-319-67431-5_21

39. Sheng, H., & Joginapelly, T. (2012). Effects of web atmospheric cues on users' emotional responses in e-commerce. *AIS Transactions on Human-Computer Interaction, 4*, 1–24. https://doi.org/10.17705/1thci.00036

40. Fischer, T., & Riedl, R. (2020). Technostress measurement in the field: A case report. In F. D. Davis, R. Riedl, J. vom Brocke, P.-M. Léger, A. B. Randolph & T. Fischer (Eds.), *Information Systems and Neuroscience: NeuroIS Retreat 2020* (Vol. 43, pp. 71–78). LNISO. Springer, Cham. https://doi.org/10.1007/978-3-030-60073-0_9

41. Adam, M. T. P., Gamer, M., Krämer, J., & Weinhardt, C. (2011). Measuring emotions in electronic markets. In *Proceedings of the 32nd International Conference on Information Systems*.

42. Adam, M. T. P., Krämer, J., & Weinhardt, C. (2012). Excitement up! Price down! Measuring emotions in Dutch auctions. *International Journal of Electronic Commerce, 17*, 7–40. https://doi.org/10.2753/JEC1086-4415170201

43. Lutz, B., Adam, M. T. P., Feuerriegel, S., Pröllochs, N., & Neumann, D. (2020). Affective information processing of fake news: Evidence from NeuroIS. In F. D. Davis, R. Riedl, J. vom Brocke, P.-M. Léger, A. B. Randolph & T. Fischer (Eds.), *Information Systems and Neuroscience: NeuroIS Retreat 2019* (Vol. 32, pp. 121–128). LNISO. Springer, Cham. https://doi.org/10.1007/978-3-030-28144-1_13

44. Lutz, B., Adam, M. T. P., Feuerriegel, S., Pröllochs, N., & Neumann, D. (2020). Identifying linguistic cues of fake news associated with cognitive and affective processing: Evidence from NeuroIS. In F. D. Davis, R. Riedl, J. vom Brocke, P.-M. Léger, A. B. Randolph & T. Fischer (Eds.), *Information Systems and Neuroscience: NeuroIS Retreat 2020* (Vol. 43, pp. 16–23). LNISO. Springer, Cham. https://doi.org/10.1007/978-3-030-60073-0_2

45. Astor, P. J., Adam, M. T. P., Jerčić, P., Schaaff, K., & Weinhardt, C. (2013). Integrating biosignals into information systems: A NeuroIS tool for improving emotion regulation. *Journal of Management Information Systems, 30*, 247–278. https://doi.org/10.2753/MIS0742-1222300309

46. Barral, O., Kosunen, I., & Jacucci, G. (2018). No need to laugh out loud: Predicting humor appraisal of comic strips based on physiological signals in a realistic environment. *ACM Transactions on Computer-Human Interaction, 24*, 1–29. https://doi.org/10.1145/3157730

47. Clayton, R. B., Leshner, G., & Almond, A. (2015). The extended iSelf: The impact of iPhone separation on cognition, emotion, and physiology. *Journal of Computer-Mediated Communication, 20*, 119–135. https://doi.org/10.1111/jcc4.12109

48. Hariharan, A., Dorner, V., & Adam, M. T. P. (2017). Impact of cognitive workload and emotional arousal on performance in cooperative and competitive interactions. In F. D. Davis, R. Riedl, J. vom Brocke, P.-M. Léger & A. B. Randolph (Eds.), *Information Systems and Neuroscience: Gmunden Retreat on NeuroIS 2016* (Vol. 16, pp. 35–42). LNISO. Springer, Cham. https://doi.org/10.1007/978-3-319-41402-7_5

49. Ortiz de Guinea, A., & Webster, J. (2013). An investigation of information systems use patterns: Technological events as triggers, the effect of time, and consequences for performance. *MIS Quarterly, 37*, 1165–1188. https://doi.org/10.25300/MISQ/2013/37.4.08

50. Shalom, J. G., Israeli, H., Markovitzky, O., & Lipsitz, J. D. (2015). Social anxiety and physiological arousal during computer mediated vs. face to face communication. *Computers in Human Behavior, 44*, 202–208. https://doi.org/10.1016/j.chb.2014.11.056

51. Teubner, T., Adam, M. T. P., & Riordan, R. (2015). The impact of computerized agents on immediate emotions, overall arousal and bidding behavior in electronic auctions. *Journal of the Association for Information Systems, 16*, 838–879. https://doi.org/10.17705/1jais.00412

52. Walla, P., & Lozovic, S. (2020). The effect of technology on human social perception: A multi-methods NeuroIS pilot investigation. In F. D. Davis, R. Riedl, J. vom Brocke, P.-M. Léger, A. B. Randolph & T. Fischer (Eds.), *Information Systems and Neuroscience: NeuroIS Retreat 2019* (Vol. 32, pp. 63–71). LNISO. Springer, Cham. https://doi.org/10.1007/978-3-030-28144-1_7

53. Buettner, R., Bachus, L., Konzmann, L., & Prohaska, S. (2019). Asking both the user's heart and its owner: Empirical evidence for substance dualism. In F. D. Davis, R. Riedl, J. vom Brocke, P.-M. Léger & A. B. Randolph (Eds.), *Information Systems and Neuroscience: NeuroIS Retreat 2018* (Vol. 29, pp. 251–257). LNISO. Springer, Cham. https://doi.org/10.1007/978-3-030-01087-4_30

54. Tozman, T., Magdas, E. S., MacDougall, H. G., & Vollmeyer, R. (2015). Understanding the psychophysiology of flow: A driving simulator experiment to investigate the relationship between flow and heart rate variability. *Computers in Human Behavior, 52*, 408–418. https://doi.org/10.1016/j.chb.2015.06.023

55. Cipresso, P., Serino, S., Gaggioli, A., Albani, G., Mauro, A., & Riva, G. (2015). Psychometric modeling of the pervasive use of Facebook through psychophysiological measures: Stress or optimal experience? *Computers in Human Behavior, 49*, 576–587. https://doi.org/10.1016/j.chb.2015.03.068

56. Kothgassner, O. D., Felnhofer, A., Hlavacs, H., Beutl, L., Palme, R., Kryspin-Exner, I., & Glenk, L. M. (2016). Salivary cortisol and cardiovascular reactivity to a public speaking task in a virtual and real-life environment. *Computers in Human Behavior, 62*, 124–135. https://doi.org/10.1016/j.chb.2016.03.081

57. Léger, P.-M., Davis, F. D., Cronan, T. P., & Perret, J. (2014). Neurophysiological correlates of cognitive absorption in an enactive training context. *Computers in Human Behavior, 34*, 273–283. https://doi.org/10.1016/j.chb.2014.02.011

58. Smith, A.-L., Owen, H., & Reynolds, K. J. (2013). Heart rate variability indices for very short-term (30 beat) analysis. Part 1: Survey and toolbox. *Journal of Clinical Monitoring and Computing, 27*, 569–576. https://doi.org/10.1007/s10877-013-9471-4

59. Bravi, A., Longtin, A., & Seely, A. J. E. (2011). Review and classification of variability analysis techniques with clinical applications. *Biomedical Engineering Online, 10*, 90. https://doi.org/10.1186/1475-925X-10-90

60. Tarvainen, M. P., Niskanen, J.-P., Lipponen, J. A., Ranta-aho, P. O., & Karjalainen, P. A. (2014). Kubios HRV—Heart rate variability analysis software. *Computer Methods and Programs in Biomedicine, 113*, 210–220. https://doi.org/10.1016/j.cmpb.2013.07.024

61. Lipponen, J. A., & Tarvainen, M. P. (2019). A robust algorithm for heart rate variability time series artefact correction using novel beat classification. *Journal of Medical Engineering & Technology, 43*, 173–181. https://doi.org/10.1080/03091902.2019.1640306

62. Niskanen, J.-P., Tarvainen, M. P., Ranta-aho, P. O., & Karjalainen, P. A. (2004). Software for advanced HRV analysis. *Computer Methods and Programs in Biomedicine, 76*, 73–81. https://doi.org/10.1016/j.cmpb.2004.03.004

63. Tarvainen, M. P., Ranta-aho, P. O., & Karjalainen, P. A. (2002). An advanced detrending method with application to HRV analysis. *IEEE Transactions on Biomedical Engineering, 49*, 172–175. https://doi.org/10.1109/10.979357

64. Baumgartner, D., Fischer, T., Riedl, R., & Dreiseitl, S. (2019). Analysis of heart rate variability (HRV) feature robustness for measuring technostress. In F. D. Davis, R. Riedl, J. vom Brocke, P.-M. Léger & A. B. Randolph (Eds.), *Information Systems and Neuroscience: NeuroIS Retreat 2018* (Vol. 29, pp. 221–228). LNISO. Springer, Cham.

65. Machado, A. V., Pereira, M. G., Souza, G. G. L., Xavier, M., Aguiar, C., de Oliveira, L., & Mocaiber, I. (2021). Association between distinct coping styles and heart rate variability changes to an acute psychosocial stress task. *Science and Reports, 11*, 24025. https://doi.org/10.1038/s41598-021-03386-6

66. Baevsky, R. M. (2002). Analysis of heart rate variability in space medicine. *Human Physiology, 28*, 202–213.

67. Baevsky, R. M., & Chernikova, A. G. (2017). Heart rate variability analysis: Physiological foundations and main methods. *Cardiometry*, 66–76. https://doi.org/10.12710/cardiometry.2017.10.6676

68. Fiske, D. W., & Fiske, S. T. (2005). Laboratory studies. In K. Kempf-Leonard (Ed.), *Encyclopedia of Social Measurement* (pp. 435–439). Elsevier. https://doi.org/10.1016/B0-12-369398-5/00407-2

69. Senior, C., Russell, T., & Gazzaniga, M. S. (2009). *Methods in mind*. The MIT Press, Cambridge.

70. Li, P., Zhao, L., Jiang, Z., Yu, M., Li, Z., Zhou, X., & Zhao, Y. (2019). A wearable and sensitive graphene-cotton based pressure sensor for human physiological signals monitoring. *Science and Reports, 9*, 14457. https://doi.org/10.1038/s41598-019-50997-1

71. Libanori, A., Chen, G., Zhao, X., Zhou, Y., & Chen, J. (2022). Smart textiles for personalized healthcare. *Nature Electronics, 5*, 142–156. https://doi.org/10.1038/s41928-022-00723-z

72. Wang, A., Nguyen, D., Sridhar, A. R., & Gollakota, S. (2021). Using smart speakers to contactlessly monitor heart rhythms. *Communications Biology, 4*, 319. https://doi.org/10.1038/s42003-021-01824-9

73. Goverdovsky, V., von Rosenberg, W., Nakamura, T., Looney, D., Sharp, D. J., Papavassiliou, C., Morrell, M. J., & Mandic, D. P. (2017). Hearables: Multimodal physiological in-ear sensing. *Science and Reports, 7*, 6948. https://doi.org/10.1038/s41598-017-06925-2

74. Riedl, R., Davis, F. D., & Hevner, A. R. (2014). Towards a NeuroIS research methodology: Intensifying the discussion on methods, tools, and measurement. *Journal of the Association for Information Systems, 15*, I–XXXV. https://doi.org/10.17705/1jais.00377

75. Pashler, H., & Harris, C. R. (2012). Is the replicability crisis overblown? Three arguments examined. *Perspectives on Psychological Science, 7*, 531–536. https://doi.org/10.1177/1745691612463401

76. Faul, F., Erdfelder, E., Buchner, A., & Lang, A.-G. (2009). Statistical power analyses using G*Power 3.1: Tests for correlation and regression analyses. *Behavior Research Methods, 41*, 1149–1160. https://doi.org/10.3758/BRM.41.4.1149

77. Maxwell, S. E., & Kelley, K. (2011). Ethics and sample size planning. In A. T. Panter & S. K. Sterba (Eds.), *Handbook of Ethics in Quantitative Methodology* (pp. 159–184). Routledge, New York.

Measurement of Heart Rate and Heart Rate Variability in NeuroIS Research: Review of Empirical Results

Fabian J. Stangl and René Riedl

Abstract Heart rate (HR) and heart rate variability (HRV) measurements are important indicators of an individual´s physiological state. In Neuro-Information-Systems (NeuroIS) research, these physiological parameters can be used to measure autonomic nervous system (ANS) activity, contributing to a better understanding of cognitive and affective processes in the Information Systems (IS) discipline. Based on a previous systematic literature analysis (Stangl and Riedl, 2022 [1]), in the present paper we review the major empirical results of NeuroIS research based on HR and HRV measurement. Thus, this review provides insights to advance the field from an empirically grounded perspective.

Keywords Heart Rate (HR) · Heart Rate Variability (HRV) · NeuroIS · Autonomic Nervous System (ANS) · Empirical Studies

1 Introduction

Heart rate (HR) and heart rate variability (HRV) are important indicators of an individual's physiological state [2], and both indicators can be used to measure autonomic nervous system (ANS) activity. HRV, defined as the variability of the time interval between heartbeats [3, 4], is affected by autonomic, respiratory, circulatory, endocrine, and mechanical influences of the heart [5]. HR and HRV measurements are critical in various Information Systems (IS) research domains, such as digital health [6–8], human–computer interaction [9–11], or technostress [12, 13]. Moreover, HR and HRV are also used in various other domains. For example, Penzel et al.

F. J. Stangl (✉) · R. Riedl
University of Applied Sciences Upper Austria, Steyr, Austria
e-mail: fabian.stangl@fh-steyr.at

R. Riedl
e-mail: rene.riedl@fh-steyr.at

R. Riedl
Johannes Kepler University Linz, Linz, Austria

© The Author(s), under exclusive license to Springer Nature Switzerland AG 2022   285
F. D. Davis et al. (eds.), *Information Systems and Neuroscience*,
Lecture Notes in Information Systems and Organisation 58,
https://doi.org/10.1007/978-3-031-13064-9_29

[14] created a model for the temporal sequence of sleep stages and wake states based on HR data. HRV is also a potential predictor of physical morbidity and mortality [3]. Hence, the dynamics of HR, along with the related HRV measure, are important indicators to assess human health, as well as physiological and psychological phenomena.

In the present paper, we provide insights on HR and HRV in NeuroIS research from an empirically grounded perspective. NeuroIS contributes to a better understanding of users' cognitive and affective processes that explain why and how certain effects occur in the use of digital technologies (e.g., [10, 11, 15, 16]). The foundation of the present paper is a recently published systematic literature review which surveyed the existing NeuroIS literature on HR and HRV with a focus on measurement based on wearable devices [1]. However, the current paper goes beyond this original review by presenting major empirical research findings. Thus, the objective is to support the mainstream IS researcher by analyzing the previous research on HR and HRV in the NeuroIS literature to develop a "big-picture" view. Based on the results of our original analysis of the NeuroIS literature published in peer-reviewed academic journals and conference proceedings [1], in the present paper we address the following research question: *What are the main research findings of existing empirical NeuroIS studies which use HR and HRV measurement?*

The remainder of this paper is structured as follows: The following Sect. 2 outlines the genesis and progress of HR measurement. The knowledge presented in this section summarizes important milestones with the aim of providing a brief overview. Section 3 describes the research methodology of this literature review. Results are presented in Sect. 4. Finally, in Sect. 5, we make concluding remarks and address implications for future IS research based on HR and HRV.

2   Genesis of Heart Rate Measurement and Opportunities for NeuroIS

The origin for HR measurement was first documented by Stephen Hales in 1733 [17], who reported variations in blood levels when he performed invasive measurements of arterial blood pressure by inserting a cannula into the carotid artery of a mare and connecting it to a glass cylinder (see Fig. 1). However, the groundwork of today's HR measurement was laid in 1906, when Cremer [18] succeeded for the first time in recording an electrocardiogram (ECG) of a fetus. In 1958, the clinical relevance of HRV was investigated for the first time by Hon [19] and later by Hon and Lee (e.g., [20, 21]), using electrodes to measure and monitor instantaneous fetal HR. One of the major conclusions was that electronically based evaluations provide a more accurate indication of changes in HR interbeat intervals in fetal distress during normal and abnormal labor than was possible with clinical methods at the time [19–21]. As a result, since the 1970s, more studies have been performed to investigate the relationship between cardiovascular parameters and pathophysiological situations.

For example, Murray et al. [22] used the electrocardiographic computer technique to examine resting HRV to detect possible autonomic nerve damage in diabetic patients. Also, Wolf et al. [23] found that patients who required treatment for acute myocardial infarction (i.e., heart attack), had lower HRs compared with patients without myocardial infarction, using a 60-s ECG rhythm strip. Since the 1980s, with the rise of computer technologies for processing biological signals, HRV measurement studies have become more established in clinical practice. Indeed, studies from this period with Holter recordings showed a correlation between a myocardial infarction and a lower HRV (e.g., [24, 25]). In addition, the possibilities of computer technologies allowed specific experiments to be performed to measure ANS activity. Akselrod et al. [26] introduced, for example, power spectral analysis of HR fluctuations in trained dogs to quantitatively assess the function of beat-to-beat cardiovascular control.

Fig. 1 Stephen Hales measuring blood pressure in a horse (*Source* [32, p. 69])

Since the introduction of the first pulse measuring device by Anastasios Filadelfeus [27] in the nineteenth century, research in the field of HR and HRV has made progress, especially due to the development of technological capabilities; see Fig. 2 for the measurement device with a description, Fig. 3 for the application of the measurement device, and Fig. 4 for exemplary recordings of the measurement device, all introduced by Anastasios Filadelfeus. Indeed, besides ECG as a measurement method, there are other methodological approaches in HR and HRV research, such as measurement with a chest strap or light-based technology for measuring blood volume pulse (please see [1] for an overview of methodological approaches identified in NeuroIS literature to perform heart-related measurements). Such a technological progress has many implications, and also important ones for IS research as it enables different research avenues. As an example, Riedl and Léger [10] outline that biological states and processes could be used in real time to develop adaptive systems that can positively impact practical and relevant outcome variables such as health, well-being, satisfaction, and productivity. Conroy et al. [28] used HR, among other indicators, to continuously monitor changing health status to aid military workforce readiness during the COVID-19 pandemic.

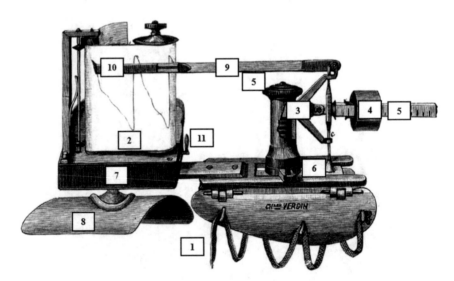

Fig. 2 Measurement device introduced by Anastasios Filadelfeus (*Adapted representation from source* [27, p. 363]). Number 1: The part to be attached to the wrist; Number 2: The recording of the measurement on the cardboard tape; Number 3: A square box made of perforated aluminum, which supports the weight of the box (Number 5) that creates the pressure on the artery; Number 4: A regulator that applies the pressure on the artery; Number 5: A slider to adjust all the above parts (Number 3–4) of the measurement device and also to adjust the levers and scales (Number 9–10) parallel through the perforation of the diagram; Number 6: A disc of woven wire of similar shape; Number 7: The measurement unit; Number 8: A curved plate that moves around the center and rests on a on the arm; Number 9 and 10: Levers and scales; Number 11: A handle for starting or stopping the measurement

Fig. 3 Application of the measurement device introduced by Anastasios Filadelfeus (*Source* [27, p. 366])

Fig. 4 Exemplary recording with the measurement device introduced by Anastasios Filadelfeus (*Source* [27, p. 365]).

From a NeuroIS perspective, it can be expected that the ongoing technological progress will enable cheaper and more mobile measurements in the future [29]. This would also lead to further field studies in the future. Indeed, Riedl et al. [30] found that empirical NeuroIS studies are predominantly conducted in the laboratory. Moreover, further technological progress would advance NeuroIS research by encouraging to collaborate with industry, especially in the context of IS design, by providing neuroscience-based advice for the design of digital products and services [31].

3 Review Methodology

The starting point for our literature analysis was a recently published review by Riedl et al. [30], which investigated the development of the NeuroIS research field during the period 2008–2017. For the subsequent literature search, we considered peer-reviewed publications included in the Senior Scholars' Basket of the Association for Information Systems (AIS), existing NeuroIS Retreat conference proceedings as well as further academic journals and AIS conferences. For our literature search, we used generic terms that represent the NeuroIS field and terms representing the various methods of measuring HR and HRV highlighted in the publications. The review process was based on existing recommendations for conducting literature searches [33–35]. In total, our literature base consists of 23 completed empirical studies which were published in the period January 2011–2021. For further details on the review methodology, please see [1].

Here, we review the major research results of completed empirical studies. The following exclusion criteria (EC) were applied to select papers for further analyses.

EC1: We excluded papers if, based on current knowledge, the used device is not recommended for HR and HRV measurement for consumer, clinical, or research purposes according to validation studies (i.e., we therefore excluded [36–39]).

EC2: We excluded papers if the HR and HRV measurement could not be ensured continuously (i.e., we therefore excluded [40] due to unexpected technical problems in the collection of HR data).

Finally, we merged overlapping papers that used the same research methodology (i.e., identical experimental design) with same research objective and overlapping research outcome in each completed empirical study. In particular, we merged Adam et al. [41, 42] as well as Lutz et al. [43, 44], considering the more comprehensive paper version for further analysis (i.e., [42, 44]). As a result of the application of the exclusion criteria and the merging process, we ended-up with 16 publications that were included in our review, including 12 peer-reviewed journal papers and 4 peer-reviewed conference proceedings papers.

4 Review Results

In this section, we present the main findings of our literature review, guided by our RQ. An overview of the content of this section related to our RQ and the corresponding metrics is provided in Table 1.

Measurement Purpose. To get an overview of the measured constructs, Table 2 shows details with the corresponding references. Specifically, we found the following results: Out of the 16 papers, the measurement purpose (i.e., investigated construct) was arousal in 7 papers (44%), anxiety (i.e., anxiety caused by separation from mobile

Table 1 Overview of research question and metrics

Research question	Metrics
What are the main research findings of existing empirical NeuroIS studies which use HR and HRV measurement?	• Measurement purpose • Insights

Table 2 Measurement purpose (Constructs) in HR and HRV papers in reviewed publications

Measurement purpose (constructs)	Reference(s)	Sum	Percentage (%)
Arousal	[42, 53–58]	7	44
Perceived anxiety	[45, 46]	2	13
Perceived stress	[47, 48]	2	13
Affective information processing	[44]	1	6
Arousal and valence	[49]	1	6
Humor appraisal	[50]	1	6
Perceived cognitive absorption	[51]	1	6
Effect of respiration	[52]	1	6

phone) in 2 papers (13%) [45, 46], stress in 2 papers (13%) [47, 48], and five further papers investigated the following constructs (6% in each case): affective information processing [44], arousal *and* valence [49], humor appraisal [50], cognitive absorption [51], and voluntary and autonomic effect of respiration [52].

Insights. Based on the analysis of N = 16 HR and HRV publications, we identified several findings. In the following, we present example findings. Our full analysis is presented in Table 3. HR can be affected by emotional stimuli. For example, Adam et al. [41] found that, on average, HR is increases during Dutch auctions if compared to a baseline condition, while HR is higher in fast auctions than in slow ones [42]. Interestingly, in an experiment by Ortiz de Guinea and Webster [55] HR decreased after an unexpected IT-mediated interruption. In another study, Cipresso et al. [52] demonstrated that HRV decreased with increasing anxiety. Further, Léger et al. [51] found that participants with lower HR and higher HRV (interpreted as "experiencing a lower level of cognitive effort when using the software", p. 279), reported a higher level of cognitive absorption than others. Table 3 summarizes the main insights of the empirical studies on HR and HRV in NeuroIS research.

5   Implications and Concluding Remarks

The aim of this article was to review the main empirical results of HR and HRV in NeuroIS research. HR and HRV are indicators of an individual's physiological state that also reveal information about ANS activity. For example, HRV as biomarker can provide significant information about the individual´s health status [59]. Due to the ongoing technological progress, a variety of research opportunities for NeuroIS can be expected in the future, and with them more opportunities for researchers in the field of HR and HRV research. As an example, Firstbeat Bodyguard 2 (BG2) from the Firstbeat Technologies Oy[1] is a device for monitoring HR beat by beat in real time [60]. The device consists of an ECG with electrodes and determines HRV with

---

[1] Firstbeat Technologies Oy, https://www.firstbeat.com (accessed on April 19, 2022).

Table 3 Insights in HR and HRV papers in reviewed publications

Metric	Effect	Description	References
HR	Increasing	In fast auctions (if compared to slow ones)	[42]
		If interaction with mobile phone is not possible	[45]
		When separated from mobile phone (partial support)	[46]
		During public speaking	[48]
		When watching and listening to a story face-to face (highest in comparison than when reading a written story or listening to a recorded audio of a person reading a story without another person present at each time)	[49]
		When processing humorous content	[50]
		During arousal	[52]
		In auction game (if compared to baseline)	[53]
		When decision performance is low	[53]
		During poor performance	[54]
		In socially anxious individuals in communication processes	[56]
		When interacting with human counterparts in auction game (if compared to computerized ones)	[57]
HR	Decreasing	During low level of perceived cognitive absorption	[51]
		During unexpected IT-mediated interruption	[55]
HRV	Increasing	When cognitive effort is low during software usage	[51]
HRV	Decreasing	During cognitive processing in reading and writing tasks	[44]
		When stress is increasing	[47]
		When sympathetic activity is increasing	[48]
		When anxiety is increasing	[52]
		When task demand is increasing	[58]

a resolution of 1 ms, resulting in accurate data compared to a clinical ECG derived HRV [61]. Such technological possibilities allow HR and HRV to be measured and analyzed in real time in everyday life, as Fig. 5 shows. From a NeuroIS perspective, such measurement devices might enable long term field studies for various measurement purposes to measure ANS activity for HR *and* HRV research.

However, in addition to the collection of accurate data, also the processing and analysis of HR and HRV data is important in NeuroIS research. Indeed, HR and HRV can be affected by various stimuli and consequently increase or decrease (see Table 3 for insights from NeuroIS research). As an example, Fig. 6 shows how respiration can affect the variation in time intervals between the successive heartbeats (i.e., *RR interval*) [62]. From an empirical perspective, NeuroIS researchers must ensure that the results of HR and HRV research are independent of the investigator and are reported in a way that allows replication [63]. Otherwise, failure to adhere to objective ideals in data collection, processing, analysis, and reporting may lead to false-positive results [64, 65]. For various empirical aspects that NeuroIS scholars need to consider

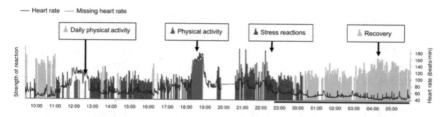

Fig. 5 HRV measurement in an individual's daily life with Firstbeat BG2 (*Source* [60, p. 1]). HR increases during physical activities and due to stress responses (i.e., HRV decreases) while HR decreases during recovery phase (i.e., HRV increases)

for objectivity, such as developing strategies to handle outliers, baseline measurements to assess study participant response, subject-experimenter interactions (e.g., greeting, attaching sensors, or announcing task instructions), or maintaining constant room temperature, humidity, and light conditions during experiments, we refer the reader to the concept of objectivity discussed by Riedl et al. [63, pp. xix–xxvi].

Based on the analysis of the results, we see two major implications for NeuroIS research on HR and HRV. *First*, HR and HRV can be used in many settings with different measurement purposes. In other words, HR and HRV can be used as physiological indicator for many psychological phenomena which are also of high relevance in IS research, including constructs such as arousal, stress, anxiety, or cognitive absorption. Specifically, our analysis of the existing NeuroIS literature on HR and HRV identified eight different measurement purposes (constructs) in 16 papers. In general, HR and HRV are important indicators of an individual's physiological state and can provide an index of ANS activity [2, 39] to objectively measure a person's ability to respond to environmental demands [66]. However, our analyses indicate that HR was predominantly investigated as indicator to measure ANS activity in NeuroIS research (see Table 3). As the research on HRV is still in a relatively nascent stage [1], we hope that the results of this review of major empirical results in NeuroIS research will encourage researchers to continue to advance this research field with empirical work and insights. Specifically, HRV as a metric for physiological measures related to ANS activity [63] is suitable for various measurement purposes (see Table 2).

Second, we found some unexpected results in HR and HRV research. Indeed, the HR surprisingly decreased after an unexpected IT-mediated interruption in the

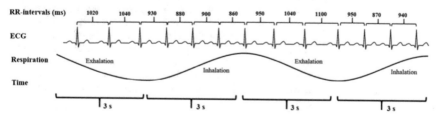

Fig. 6 Influence of respiration on HR measurement (*Source* [60, p. 3]). HR increases during inhalation (HRV decreases), while HR decreases during exhalation (HRV increases)

experiment by Ortiz de Guinea and Webster [55]. Stress-related disturbances such as unexpected IT-mediated interruptions typically lead to an increase in HR and a decrease in HRV [67], as the response towards stressors activates specific cognitive and affective processes and underlying brain mechanisms [68] which come along with activation of the sympathetic division of the ANS, including the release of stress hormones such as adrenaline, noradrenaline, and cortisol [13, 68]. In this context, it is critical for IS scholars to consider research findings in other disciplines. As an example, Moghtadaei et al. [69] indicate that the regulation of hormones plays a crucial role in cardiovascular functions in both normal and diseased states. From an IS perspective, such research findings are important because they can advance the understanding of IS constructs (e.g., arousal, stress) and the development of IS and corresponding user interfaces that improve the efficiency of human–computer interaction [9] and contribute to consideration of humanistic values such as health, well-being, and satisfaction [10, 11].

Table 4 outlines our review results for our research question: *What are the main research findings of existing empirical NeuroIS studies which use on HR and HRV measurement?* In this paper, we reviewed major research findings of completed empirical studies, focusing on the measurement purpose (i.e., the investigated constructs and phenomena) and concrete insights. Together with our recently published systematic literature review [1], the aim of this review was to contribute to HR and HRV research in NeuroIS by advancing the field from an empirically grounded perspective.

Acknowledgements This research was funded by the Austrian Science Fund (FWF) as part of the project "Technostress in Organizations" (project number: P 30865) at the University of Applied Sciences Upper Austria.

Table 4 Overview of implications for research

Metric	Implications
Measurement purpose	Our review indicates that that arousal as a construct has been studied primarily in NeuroIS research on HR and HRV (see Table 2). HR as a measure, along with the related HRV measure, contributes to a better understanding of the various ANS activities, which are of great importance for IS research, such as perceived anxiety or stress
Insights	Our analysis of insights from HR and HRV papers in reviewed publications revealed that HR has been predominantly studied as an indicator to measure ANS activity in NeuroIS research (see Table 3). However, with the rise of computer technologies, the importance of HRV as a measure in clinical practice has become widely established since its first recognition in 1958. We suspect that technological progress will enable long-term field studies for various measurement purposes to measure HR and HRV as an indicator of ANS activity for NeuroIS research

References

1. Stangl, F. J., & Riedl, R. (2022). Measurement of heart rate and heart rate variability with wearable devices: A systematic review. In *Proceedings of the 17th International Conference on Wirtschaftsinformatik.*
2. Patron, E., Messerotti Benvenuti, S., Favretto, G., Gasparotto, R., & Palomba, D. (2014). Depression and reduced heart rate variability after cardiac surgery: The mediating role of emotion regulation. *Autonomic Neuroscience, 180,* 53–58. https://doi.org/10.1016/j.autneu. 2013.11.004
3. Pizzoli, S. F. M., Marzorati, C., Gatti, D., Monzani, D., Mazzocco, K., & Pravettoni, G. (2021). A meta-analysis on heart rate variability biofeedback and depressive symptoms. *Science and Reports, 11,* 6650. https://doi.org/10.1038/s41598-021-86149-7
4. Agliari, E., Barra, A., Barra, O. A., Fachechi, A., Franceschi Vento, L., & Moretti, L. (2020). Detecting cardiac pathologies via machine learning on heart-rate variability time series and related markers. *Science and Reports, 10,* 8845. https://doi.org/10.1038/s41598-020-64083-4
5. Quintana, D. S., Alvares, G. A., & Heathers, J. A. J. (2016). Guidelines for reporting articles on psychiatry and heart rate variability (GRAPH): Recommendations to advance research communication. *Translational Psychiatry, 6,* e803–e803. https://doi.org/10.1038/tp.2016.73
6. Alavi, A., Bogu, G. K., Wang, M., Rangan, E. S., Brooks, A. W., Wang, Q., Higgs, E., Celli, A., Mishra, T., Metwally, A. A., Cha, K., Knowles, P., Alavi, A. A., Bhasin, R., Panchamukhi, S., Celis, D., Aditya, T., Honkala, A., Rolnik, B., ... Snyder, M. P. (2022). Real-time alerting system for COVID-19 and other stress events using wearable data. *Nature Medicine, 28,* 175–184. https://doi.org/10.1038/s41591-021-01593-2
7. Bayoumy, K., Gaber, M., Elshafeey, A., Mhaimeed, O., Dineen, E. H., Marvel, F. A., Martin, S. S., Muse, E. D., Turakhia, M. P., Tarakji, K. G., & Elshazly, M. B. (2021). Smart wearable devices in cardiovascular care: Where we are and how to move forward. *Nature Reviews Cardiology, 18,* 581–599. https://doi.org/10.1038/s41569-021-00522-7
8. Dunn, J., Kidzinski, L., Runge, R., Witt, D., Hicks, J. L., Schüssler-Fiorenza Rose, S. M., Li, X., Bahmani, A., Delp, S. L., Hastie, T., & Snyder, M. P. (2021). Wearable sensors enable personalized predictions of clinical laboratory measurements. *Nature Medicine, 27,* 1105–1112. https://doi.org/10.1038/s41591-021-01339-0
9. Ayaz, H., Shewokis, P. A., Bunce, S., Izzetoglu, K., Willems, B., & Onaral, B. (2012). Optical brain monitoring for operator training and mental workload assessment. *NeuroImage, 59,* 36–47. https://doi.org/10.1016/j.neuroimage.2011.06.023
10. Riedl, R., & Léger, P.-M. (2016). *Fundamentals of NeuroIS: Information systems and the brain.* Springer, Heidelberg. https://doi.org/10.1007/978-3-662-45091-8
11. Riedl, R., Banker, R. D., Benbasat, I., Davis, F. D., Dennis, A. R., Dimoka, A., Gefen, D., Gupta, A., Ischebeck, A., Kenning, P. H., Müller-Putz, G. R., Pavlou, P. A., Straub, D. W., vom Brocke, J., & Weber, B. (2010). On the foundations of NeuroIS: Reflections on the Gmunden Retreat 2009. *Communications of the Association for Information Systems, 27,* 243–264. https://doi.org/10.17705/1CAIS.02715
12. Baumgartner, D., Fischer, T., Riedl, R., & Dreiseitl, S. (2019). Analysis of heart rate variability (HRV) feature robustness for measuring technostress. In F. D. Davis, R. Riedl, J. vom Brocke, P.-M. Léger & A. B. Randolph (Eds.), *Information Systems and Neuroscience: NeuroIS Retreat 2018* (Vol. 29, pp. 221–228). LNISO. Springer. https://doi.org/10.1007/978-3-030-01087-4_27
13. Riedl, R. (2013). On the biology of technostress: Literature review and research agenda. *ACM SIGMIS Database: The DATA BASE for Advances in Information Systems, 44,* 18–55. https://doi.org/10.1145/2436239.2436242
14. Penzel, T., Kantelhardt, J. W., Lo, C.-C., Voigt, K., & Vogelmeier, C. (2003). Dynamics of heart rate and sleep stages in normals and patients with sleep apnea. *Neuropsychopharmacology, 28,* S48–S53. https://doi.org/10.1038/sj.npp.1300146
15. Riedl, R., Davis, F. D., Banker, R. D., & Kenning, P. H. (2017). *Neuroscience in information systems research: Applying knowledge of brain functionality without neuroscience tools.* Springer, Cham. https://doi.org/10.1007/978-3-319-48755-7

16. Dimoka, A., Davis, F. D., Gupta, A., Pavlou, P. A., Banker, R. D., Dennis, A. R., Ischebeck, A., Müller-Putz Cham, G. R., Benbasat, I., Gefen, D., Kenning, P. H., Riedl, R., vom Brocke, J., & Weber, B. (2012). On the use of neurophysiological tools in IS research: Developing a research agenda for NeuroIS. *MIS Quarterly, 36*, 679–702. https://doi.org/10.2307/41703475

17. Hales, S. (1733). *Statical essays: Containing haemastaticks; or, an account of some hydraulick and hydrostatical experiments made on the blood and blood vessels of animals. Also an account of some experiments on stones in the kidneys and bladder; with an enquiry into the nature of those anomalous concretions. To which is added, an appendix, containing observations and experiments relating to several subjects in the first volume, the greater part of which were read at several meetings before the Royal Society. With an index to both volumes. Vol. II.* Printed for W. Innys and R. Manby, at the west-end of St. Paul's, and T. Woodward, at the Half-Moon between Temple-Gate, Fleetstreet, London. https://doi.org/10.5962/bhl.title.106596

18. Cremer, M. (1906). Über die direkte Ableitung der Aktionsströme des menschlichen Herzens vom Oesophagus und über das Elektrokardiogramm des Fötus. *Münchener Medizinische Wochenschrift., 53*, 811–813.

19. Hon, E. H. (1958). The electronic evaluation of the fetal heart rate: Preliminary report. *American Journal of Obstetrics and Gynecology, 75*, 1215–1230. https://doi.org/10.1016/0002-937 8(58)90707-5

20. Hon, E. H., & Lee, S. T. (1963). Electronic evaluations of the fetal heart rate. VIII. Patterns preceding fetal death, further observations. *American Journal of Obstetrics and Gynecology, 87*, 814–826.

21. Hon, E. H., & Lee, S. T. (1964). Averaging techniques in fetal electrocardiography. *Medical Electronics & Biological Engineering, 2*, 71–76. https://doi.org/10.1007/BF02474362

22. Murray, A., Ewing, D. J., Campbell, I. W., Neilson, J. M., & Clarke, B. F. (1975). RR interval variations in young male diabetics. *Heart, 37*, 882–885. https://doi.org/10.1136/hrt.37.8.882

23. Wolf, M. M., Varigos, G. A., Hunt, D., & Sloman, J. G. (1978). Sinus arrhythmia in acute myocardial infarction. *Medical Journal of Australia, 2*, 52–53. https://doi.org/10.5694/j.1326-5377.1978.tb131339.x

24. Kleiger, R. E., Miller, J. P., Bigger, J. T., & Moss, A. J. (1987). Decreased heart rate variability and its association with increased mortality after acute myocardial infarction. *American Journal of Cardiology, 59*, 256–262. https://doi.org/10.1016/0002-9149(87)90795-8

25. Malik, M., Farrell, T., Cripps, T., & Camm, A. J. (1989). Heart rate variability in relation to prognosis after myocardial infarction: Selection of optimal processing techniques. *European Heart Journal, 10*, 1060–1074. https://doi.org/10.1093/oxfordjournals.eurheartj.a059428

26. Akselrod, S., Gordon, D., Ubel, F. A., Shannon, D. C., Berger, A. C., & Cohen, R. J. (1981). Power spectrum analysis of heart rate fluctuation: A quantitative probe of beat-to-beat cardiovascular control. *Science, 213*, 220–222. https://doi.org/10.1126/science.6166045

27. Filadelfeus, A. (1899). Ανακάλυψη νέου Σφυγμομετρογραφοσ. In Arsenis I. A. (Ed.), *Ποικίλη Στοά* (Vol. 14, pp. 363–366). University of Cyprus, Athen. http://hdl.handle.net/10797/26181

28. Conroy, B., Silva, I., Mehraei, G., Damiano, R., Gross, B., Salvati, E., Feng, T., Schneider, J., Olson, N., Rizzo, A. G., Curtin, C. M., Frassica, J., & McFarlane, D. C. (2022). Real-time infection prediction with wearable physiological monitoring and AI to aid military workforce readiness during COVID-19. *Science and Reports, 12*, 3797. https://doi.org/10.1038/s41598-022-07764-6

29. Loos, P., Riedl, R., Müller-Putz, G. R., vom Brocke, J., Davis, F. D., Banker, R. D., & Léger, P.-M. (2010). NeuroIS: Neuroscientific approaches in the investigation and development of information systems. *Business & Information Systems Engineering, 2*, 395–401. https://doi.org/10.1007/s12599-010-0130-8

30. Riedl, R., Fischer, T., Léger, P.-M., & Davis, F. D.: A decade of NeuroIS research: Progress, challenges, and future directions. *ACM SIGMIS Database: The DATA BASE for Advances in Information Systems, 51*, 13–54. https://doi.org/10.1145/3410977.3410980

31. vom Brocke, J., Hevner, A., Léger, P. M., Walla, P., & Riedl, R. (2020). Advancing a NeuroIS research agenda with four areas of societal contributions. *European Journal of Information Systems, 29*, 9–24. https://doi.org/10.1080/0960085X.2019.1708218

32. Buchanan, J. W. (2013). The history of veterinary cardiology. *Journal of Veterinary Cardiology, 15*, 65–85. https://doi.org/10.1016/j.jvc.2012.12.002
33. Webster, J., & Watson, R. T. (2002). Analyzing the past to prepare for the future: Writing a literature review. *MIS Quarterly, 26*, xiii–xxiii.
34. Kitchenham, B., & Charters, S. (2007). *Guidelines for performing systematic literature reviews in software engineering version 2.3* (EBSE Technical Report EBSE-2007–01). Keele University and University of Durham.
35. vom Brocke, J., Simons, A., Niehaves, B., Riemer, K., Plattfaut, R., & Cleven, A. (2009) Reconstructing the giant: On the importance of rigour in documenting the literature search process. In S. Newell, E. A. Whitley, N. Pouloudi, J. Wareham & L. Mathiassen (Eds.), *Proceedings of the 17th European Conference on Information Systems* (pp. 2206–2217).
36. Gaskin, J., Jenkins, J., Meservy, T., Steffen, J., & Payne, K. (2017). Using wearable devices for non-invasive, inexpensive physiological data collection. In *Proceedings of the 50th Hawaii International Conference on System Sciences* (pp. 597–605). https://doi.org/10.24251/HICSS.2017.072
37. Jensen, M., Piercy, C., Elzondo, J., Twyman, N., Valacich, J., Miller, C., Lee, Y.-H., Dunbar, N., Bessarabova, E., Burgoon, J., Adame, B., & Wilson, S. (2016). Exploring failure and engagement in a complex digital training game: A multi-method examination. *AIS Transactions on Human-Computer Interaction, 8*, 1–20. https://doi.org/10.17705/1thci.08102
38. Öksüz, N., Biswas, R., Shcherbatyi, I., & Maass, W. (2018). Measuring biosignals of overweight and obese children for real-time feedback and predicting performance. In F. D. Davis, R. Riedl, J. vom Brocke, P.-M. Léger & A. B. Randolph (Eds.), *Information Systems and Neuroscience: Gmunden Retreat on NeuroIS 2017* (Vol. 25, pp. 185–193). LNISO. Springer, Cham. https://doi.org/10.1007/978-3-319-67431-5_21
39. Sheng, H., & Joginapelly, T. (2012) Effects of web atmospheric cues on users' emotional responses in e-commerce. *AIS Transactions on Human-Computer Interaction, 4*, 1–24 (2012). https://doi.org/10.17705/1thci.00036
40. Fischer, T., & Riedl, R. (2020). Technostress measurement in the field: A case report. In F. D. Davis, R. Riedl, J. vom Brocke, P.-M. Léger, A. B. Randolph & T. Fischer (Eds.), *Information Systems and Neuroscience: NeuroIS Retreat 2020* (Vol. 43, pp. 71–78). LNISO. Springer, Cham. https://doi.org/10.1007/978-3-030-60073-0_9
41. Adam, M. T. P., Gamer, M., Krämer, J., & Weinhardt, C. (2011). Measuring Emotions in Electronic Markets. In *Proceedings of the 32nd International Conference on Information Systems*.
42. Adam, M. T. P., Krämer, J., & Weinhardt, C. (2012). Excitement up! Price down! Measuring emotions in Dutch auctions. *International Journal of Electronic Commerce, 17*, 7–40. https://doi.org/10.2753/JEC1086-4415170201
43. Lutz, B., Adam, M. T. P., Feuerriegel, S., Pröllochs, N., & Neumann, D. (2019). Affective information processing of fake news: Evidence from NeuroIS. In F. D. Davis, R. Riedl, J. vom Brocke, P.-M. Léger, A. B. Randolph & Fischer, T. (Eds.), *Information Systems and Neuroscience: NeuroIS Retreat 2019* (Vol. 32, pp. 121–128). LNISO. Springer, Cham. https://doi.org/10.1007/978-3-030-28144-1_13
44. Lutz, B., Adam, M. T. P., Feuerriegel, S., Pröllochs, N., & Neumann, D. (2020). Identifying linguistic cues of fake news associated with cognitive and affective processing: Evidence from NeuroIS. In F. D. Davis, R. Riedl, J. vom Brocke, P.-M. Léger, A. B. Randolph & T. Fischer (Eds.), *Information Systems and Neuroscience: NeuroIS Retreat 2020* (Vol. 43, pp. 16–23). LNISO. Springer, Cham. https://doi.org/10.1007/978-3-030-60073-0_2
45. Clayton, R. B., Leshner, G., & Almond, A. (2015). The extended iSelf: The impact of iPhone separation on cognition, emotion, and physiology. *Journal Computer Communications, 20*, 119–135. https://doi.org/10.1111/jcc4.12109
46. Konok, V., Pogány, Á., & Miklósi, Á. (2017). Mobile attachment: Separation from the mobile phone induces physiological and behavioural stress and attentional bias to separation-related stimuli. *Computers in Human Behavior, 71*, 228–239. https://doi.org/10.1016/j.chb.2017.02.002

47. Buettner, R., Bachus, L., Konzmann, L., & Prohaska, S. (2018). Asking both the user's heart and its owner: Empirical evidence for substance dualism. In F. D. Davis, R. Riedl, J. vom Brocke, P.-M. Léger & A. B. Randolph (Eds.), *Information Systems and Neuroscience: NeuroIS Retreat 2018* (Vol. 29, pp. 251–257). LNISO. Springer, Cham. https://doi.org/10.1007/978-3-030-01087-4_30
48. Kothgassner, O. D., Felnhofer, A., Hlavacs, H., Beutl, L., Palme, R., Kryspin-Exner, I., & Glenk, L. M. (2016). Salivary cortisol and cardiovascular reactivity to a public speaking task in a virtual and real-life environment. *Computers in Human Behavior, 62*, 124–135. https://doi.org/10.1016/j.chb.2016.03.081
49. Walla, P., & Lozovic, S. (2020). The effect of technology on human social perception: A multi-methods NeuroIS pilot investigation. In F. D. Davis, R. Riedl, J. vom Brocke, P.-M. Léger, A. B. Randolph & T. Fischer (Eds.), *Information Systems and Neuroscience: NeuroIS Retreat 2019* (Vol. 32, pp. 63–71). LNISO. Springer, Cham. https://doi.org/10.1007/978-3-030-28144-1_7
50. Barral, O., Kosunen, I., & Jacucci, G. (2018). No need to laugh out loud: Predicting humor appraisal of comic strips based on physiological signals in a realistic environment. *ACM Transactions on Computer–Human Interaction, 24*, 1–29. https://doi.org/10.1145/3157730
51. Léger, P.-M., Davis, F. D., Cronan, T. P., & Perret, J. (2014). Neurophysiological correlates of cognitive absorption in an enactive training context. *Computers in Human Behavior, 34*, 273–283. https://doi.org/10.1016/j.chb.2014.02.011
52. Cipresso, P., Serino, S., Gaggioli, A., Albani, G., Mauro, A., & Riva, G. (2015). Psychometric modeling of the pervasive use of Facebook through psychophysiological measures: Stress or optimal experience? *Computers in Human Behavior, 49*, 576–587. https://doi.org/10.1016/j.chb.2015.03.068
53. Astor, P. J., Adam, M. T. P., Jerčić, P., Schaaff, K., & Weinhardt, C. (2013). Integrating biosignals into information systems: A NeuroIS tool for improving emotion regulation. *Journal of Management Information Systems, 30*, 247–278. https://doi.org/10.2753/MIS0742-1222300309
54. Hariharan, A., Dorner, V., & Adam, M. T. P. (2017). Impact of cognitive workload and emotional arousal on performance in cooperative and competitive interactions. In F. D. Davis, R. Riedl, J. vom Brocke, P.-M. Léger & A. B. Randolph (Eds.), *Information Systems and Neuroscience: Gmunden Retreat on NeuroIS 2016* (Vol. 16, pp. 35–42). LNISO. Springer, Cham. https://doi.org/10.1007/978-3-319-41402-7_5
55. Ortiz de Guinea, A., & Webster, J. (2013). An investigation of information systems use patterns: Technological events as triggers, the effect of time, and consequences for performance. *MIS Quarterly, 37*, 1165–1188. https://doi.org/10.25300/MISQ/2013/37.4.08
56. Shalom, J. G., Israeli, H., Markovitzky, O., & Lipsitz, J. D.: Social anxiety and physiological arousal during computer mediated versus face to face communication. *Computers in Human Behavior, 44*, 202–208. https://doi.org/10.1016/j.chb.2014.11.056
57. Teubner, T., Adam, M. T. P., & Riordan, R. (2015). The impact of computerized agents on immediate emotions, overall arousal and bidding behavior in electronic auctions. *Journal of the Association for Information Systems, 16*, 838–879. https://doi.org/10.17705/1jais.00412
58. Tozman, T., Magdas, E. S., MacDougall, H. G., & Vollmeyer, R. (2015). Understanding the psychophysiology of flow: A driving simulator experiment to investigate the relationship between flow and heart rate variability. *Computers in Human Behavior, 52*, 408–418. https://doi.org/10.1016/j.chb.2015.06.023
59. Task Force of the European Society of Cardiology the North American Society of Pacing Electrophysiology. (1996). Heart rate variability: Standards of measurement, physiological interpretation, and clinical use. *Circulation, 93*, 1043–1065. https://doi.org/10.1161/01.CIR.93.5.1043
60. Firstbeat Technologies Ltd. (2014). Stress and recovery analysis method based on 24-hour heart rate variability.
61. Parak, J., & Korhonen, I. (2013). Accuracy of Firstbeat Bodyguard 2 beat-to-beat heart rate monitor.

62. Cowan, M. J. (1995). Measurement of heart rate variability. *Western Journal of Nursing Research, 17*, 32–48. https://doi.org/10.1177/019394599501700104

63. Riedl, R., Davis, F. D., & Hevner, A. R. (2014). Towards a NeuroIS research methodology: Intensifying the discussion on methods, tools, and measurement. *Journal of the Association for Information System, 15*, I–XXXV. https://doi.org/10.17705/1jais.00377

64. Simmons, J. P., Nelson, L. D., & Simonsohn, U. (2011). False-positive psychology: Undisclosed flexibility in data collection and analysis allows presenting anything as significant. *Psychological Science, 22*, 1359–1366. https://doi.org/10.1177/0956797611417632

65. Wiggins, B. J., & Christopherson, C. D. (2019). The replication crisis in psychology: An overview for theoretical and philosophical psychology. *Journal of Theoretical and Philosophical Psychology, 39*, 202–217. https://doi.org/10.1037/teo0000137

66. Francis, H. M., Penglis, K. M., & McDonald, S. (2016). Manipulation of heart rate variability can modify response to anger-inducing stimuli. *Social Neuroscience, 11*, 545–552. https://doi.org/10.1080/17470919.2015.1115777

67. Yates, D. (2021). Heightening the threat. *Nature Reviews Neuroscience, 22*, 4–5. https://doi.org/10.1038/s41583-020-00417-5

68. Riedl, R., Kindermann, H., Auinger, A., & Javor, A. (2012). Technostress from a neurobiological perspective: System breakdown increases the stress hormone cortisol in computer users. *Business & Information Systems Engineering, 4*, 61–69. https://doi.org/10.1007/s12599-012-0207-7

69. Moghtadaei, M., Langille, E., Rafferty, S. A., Bogachev, O., & Rose, R. A. (2017). Altered heart rate regulation by the autonomic nervous system in mice lacking natriuretic peptide receptor C (NPR-C). *Science and Reports, 7*, 17564. https://doi.org/10.1038/s41598-017-17690-7

Investigating Mind-Wandering Episodes While Using Digital Technologies: An Experimental Approach Based on Mixed-Methods

Caroline Reßing, Frederike M. Oschinsky, Michael Klesel, Björn Niehaves, René Riedl, Patrick Suwandjieff, Selina C. Wriessnegger, and Gernot R. Müller-Putz

Abstract In today's fast-paced world, our brain spends almost half of our waking hours distracted from current environmental stimuli, often referred to as mind wandering in the scientific literature. At the same time, people frequently have several hours daily screen time, signifying the ubiquity of digital technologies. Here, we investigate mind wandering while using digital technologies. Measuring mind wandering (i.e., off-task thought), however, comes with challenges. In this research-in-progress paper, we present an experimental approach based on EEG, eye-tracking,

C. Reßing (✉) · F. M. Oschinsky · B. Niehaves
University of Siegen, Siegen, Germany
e-mail: caroline.ressing@uni-siegen.de

F. M. Oschinsky
e-mail: frederike.oschinsky@uni-siegen.de

B. Niehaves
e-mail: bjoern.niehaves@uni-siegen.de

M. Klesel
University of Twente, Enschede, The Netherlands
e-mail: m.klesel@utwente.nl

R. Riedl
University of Applied Sciences Upper Austria, Steyr, Austria
e-mail: rene.riedl@fh-steyr.at

Johannes Kepler University Linz, Linz, Austria

P. Suwandjieff · S. C. Wriessnegger · G. R. Müller-Putz
Graz University of Technology, Graz, Austria
e-mail: suwandjieff@tugraz.at

S. C. Wriessnegger
e-mail: s.wriessnegger@tugraz.at

G. R. Müller-Putz
e-mail: gernot.mueller@tugraz.at

S. C. Wriessnegger · G. R. Müller-Putz
BioTechMed Graz, Graz, Austria

© The Author(s), under exclusive license to Springer Nature Switzerland AG 2022   301
F. D. Davis et al. (eds.), *Information Systems and Neuroscience*,
Lecture Notes in Information Systems and Organisation 58,
https://doi.org/10.1007/978-3-031-13064-9_30

questionnaires, and performance data to measure wind wandering. Our work draws upon the Unusual Uses Task, a widely used task to measure divergent thinking (as a proxy for mind wandering). We describe the experimental setup and discuss initial findings.

Keywords Mind-wandering · EEG · Eye-tracking · Unusual uses task · NeuroIS

1 Introduction

Mind-wandering episodes are described as a shift of attention away from a primary task and toward dynamic, unconstrained, spontaneous thoughts [1]. In times of constant use of digital technologies (hereafter: use of technology), it is almost impossible not to get distracted. Our thoughts frequently and automatically wander back and forth [2]. Killingsworth and Gilbert showed that our minds wander nearly half of our waking time (46.9%), with a tendency to wander to negative content and get stuck in rumination [3]. In contrast to initial research that stresses the negative effects, the more current literature refers to mind wandering as a positive feature and as something purposeful we all experience [4]. Specifically, mind wandering can promote problem-solving and creative skills [5, 6]. Such skills are highly relevant Information Systems (IS) phenomena because creativity is a crucial characteristic for knowledge workers [7].

So-called "Aha moments" let ideas appear true, satisfying, and valuable [8]. Although breakthrough ideas often seem to happen when inventors think about completely unrelated things, there is a paucity of research on technological tools to foster mind-wandering episodes and thus "Aha moments" [9]. Consequently, discovering and manipulating mind wandering while using technology is a critical endeavor. In particular, the development of neuroadaptive systems [10–15] that could induce mind wandering would constitute a breakthrough in research and innovation. Yet, to accomplish this goal, it is necessary to measure mind-wandering episodes in a valid manner.

So far, mind-wandering episodes are mainly studied with experience samples and questionnaires [16]. Given the potential shortcomings of self-reports (e.g., social desirability), we pursue a triangulation approach to achieve more valid measurement. Validity refers to "the extent to which a measurement instrument measures the construct [mind wandering] that it purports to measure" [17, p. 14]. In this context, triangulation is used to increase the robustness and validity of our research findings. The goal is to measure mind-wandering episodes while using technology more accurately compared to the extant literature.

The measurements involved include electroencephalography (EEG), eye-tracking, questionnaires, and creative performance. First, EEG is an established brain imaging tool that non-invasively assesses mind-wandering episodes without interfering with a task [18]. It allows temporal inferences at the millisecond level. Second, eye-tracking acts as a reliable time-critical indicator of visual attention by detecting changes in eye behavior [19]. Three visual mechanisms were observed in the present study: visual uncoupling, perceptual uncoupling, and internal coupling. Third, we present questionnaires to investigate spontaneous as well as deliberate mind-wandering episodes while solving a task [20]. Fourth, creative performance is operationalized based on the number of new and useful answers to a divergent thinking task.

The present work is adapted to a technology-related environment because all steps can only be carried out using a computer and involve a specific reference to technological hardware. The experiment is based on the Unusual Uses Task (UUT) [21, 22] and the work by Baird et al. [23]. The UUT measures divergent thinking. The participants are asked to generate as many unusual uses for everyday life objects as possible, for example, newspapers or headsets. The number and originality of the responses are taken as an index of creative thinking. All in all, the combination of brain data, eye-tracking data, self-reports, and behavioral data promises to shed light on mind-wandering episodes while using technology in an innovative way. Our overall expectation is that if we can design technologies that allow users to let their mind wander, we can increase the likelihood of "Aha moments" and overall creativity.

2 Theoretical Background

2.1 Detecting Mind Wandering with EEG

EEG is a widely used tool for the non-invasive measurement of bioelectrical brain ac-activity, in many areas, from basic sciences to diagnosis and treatment [24]. Here, we use EEG for the detection of mind-wandering episodes [18, 25]. The method enables recording of cognitive processes underlying perception, memory, and attention, among others, by measuring electrical signals generated by the brain (e.g., ideas) on the scalp. The most significant advantage of EEG is its temporal resolution, allowing for detection of complex patterns of neural activity which are a consequence of stimulus perception on a millisecond level [24]. For a detailed comparison of EEG with two other major brain imaging tools (fMRI, fNIRS), please see Table 1 in [26].

Regarding EEG, the first indicator we use is the event-related potential (ERP), which is "a waveform complex resulting from an external stimulus or an event." [24, p. 932]. A participant is repeatedly exposed to a defined number of stimuli (i.e., visual information on a screen), which allows for the measurement of such an evoked potential (EP). This includes amplitudes, positive or negative polarity (P and N), millisecond latency, and scalp distribution. The P-N appears to be attenuated during mind-wandering episodes due to perceptual decoupling [27–29].

The second indicator is to detect oscillatory EEG components varying in different frequency bands. According to Müller-Putz et al. [24, p. 918], these are defined as alpha (8–13 Hz), beta (13–25 Hz), gamma (25–200 Hz), delta (1–4 Hz) and theta (4–8 Hz) bands (note that slight deviations from these bands can be observed in the scientific literature, hence the bands are rough specifications). Studies that examined mind wandering concluded that considering alpha waves is crucial [30]. Alpha waves are primarily found during relaxation [24] and low stimulation [31]. They tend to increase during mental imagery [32] and before solving a creative problem with insight [33]. In addition, higher internal processing demands were found to increase EEG alpha activity in posterior brain regions and the parietal/occipital alpha power could be a neural correlate of mind wandering [34–36].

2.2 Detecting Mind Wandering with Eye-Tracking

Eye-tracking can be used to detect a possible daydream and to draw conclusions about the onset of relevant time points in the EEG signal. Eye-tracking is a consumer-friendly tool for detecting mind-wandering episodes. It can be considered a "time-critical [indicator] of internal attentional demands" that captures eye behavior [19, p. 1]. Mind wandering can increase blink rates and may come along with longer blink durations [37, 38]. Similarly, the number of saccades decreases, indicating less rapid, simultaneous, and voluntary eye movements. In addition, saccade amplitude increases [19] and pupil diameter (PD) increases [39]. In this context, gaze aversion is reduced during cognitive load (i.e., demanding activities) by decreasing the processing of distracting external stimuli to shield internal processes [40, 41]. During mind-wandering episodes, pupillary activity tends to be less guided and more spontaneous [42]. Eye-closing measures are shown to promote performance in creativity tasks [10].

3 Research Design

Experimental design. The participants' brain activity and eye movements were measured while sitting in front of a computer working on a creative task. In addition, a short questionnaire was presented. The experiment took place in the EEG laboratory at Graz University of Technology, which guaranteed freedom from interference. The procedure is non-invasive, painless, and injury-free. The local ethics committee at Medical University Graz approved the study.

Following Baird et al. [23], the experiment was conducted with three groups (rest, demanding task, undemanding task) [22]. The instruction for all participants was to list as many unusual uses as possible for each stimulus (i.e., four objects, for details see next paragraph) and to solve UUT problems. Before the experiment, participants were equipped with EEG (BrainProducts GmbH, Germany) and an eye-tracking device (Pupil Labs, Germany). An ultra-sound sensor system (Zebris) was used to record the thirty-two electrode positions on the participants' head to allow and facilitate later analyses. The impedance of the electrodes was set at 15 k Ohm. The pupil headset was calibrated with a baseline before onset. The participants were instructed to limit movement. Instructions for the measurement procedure were given once participants were seated.

After EEG and eye-tracking montage the experiment started, and the participants solved four UUT tasks (two minutes each). Two images showed non-technological artifacts (bricks, newspaper), and two images showed technological artifacts (headset, computer mouse). The participants verbally expressed their responses, which were then recorded by sound and video. This method reduced motor distortions in the EEG to a minimum.

After completing the UUT problems, participants were randomly assigned to one of three groups (rest, demanding task, undemanding task) following a between-subjects design. Participants in the demanding-task condition performed a 3-back task, while those in the undemanding-task condition performed a 1-back task [11]. In the 1 or 3-back task, participants had to remember whether a given number was the same as 1 or 3 numbers before and confirm or deny this by pressing the left or right key on a keyboard. In the rest condition, participants were asked to sit quietly and look at a grey screen. The intervention lasted twelve minutes. After that, all participants answered a short questionnaire about spontaneous mind wandering during the UUTs [19]. Next, the participants completed the same four UUTs as before (two minutes each). After that, they were released from the EEG and eye-tracking tools. They were asked to complete a longer questionnaire about spontaneous and deliberate mind-wandering episodes in their everyday life. Finally, they were thanked and debriefed.

In sum, the participants' cognitive processes while using technology were recorded using an EEG and eye-tracking device. The high temporal resolution of the tools (millisecond level) made it possible to determine thought patterns and highlighted the typical course of thought. Self-reports on perceived mind wandering were also collected. Creative performance was analyzed by two independent raters.

Participants. In this study, a total of 45 healthy volunteers aged between 21 to 47 years (average 29.2 years) took part (24 females, 21 males). All participants gave written consent, were informed by the experimenters regarding the aim of the study, its content, and how the investigation will be conducted. The participants could terminate or discontinue their participation at any time without giving any reason and without any consequences. The recorded data of each participant was anonymized.

Data analysis. A major indicator for possible mind-wandering episodes is the gaze, more specifically, gaze fixations. Thus, the first task was to find a suiting value for the minimum time a fixation lasts. We analyzed the eye-tracking data with a fixation detection algorithm and placed our minimum fixation time at one second. This was done, on the one hand, to find a value that was small enough to show possible mind-wandering episodes and, on the other hand, to limit the number of epochs to be analyzed later. The logic is as follows: If a person engages in mind wandering, it is highly unlikely that this person—at the same time—moves gaze from one to another point on the screen. To clean the EEG data, we used an independent component analysis (ICA). Independent components like eye movements or muscle artifacts were rejected based on visual inspection. Thereafter, the data was bandpass filtered from 0.5 to 40 Hz. The fixation timings delivered from the eye-tracker were then used

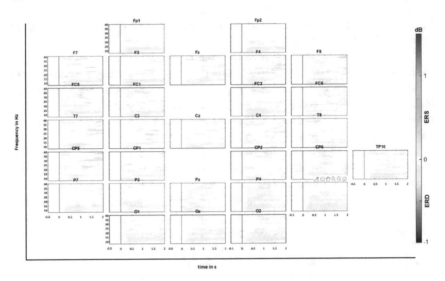

Fig. 1 Grand average ERD/ERS topographical map of 45 participants during mind-wandering episodes during the first block of UUTs (total of 3336 trials). x-axes: shows the time: [−0.5, 0.05, 2]s with the reference period from [−0.5, 0]s. y-axes show the frequency [8, 40]Hz in overlapping 2-Hz bands. Scale −100 to +150. Bootstrap significance test ($\alpha = 0.05$)

to calculate and plot an event-related desynchronization/synchronization (ERD/S) map in overlapping 2 Hz bands. EEG epochs around the gaze fixation time points [−0.5, 2]s during the four UUTs of all 45 participants were used. As a reference interval, we defined [−0.5, 0]s. A bootstrapping method was then used to test for significant ERD/ERS (0.05) values.

Preliminary results. Preliminary findings can already be drawn from the ERD analysis of the whole group during mind-wandering episodes detected through gaze fixation during the initial four UUT problems. Figure 1 shows only significant changes (alpha = 5%) in band power compared to the reference. Generally, an alpha band ERD can be observed on almost all channels, which is more pronounced in the occipital region (O1, Oz, O2). This ERD starts immediately as eye gaze is fixated (second 0) and could be an indication of visual processing during mind wandering. The next steps are ERD analyses group specific for the second block of UUTs after the intervention of rest, 1-back and 3-back tests. In later studies, the data will be analyzed separately, with particular attention to alpha power and the parietal/occipital regions.

4 Future Work

For in-depth statistical analysis, we will use correlation, regression analysis, analysis of variance, and multivariate analysis. This far, our experimental approach demonstrated how and whether neurophysiological data (brain, eye-tracking) can be used to detect mind wandering while using technology. In the future, we seek a triangulation of the data. This analysis will be based on EEG, eye-tracking, questionnaires, and creative performance. Specifically, we will compare group-wise differences using perceived measures with more objective neurophysiological measures.

Our work contributes to the current literature by identifying and measuring a crucial cognitive process while using technology that can foster creativity, namely mind wandering. This can pave the way for designing innovative neuroadaptive systems [12–15, 43, 44]. Our study presents a first step towards automatic observation and interpretation of mind-wandering episodes while using digital technology, which could help to design human–computer interaction tasks and human-centered technological artifacts in the future. We strongly believe that neuroadaptive systems offer significant theoretical and practical potential and that this study contributes to related NeuroIS research.

Acknowledgements This work was supported by the Eureka project funded by Volkswagen Foundation (grant: 96982).

References

1. Andrews-Hanna, J. R., Irving, Z. C., Fox, K. C., Spreng, R. N., & Christoff, K. (2018). The neuroscience of spontaneous thought: An evolving interdisciplinary field. In K. C. R. Fox & K. Christoff (Eds.), *The Oxford Handbook of Spontaneous Thought: Mind-Wandering, Creativity, and Dreaming* (pp. 143–164). Oxford University Press, Oxford.
2. Christoff, K., Irving, Z. C., Fox, K. C., Spreng, R. N., & Andrews-Hanna, J. R. (2016). Mind-wandering as spontaneous thought: A dynamic framework. *Nature Reviews Neuroscience, 17*, 718–731.
3. Killingsworth, M. A., & Gilbert, D. T. (2010). A wandering mind is an unhappy mind. *Science, 330*, 932.
4. Irving, Z. C. (2016). Mind-wandering is unguided attention: Accounting for the "purposeful" wanderer. *Philosophical Studies, 173*, 547–571.
5. Oschinsky, F. M., Niehaves, B., Riedl, R., Klesel, M., Wriessnegger, S. C., & Mueller-Putz, G. R. (2021). On how mind wandering facilitates creative incubation while using information technology: A research agenda for robust triangulation. In F. D. Davis, R. Riedl, J. vom Brocke, P.-M. Léger, A. B. Randolph & G. R. Müller-Putz (Eds.), *Information Systems and Neuroscience: NeuroIS Retreat 2021* (Vol. 52, pp. 139–147). LNISO. Springer, Cham.
6. Preiss, D. D., Cosmelli, D., Grau, V., & Ortiz, D. (2016). Examining the influence of mind wandering and metacognition on creativity in university and vocational students. *Learning and Individual Differences, 51*, 417–426.
7. Sokół, A., & Figurska, I. (2021). The importance of creative knowledge workers in creative organization. *Energies, 14*, 6751.
8. Laukkonen, R. E., Kaveladze, B. T., Tangen, J. M., & Schooler, J. W. (2020). The dark side of Eureka: Artificially induced Aha moments make facts feel true. *Cognition, 196*, 104122.
9. Napier, N. K., Bahnson, P. R., Glen, R., Maille, C. J., Smith, K., & White, H. (2009). When "Aha moments" make all the difference. *Journal of Management Inquiry, 18*, 64–76.
10. Ritter, S. M., Abbing, J., & Van Schie, H. T. (2018). Eye-closure enhances creative performance on divergent and convergent creativity tasks. *Frontiers in Psychology*, 1315.

11. Kane, M. J., Conway, A. R. A., Miura, T. K., & Colflesh, G. J. H. (2007). Working memory, attention control, and the n-back task: A question of construct validity. *Journal of Experimental Psychology: Learning, Memory, and Cognition, 33*, 615–622.
12. vom Brocke, J., Hevner, A., Léger, P.-M., Walla, P., & Riedl, R. (2020). Advancing a NeuroIS research agenda with four areas of societal contributions. *European Journal of Information Systems, 29*, 9–24.
13. vom Brocke, J., Riedl, R., & Léger, P.-M. (2013). Application strategies for neuroscience in information systems design science research. *Journal of Computer Information Systems, 53*, 1–13.
14. Loos, P., Riedl, R., Müller-Putz, G. R., Vom Brocke, J., Davis, F. D., Banker, R. D., & Léger, P.-M. (2010). NeuroIS: Neuroscientific approaches in the investigation and development of information systems. *Business & Information Systems Engineering, 2*, 395–401.
15. Riedl, R., & Léger, P.-M. (2016). *Fundamentals of NeuroIS: Information systems and the brain.* Heidelberg: Springer.
16. Klesel, M., Oschinsky, F. M., Conrad, C., & Niehaves, B. (2021). Does the type of mind wandering matter? Extending the inquiry about the role of mind wandering in the IT use experience. *Internet Research, 31*, 1018–1039.
17. Riedl, R., Davis, F. D., & Hevner, A. R. (2014). Towards a NeuroIS research methodology: Intensifying the discussion on methods, tools, and measurement. *Journal of the Association for Information Systems, 15*, xiv.
18. Conrad, C., & Newman, A. (2021). Measuring mind wandering during online lectures assessed with EEG. *Frontiers in Human Neuroscience, 15*, 697532.
19. Ceh, S. M., Annerer-Walcher, S., Körner, C., Rominger, C., Kober, S. E., Fink, A., & Benedek, M. (2020). Neurophysiological indicators of internal attention: An electroencephalography–eye-tracking coregistration study. *Brain and Behavior, 10*, 01790.
20. Oschinsky, F. M., Klesel, M., & Niehaves, B. (2022). *Mind wandering in information technology use-Scale development and cross-validation.* ACM SIGMIS Database: The DATA BASE for Advances in Information Systems.
21. Milgram, R. M., & Milgram, N. A. (1976). Creative thinking and creative performance in Israeli students. *Journal of Educational Psychology, 68*, 255.
22. Wallach, M. A., & Kogan, N. (1965). *Modes of thinking in young children; a study of the creativity-intelligence distinction.* Holt, Rinehart and Winston.
23. Baird, B., Smallwood, J., Mrazek, M. D., Kam, J. W., Franklin, M. S., & Schooler, J. W. (2012). Inspired by distraction: Mind wandering facilitates creative incubation. *Psychological Science, 23*, 1117–1122.
24. Müller-Putz, G. R., Riedl, R., & Wriessnegger, S. C. (2015). Electroencephalography (EEG) as a research tool in the information systems discipline: Foundations, measurement, and applications. *Communications of the Association for Information Systems, 37*, 911–948.
25. Arnau, S., Löffler, C., Rummel, J., Hagemann, D., Wascher, E., & Schubert, A. (2020). Inter-trial alpha power indicates mind wandering. *Psychophysiology, 57*.
26. Weber, B., Fischer, T., & Riedl, R. (2021). Brain and autonomic nervous system activity measurement in software engineering: A systematic literature review. *Journal of Systems & Software, 178*, 110946.
27. Baird, B., Smallwood, J., Lutz, A., & Schooler, J. W. (2014). The decoupled mind: Mind-wandering disrupts cortical phase-locking to perceptual events. *Journal of Cognitive Neuroscience, 26*, 2596–2607.
28. Broadway, J. M., Franklin, M. S., & Schooler, J. W. (2015). Early event-related brain potentials and hemispheric asymmetries reveal mind-wandering while reading and predict comprehension. *Biological Psychology, 107*, 31–43.
29. Kam, J. W., & Handy, T. C. (2013). The neurocognitive consequences of the wandering mind: A mechanistic account of sensory-motor decoupling. *Frontiers in Psychology, 4*.
30. Compton, R. J., Gearinger, D., & Wild, H. (2019). The wandering mind oscillates: EEG alpha power is enhanced during moments of mind-wandering. *Cognitive, Affective, & Behavioral Neuroscience, 19*, 1184–1191.

31. Thut, G., Nietzel, A., Brandt, S. A., & Pascual-Leone, A. (2006). α-Band electroencephalographic activity over occipital cortex indexes visuospatial attention bias and predicts visual target detection. *Journal of Neuroscience, 26*, 9494–9502.
32. Cooper, N. R., Croft, R. J., Dominey, S. J., Burgess, A. P., & Gruzelier, J. H. (2003). Paradox lost? Exploring the role of alpha oscillations during externally vs. internally directed attention and the implications for idling and inhibition hypotheses. *International Journal of Psychophysiology, 47*, 65–74.
33. Jung-Beeman, M., Bowden, E. M., Haberman, J., Frymiare, J. L., Arambel-Liu, S., Greenblatt, R., Reber, P. J., Kounios, J., & Dehaene, S. (2004). Neural activity when people solve verbal problems with insight. *PLoS Biology, 2*, 97.
34. Agnoli, S., Zanon, M., Mastria, S., Avenanti, A., & Corazza, G. E. (2020). Predicting response originality through brain activity: An analysis of changes in EEG alpha power during the generation of alternative ideas. *NeuroImage, 207*, 116385.
35. Fink, A., & Benedek, M. (2014). EEG alpha power and creative ideation. *Neuroscience & Biobehavioral Reviews, 44*, 111–123.
36. Stevens, C. E., Jr., & Zabelina, D. L. (2019). Creativity comes in waves: An EEG-focused exploration of the creative brain. *Current Opinion in Behavioral Sciences, 27*, 154–162.
37. Annerer-Walcher, S., Körner, C., & Benedek, M. (2018). Eye behavior does not adapt to expected visual distraction during internally directed cognition. *PLoS ONE, 13*, 0204963.
38. Smilek, D., Carriere, J. S., & Cheyne, J. A. (2010). Out of mind, out of sight: Eye blinking as indicator and embodiment of mind wandering. *Psychological Science, 21*, 786–789.
39. Doherty-Sneddon, G., & Phelps, F. G. (2005). Gaze aversion: A response to cognitive or social difficulty? *Memory and Cognition, 33*, 727–733.
40. Markson, L., & Paterson, K. B. (2009). Effects of gaze-aversion on visual-spatial imagination. *British Journal of Psychology, 100*, 553–563.
41. Smallwood, J., & Schooler, J. W. (2015). The science of mind wandering: Empirically navigating the stream of consciousness. *Annual Review of Psychology, 66*, 487–518.
42. Franklin, M. S., Broadway, J. M., Mrazek, M. D., Smallwood, J., & Schooler, J. W. (2013). Window to the wandering mind: Pupillometry of spontaneous thought while reading. *Quarterly Journal of Experimental Psychology, 66*, 2289–2294.
43. Demazure, T., Karran, A., Léger, P.-M., Labonté-LeMoyne, É., Sénécal, S., Fredette, M., & Babin, G. (2021). Enhancing sustained attention: A pilot study on the integration of a brain-computer interface with an enterprise information system. *Business & Information Systems Engineering, 63*, 653–668.
44. Astor, P. J., Adam, M. T., Jerčić, P., Schaaff, K., & Weinhardt, C. (2013). Integrating biosignals into information systems: A NeuroIS tool for improving emotion regulation. *Journal of Management Information Systems, 30*, 247–278.

The Effects of Artificial Intelligence (AI) Enabled Personality Assessments During Team Formation on Team Cohesion

Nicolette Gordon and Kimberly Weston Moore

Abstract Studies across multiple industries have shown the importance of team cohesion and its diverse application [1, 44, 46]. This study reviews the literature regarding the types of factors that affect team cohesion. Within the literature review we examine the question; Can Artificial Intelligence (AI) be a tool to assist in accurately determining teams using employee self-reported preferences and capabilities? Opportunities exist within the literature for areas of practice where AI enabled personality tests are used for the purposes of creating teams. The purpose of this paper is to examine the possible relationship between AI enabled personality assessments results and their ability to garner team cohesion.

Keywords Project management · Artificial intelligence · Scope creep · Personality assessment · Team cohesion · Five factor model · Facial expression recognition

1 Introduction

In 2018 the Project Management Institute (PMI) conducted the Pulse of the Profession survey which uncovered that organizations around the world waste $1 million every 20 seconds equating to $2 trillion dollars a year [2]. From the survey five critical factors were identified and they were executive sponsored engagement, lack of connection between strategy and design, strategy implementation investment, disruption management and scope creep. This review will focus on the scope creep critical factor.

Larson and Larson [32] defined scope creep as the addition of unauthorized functionality including features, new product, requirements, or work. Studies show scope

N. Gordon (✉) · K. W. Moore
Kennesaw State University, Coles College of Business, Kennesaw, GA, USA
e-mail: ngordo20@students.kennesaw.edu

K. W. Moore
e-mail: kmoor228@students.kennesaw.edu

© The Author(s), under exclusive license to Springer Nature Switzerland AG 2022
F. D. Davis et al. (eds.), *Information Systems and Neuroscience*,
Lecture Notes in Information Systems and Organisation 58,
https://doi.org/10.1007/978-3-031-13064-9_31

creep as one cause for diminished team performance with one reason being personality differences [3, 53]. Cohesive teams offer many benefits such as performance
improvement. When team members have issues such as distrust, dislike and disin-
terest team motivation decline and the likelihood they will display behaviors related
to teams that result in a positive effect are diminished [47]. Since scope creep has
been linked to personality differences the purpose of this paper is to examine literature that could link the effects of artificial intelligence used when administering
personality assessments. Next, we examine the use of personality assessment results
in team selection.

In the 1930s, the social sciences recognized personality psychology as an identifiable discipline [37]. For years managers have utilized assessments for human
resource planning, employee opinion surveys, and program evaluation [44]. Team
performance is influenced by both team cohesion and personality [1, 8, 21, 26, 42,
48] and the variable cohesion has been considered the most important for small teams
[1, 14, 33]. Cohesion is also considered a critical variable in models of effective work
teams [12, 19, 27].

In the literature we found a limited number of articles regarding the use of AI
in team selection. The articles we reviewed focused on AI before team selection
and the use of AI after the team was formed. We suggest a more technical method
for project team selection to encourage the chance of better team cohesion. Further
we hope to encourage research for investigating the effects of the addition of AI
enabled personality testing. Through the lens of Tuckman's theory [52], we address
the following question: **Does the use of AI-enabled personality assessments during
team formation improve team cohesion?**

The objective of this study is to provide a literature review on the effects of
Artificial Intelligence enabled personality assessments on team cohesion. There are
numerous articles on personality assessment and team cohesion as it relates to the use
of AI after the team has been created [35, 50, 55]. However, studies of AI for team
creation were minimal. Since it is possible to distort a personality assessment by the
respondent's environment or their ability to answer based on what the respondent
perceives as the correct answer [20], we question if AI can help mitigate this risk
providing a more accurate assessment.

2 Literature Review

2.1 Team Cohesion

Cohesion is considered a valuable small group variable [33] because of the relationship between cohesion and positive group outcomes [9]. Teams provide numerous
advantages. These advantages are realized to an extent only in cohesive teams [47].

When teams are cohesive, they are more likely to be motivated to achieve the established goals [1]. Thus, when groups are studied, cohesion is a primary concern in numerous disciplines [14].

Cohesion is defined in diverse ways across literature which indicates the distinction of how researchers interpret the word cohesion. Moreover, some scholars define cohesion as unidimensional for example Seashore [49] defines cohesion as members attraction to the group or resistance to leaving. And some multidimensional such as Festinger [22], who provides a historical definition for cohesion, states it is, "resultant of all the forces acting on members to remain in the group". He created a list of the five most common subdimensions and their definitions.

The subdimensions of cohesion according to Festinger [22] are task, social, belongingness, group pride, and morale. The subdimension task is defined as, an attraction or bonding between group members that is based on a shared commitment to achieving the groups goals and objectives; social cohesion, a closeness and attraction within the group that is based on social relationships within the group [13]. Belongingness is the degree to which members of a group are attracted to each other [49]. Group pride is the extent to which group members exhibit liking for the status or the ideologies that the group supports or represents, or the shared importance of being a member of the group [9]. Finally, moral is individuals' high degree of loyalty to fellow group members and their willingness to endure frustration for the group [17]. Offering a unique perspective, another historical definition by Gross and Martin [25], refers to cohesion as "the resistance of the group to disruptive forces".

Cohesion is often recognized as an individual team member's desire to stick together within a team [1]. Specifically, Carron [13] defines cohesion as a "dynamic process which is reflected in the tendency of the group to stick together and remain united in the pursuit of its goals and objectives." Casey et al. [15] further defines cohesion as the shared bond or attraction that drives team members to stay together and to want to work together which is essential for teams [9, 18].

Many researchers have utilized the position introduced by Festinger et al. [22] that the cohesiveness of a group is the desire of individuals to maintain their membership in a group, is contributed to by several independent forces, but most investigations have focused on one force, inter-member attraction [33].

2.2 AI Enabled Personality Assessment

Personality testing has been an intrinsic aspect of industrial-organizational (I-O) and vocational psychology for the last 85 years [23]. During this time inventories were developed to measure personality traits such as introversion, extraversion, and neuroticism [11]. We define personality as an individual's psychophysical systems that determine their characteristic behavior and thought [4]. Major personality theorists created personality theories within one of five perspectives: social cognitive perspective, humanistic perspective, psychoanalytic, or trait perspective [10].

The social cognitive theory approaches self-development, adaptation, and change from an agentic standpoint [6]. The humanistic theory of personality and human development focuses on the basic human needs and the idea that people strive toward self-actualization in a positive climate [36]. Psychoanalytic theory's view of personality consists of sexual pleasures known as the id, a reality-oriented executive known as the ego, and an internalized set of ideals known as the superego [10]. Trait theory is a scientific study of traits that encompass significant dimensions of personality [43]. There are several different personality assessments used to identify specific traits.

Personality assessments are used to predict whether an employee will engage in undesirable job behaviors and whether they are trustworthy [10]. Several assessments were reviewed. Cattell's 16-PF included a set of 15 personality trait scales and one scale to assess intelligence which were designed to assess the full range of normal personality functioning [16]. The next assessment is the California Psychological Inventory or CPI, published in 1956 [24], containing 489 items. The 18 scales within CPI scales are designed to assess personality characteristics important from a social interaction point of view [24]. The NEO Personality Inventory (NEO PI) was developed by Paul Costa and Robert McCrae to assess personality dimension that is based on the "Big Five" or Five Factor Model of personality [38]. The NEO was published in 1985 as a measure of the major personality dimensions in normal personality to assess: openness, agreeableness, neuroticism, extraversion, and conscientiousness [38].

For this study we will focus on the five-factor model (FFM), or the Big Five which emerged in the 1980s [29, 56]. According to Mount and Barrick [41], many personality psychologists appear to have concluded that five personality categories, known as the Big Five, are both essential and adequate to define the core characteristics of normal personality. This personality assessment provides a comprehensive framework to examine personality and its relationship to team process [1]. The five-factor model has five traits that have convergent and discriminate validity [39] are described as: "**Conscientiousness** which is defined as the extent to which people are efficient, organized, planful, reliable responsible and thorough; **extraversion** which is defined as the extent to which people are active, assertive, energetic, enthusiastic, outgoing and talkative; **agreeableness**: it is defined as the extent to which people are appreciative, forgiving, generous, kind, sympathetic and trusting; **neuroticism**: it is defined as the extent to which people are anxious, self-pitying, tense, touchy, unstable and worrying; **openness to experience is** defined as the extent to which people are artistic, curious, imaginative, insightful, original and have wide interests." Although personality assessments have existed for several decades, a gap exists between respondents' true nature and perceived predictor scores which are frequently assumed to be attributable to deliberate response distortion or faking [20].

2.3 Artificial Intelligence

Artificial Intelligence (AI) has been around for many years. The phrase was coined in 1956 by John McCarthy. "McCarthy used the term "artificial intelligence" for a conference he was organizing, along with Marvin Minsky, Nat Rochester, and Claude Shannon—the Dartmouth Summer Research Project on Artificial Intelligence, funded by the Rockefeller Foundation" [5].

Through the years there have been many definitions of AI. Simmons and Chappell [51] defined AI as a term used to describe human like behaviors of a machine that would be considered intelligent. In 2009, it was suggested that the definition of AI not be limited and that most of the definitions at that time would fit into four categories; systems that think like humans, act like humans, think rationally and act rationally [31]. Wang [54] concluded the definition of AI depends on the field and that it is the researcher's responsibility to define their interpretation and how it will be used for their study. In this study we define AI as a machine's ability to interpret human behavior.

During our review we found studies that focused on the effect of AI on team cohesion after the teams were developed. Webber et al. [55] researched the challenges of developing effective teams and the possible use of AI to help with team development. Seeber et al. [50] studied the use of AI machines as a team member. Bansal et al. [7] studied the benefits of AI being a complementary team member concluding that when the AI was correct, team performance increased and vice versa. Hickman et al. [28] studied the use of automated video interviews and concluded that using personality assessments in an applied setting holds promise. However, we did not find in the literature research on the use of AI in formulating teams which we have identified as a gap.

2.4 Facial Expression Recognition

In this paper, the crux of the proposed use of AI is the ability to interpret facial expressions for accuracy in personality test taking. In 2019, Kim et al. [30] proposed a new recognition algorithm based on hierarchical deep neural network structure for facial expression recognition (FER). The study used two datasets CK+ (Extended Cohn-Kanade) and JAFFE (Japanese female facial expression) comprising of 150 images for each of the six emotions: anger, disgust, fear, happy, sad, and surprised. By combining geometric and appearance features algorithms a fusion of the two was created. The researchers quantitatively determined their results by using the tenfold cross validation method. The results were a greater than 91% accuracy and more than 1.3% increase in comparison to datasets. Further, Lui et al. [34] studied the use of video clips for facial recognition proposing the use of clip-aware emotion-rich feature learning network (CEFLNet) as a means of evaluating FER using expressions of emotions in short clips in videos. Since we are interested in video recorded testing and the many types of analysis, this is another area that will have to be researched further.

3 Conclusion

Personality assessments with self-report measures are one of the most widely used assessment techniques in modern psychology. There is a valid risk that clients may purposefully misrepresent their results to attain the goal of the testing entity [40]. With AI-enabled personality test, we believe there is an opportunity for capturing and quantitatively analyzing facial expression data to increase test result accuracy thereby providing a tool that could help with team formation resulting in greater team cohesion. Hickman et al. [28] studied the use of automated video interviews capturing data to be analyzed by AI and comparing it to data obtained by the interviewer. Similar tests could be performed using AI to register behaviors and environment to help in assessing an individual's team fit. We hope to research this topic further and perhaps suggest another method of team selection. In 2007, Riedl and Roithmayr [45] posit that with the help of neuroscience technology, we would be able to measure feelings and thoughts directly. Today, we know this to be true.

References

1. Aeron, S., & Pathak, S. (2012). Relationship between team member personality and team cohesion: An exploratory study in IT industry. *Management and Labour Studies, 37*(3), 267–282.
2. Alcantara, R., & Squibb, J. (2018, February 15). $1 million wasted every 20 seconds by organizations around the world. *Business Wire.* https://www.businesswire.com/news/home/201 80215005610/en/1-Million-Wasted-Every-20-Seconds-by-Organizations-around-the-World
3. Ajmal, M., Khan, M., & Al-Yafei, H. (2019). Exploring factors behind project scope creep–stakeholders' perspective. *International Journal of Managing Projects in Business.*
4. Allport, G. W. (1961). *Pattern and growth in personality.* Holt, Rinehart & Winston.
5. Andresen, S. L. (2002). John McCarthy: Father of AI. *IEEE Intelligent Systems, 17*(5), 84–85.
6. Bandura, A. (2001). *Social cognitive theory: An agentic perspective, Annual Review of Psychology, 52:1–26.* Annual Reviews Inc.
7. Bansal, G., Wu, T., Zhou, J., Fok, R., Nushi, B., Kamar, E., Weld, D., et al. (2021, May). Does the whole exceed its parts? The effect of AI explanations on complementary team performance. In *Proceedings of the 2021 CHI Conference on Human Factors in Computing Systems* (pp. 1–16).
8. Barrick, M. R., & Mount, M. K. (1991). The Big Five personality dimensions and job performance: A meta-analysis. *Personnel Psychology, 44*(1), 1–26.
9. Beal, D. J., Cohen, R. R., Burke, M. J., & McLendon, C. L. (2003). Cohesion and performance in groups: A meta-analytic clarification of construct relations. *Journal of Applied Psychology, 88*(6), 989.
10. Boyle, G. J. (2008). Critique of the five-factor model of personality. *Humanities & Social Sciences Papers, Paper 297.*
11. Butcher, J. N. (2009). Clinical personality assessment: History, evolution, contemporary models, and practical applications. *Oxford Handbook of Personality Assessment,* 5–21.

12. Carless, S. A., & De Paola, C. (2000). The measurement of cohesion in work teams. *Small Group Research, 31*(1), 71–88.
13. Carron, A. V. (1982). Cohesiveness in sport groups: Interpretations and considerations. *Journal of Sport Psychology, 4*, 123–138.
14. Carron, A. V., & Brawley, L. R. (2000). Cohesion: Conceptual and measurement issues. *Small Group Research, 31*(1), 89–106.
15. Casey-Campbell, M., & Martens, M. L. (2009). Sticking it all together: A critical assessment of the group cohesion–performance literature. *International Journal of Management Reviews, 11*(2), 223–246.
16. Cattell, H. E., & Mead, A. D. (2008). *The sixteen personality factor questionnaire (16PF).*
17. Cartwright, D., & Zander, A. (1960). Group cohesiveness: Introduction. In D. Cartwright & A. Zander (Eds.), *Group dynamics: Research and theory* (2nd ed., pp. 69–94). Harper Ro.
18. Chiocchio, F., & Essiembre, H. (2009). Cohesion and performance: A meta-analytic review of disparities between project teams, production teams, and service teams. *Small Group Research, 40*(4), 382–420.
19. Cohen, S. G., & Bailey, D. E. (1997). What makes teams work: Group effectiveness research from the shop floor to the executive suite. *Journal of Management, 23*(3), 239–290.
20. Dilchert, S., Ones, D. S., Viswesvaran, C., & Deller, J. (2006). Response distortion in personality measurement: Born to deceive, yet capable of providing valid self-assessments? *Psychology Science, 48*(3), 209.
21. Evans, C. R., & Dion, K. L. (1991). Group cohesion and performance: A meta-analysis. *Small Group Research, 22*(2), 175–186.
22. Festinger, L. (1950). Informal social communication. *Psychological Review, 57*(5), 271.
23. Gibby, R., & Zickar, M. (2008). A history of the early days of personality testing in American industry: An obsession with adjustment. *History of Psychology, 11*, 164–184. https://doi.org/10.1037/a0013041
24. Gough, H. G. (1956). *California psychological inventory.* Consulting Psychologists Press.
25. Gross, N., & Martin, W. E. (1952). On group cohesiveness. *The American Journal of Sociology, 57*, 546–564.
26. Gully, S. M., Devine, D. J., & Whitney, D. J. (1995). A meta-analysis of cohesion and performance: Effects of levels of analysis and task interdependence. *Small Group Research, 26*(4), 497–520.
27. Hackman, L., & Warnow-Blewett, J. (1987). The documentation strategy process: A model and a case study. *The American Archivist, 50*(1), 12–47.
28. Hickman, L., Bosch, N., Ng, V., Saef, R., Tay, L., & Woo, S. E. (2021). Automated video interview personality assessments: Reliability, validity, and generalizability investigations. *Journal of Applied Psychology.*
29. Hough, L. M. (1992). The Big Five personality variables—construct confusion: Description versus prediction. *Human Performance, 5*(1–2), 139–155.
30. Kim, J. H., Kim, B. G., Roy, P. P., & Jeong, D. M. (2019). Efficient facial expression recognition algorithm based on hierarchical deep neural network structure. *IEEE Access, 7*, 41273–41285.
31. Kok, J. N., Boers, E. J., Kosters, W. A., Van der Putten, P., & Poel, M. (2009). Artificial intelligence: Definition, trends, techniques, and cases. *Artificial Intelligence, 1*, 270–299.
32. Larson, R., & Larson, E. (2009). *Top five causes of scope creep ... and what to do about them.* Paper presented at PMI® Global Congress 2009—North America, Orlando, FL. Project Management Institute.
33. Lott, A. J., & Lott, B. E. (1965). Group cohesiveness as interpersonal attraction: A review of relationships with antecedent and consequent variables. *Psychological Bulletin, 64*(4), 259.
34. Liu, Y., Feng, C., Yuan, X., Zhou, L., Wang, W., Qin, J., & Luo, Z. (2022). Clip-aware expressive feature learning for video-based facial expression recognition. *Information Sciences.*
35. MacLean, T., & Thomas, D. (2019). Team challenges: Is artificial intelligence the solution? *Business Horizons.*
36. Mahrer, A. R. (1978). *Experiencing: A humanistic theory of psychology and psychiatry.* Brunner/Mazel.

37. McAdams, D. P. (1997). A conceptual history of personality psychology. In *Handbook of personality psychology* (pp. 3–39). Academic Press.
38. McCrae, R. R., & Costa, P. T. (1985). Updating Norman's "adequacy taxonomy": Intelligence and personality dimensions in natural language and in questionnaires. *Journal of Personality and Social Psychology, 49*(3), 710.
39. McCrae, R. R., & John, O. P. (1992). An introduction to the five-factor model and its applications. *Journal of Personality, 60*(2), 175–215.
40. Meyer, J. K., Hong, S. H., & Morey, L. C. (2015). Evaluating the validity indices of the personality assessment inventory-adolescent version. *Assessment, 22*, 490–496. https://doi.org/10.1177/1073191114550478
41. Mount, M. K., & Barrick, M. R. (1995). The big five personality dimensions: Implications for research and practice in human resource management. In K. M. Rowland & G. R. Ferris (Eds.), *Research in personnel and human resources management* (Vol. 13, pp. 153–200). JAI Press.
42. Mount, M. K., & Barrick, M. R. (1998). Five reasons why the "big five" article has been frequently cited: The big five personality dimensions and job performance: A meta-analysis. *Personnel Psychology, 51*(4), 849–857.
43. Pervin, L. A. (1994). A critical analysis of current trait theory. *Psychological Inquiry, 5*(2), 103–113.
44. Podsakoff, P. M., & Organ, D. W. (1986). Self-reports in organizational research: Problems and prospects. *Journal of Management, 12*(4), 531–544.
45. Riedl, R., & Roithmayr, F. (2007). Human-computer interaction and neuroscience: Science or science fiction? In *SIGHCI 2007 Proceedings* (p. 7).
46. Ronen, S., & Mikulincer, M. (2009). Attachment orientations and job burnout: The mediating roles of team cohesion and organizational fairness. *Journal of Social and Personal Relationships, 26*(4), 549–567.
47. Salas, E., Grossman, R., Hughes, A. M., & Coultas, C. W. (2015). Measuring team cohesion: Observations from the science. *Human Factors, 57*(3), 365–374.
48. Salgado, J. F. (1997). The five factor model of personality and job performance in the European community. *Journal of Applied Psychology, 82*(1), 36–43.
49. Seashore, S. E. (1954). *Group cohesiveness in the industrial work group.* University of Michigan.
50. Seeber, I., Bittner, E., Briggs, R. O., De Vreede, T., De Vreede, G. J., Elkins, A., & Söllner, M. (2020). Machines as teammates: A research agenda on AI in team collaboration. *Information & Management, 57*(2), 103174.
51. Simmons, A. B., & Chappell, S. G. (1988). Artificial intelligence-definition and practice. *IEEE Journal of Oceanic Engineering, 13*(2), 14–42.
52. Tuckman, B. (1965). *Forming, storming, norming, & performing team development model.*
53. Wagner, H. T., Beimborn, D., & Weitzel, T. (2014). How social capital among information technology and business units drives operational alignment and IT business value. *Information Systems, 31*(1), 241–272. https://doi.org/10.2753/MIS0742-1222310110
54. Wang, P. (2019). On defining artificial intelligence. *Journal of Artificial General Intelligence, 10*(2), 1–37.
55. Webber, S. S., Detjen, J., MacLean, T. L., & Thomas, D. (2019). Team challenges: Is artificial intelligence the solution? *Business Horizons, 62*(6), 741–750.
56. Wiggins, J. S. (Ed.). (1996). *The five-factor model of personality: Theoretical perspectives.* Guilford Press.

Age-Related Differences on Mind Wandering While Using Technology: A Proposal for an Experimental Study

Anna Zeuge, Frederike Marie Oschinsky, Michael Klesel, Caroline Reßing, and Bjoern Niehaves

Abstract Mind wandering (MW) is a mental activity in which our thoughts drift away and turn into internal notions and feelings. Research suggests that individuals spend up to one half of their waking hours thinking about task-unrelated things. Being the opposite of goal-directed thinking, empirical evidence suggests that MW can forester creativity and problem solving. However, and despite growing efforts to understand the role of MW in technology-related settings, the role of individual differences remains unclear. We address this gap by proposing a research model that seeks to shed further light on age-related differences in MW while using different types of technology (i.e., hedonic and utilitarian systems). Thereby, we provide a point of departure for further research on how individual characteristics influence MW while using technology.

Keywords Mind wandering · Technology use · Age · Hedonic and utilitarian systems

A. Zeuge (✉) · F. M. Oschinsky · C. Reßing · B. Niehaves
University of Siegen, Siegen, Germany
e-mail: anna.zeuge@uni-siegen.de

F. M. Oschinsky
e-mail: frederike.oschinsky@uni-siegen.de

C. Reßing
e-mail: caroline.ressing@uni-siegen.de

B. Niehaves
e-mail: bjoern.niehaves@uni-siegen.de

M. Klesel
University of Twente, Enschede, The Netherlands
e-mail: michael.klesel@utwente.nl

© The Author(s), under exclusive license to Springer Nature Switzerland AG 2022
F. D. Davis et al. (eds.), *Information Systems and Neuroscience*,
Lecture Notes in Information Systems and Organisation 58,
https://doi.org/10.1007/978-3-031-13064-9_32

1 Introduction

Mind wandering (MW) is one of the most ubiquitous mental activities [1] and happens up to 50% of our waking time [2]. MW occurs when the mind stops being focused on the present and instead starts pondering about task-unrelated things [3]. Literature has shown that MW can be related to both negative job-output (e.g., reduced performance) and positive job-output (e.g., increased creativity) [4, 5]. Due to its complexity, the investigation of MW is important to further understand how it affects human behavior.

Since MW is a ubiquitous experience, it is most likely that our minds frequently wander when using technology. In fact, there is initial evidence that the degree of MW varies among different types of systems, i.e., hedonic and utilitarian systems [6]. Hedonic systems aim to provide self-fulfilling value, while utilitarian systems aim to provide instrumental value [7]. Sullivan et al. [3] suggest that MW, while using technology, has a notable impact on creativity [8]. To this end, we argue that MW is increasingly important in the context of technology use but needs further clarification.

Despite valuable first efforts to understand MW as a subject of information system (IS) research, little is known about individual differences in terms of MW and technology use so far. This gap is critical because literature on MW has stressed the role of individual differences [9–11]. Moreover, research concerning technology-related phenomena put further emphasis on them [12–14]. Studies demonstrate that older people's minds wander less in their daily life compared to younger people [9, 15], because cognitive abilities decrease with age [16–18]. As cognitive ability influences how technology is used [19], it is important to understand how IS artifacts need to be adapted to support human computer interaction.

Our paper aims to investigate the relationship between MW and age by raising the following research question: *Is there an age-related difference on MW while using different types of systems (i.e., hedonic, utilitarian)?* We contribute to a more holistic understanding of how humans of different ages use technology when their minds trail off.

2 Theoretical Background

Christoff et al. [18, p. 719] define MW as "a mental state, or a sequence of mental states, that arises relatively freely due to an absence of strong constraints on the contents of each state and on the transitions from one mental state to another". Psychology and neuroscience research demonstrates that MW predominately occurs in non-demanding circumstances and during task-free activity, e.g., during reading or driving [20–22].

MW has been associated with negative and positive consequences: Since thoughts wander from topic to topic, MW induces a lack of awareness and is seen as a cause of poor performance, errors, disruption, disengagement, and carelessness [4, 23, 24].

Moreover, MW is perceived as adverse, as it is enhanced by stress, unhappiness, and substance abuse [25–27]. However, besides its negative effects, studies suggest that MW offers unique benefits [1]. MW can lead to an increased ability to solve problems and positive predicts creative performance [3, 5, 11]. Moreover, MW is useful as it provides mental breaks to reduce boredom from monotonous activities [11].

In general, two types of MW can be distinguished: Deliberate and spontaneous MW [28]. This differentiation goes back to Giambra [20, 29]. Deliberate MW is characterized by intentional internal thoughts such as planning the weekend while driving to work. In contrast, spontaneous MW is unintentional, for example, when drifting away during a conversation [28]. Agnoli et al. [5] demonstrate that this distinction has indeed an effect as deliberate MW is a positive predictor of creative performance, whereas spontaneous MW is a negative predictor of creative performance. Moreover, MW can occur both as a state in specific situations or as trait in everyday life [30].

IS researchers acknowledge the relevance of MW [3, 6, 31, 32]. Sullivan, Davis and Koh [3] showed that MW while using technology influences creativity and knowledge retention. The authors came up with a domain-specific definition for technology-related MW: "task unrelated thought which occurs spontaneously, and the content is related to the aspects of computer systems" [3, p. 4]. Moreover, it has been shown that using different types of IS (i.e., hedonic, or utilitarian systems) relates to the degree of MW [6]. The use of hedonic systems indicates a higher level of MW compared to the use of utilitarian systems. Despite growing efforts to investigate MW in IS research, several questions remain unanswered. Most notably, the influence of individual characteristics on MW while using technology have not been investigated so far.

This gap is critical because individuals differ in the frequently and intentionality of their MW [9–11]. For example, Maillet et al. [9] assessed age-related differences in (1) MW frequency, (2) the relationship between affect and MW and (3) content of MW. The authors suggest that older people wander less in their daily life compared to younger people. Moreover, the authors showed that older people report their off-task thoughts were more "pleasant, interesting, and clear", while the thoughts of younger people were more "dreamlike, novel, strange, and racing" [9, p. 643]. Moreover, it has been shown that impairments can affect individuals MW. For instance, attention deficit and hyperactivity disorder symptomatology positively correlate with spontaneous MW frequency and lack of awareness of MW engagement [10, 33]. Christian et al. [34] suggest that individuals' gender and culture has an impact on the visual perspective while MW. They found out that females and residents from western nations most frequently adopted a first-person point of view, whereas a third-person perspective was more common among residents from eastern countries. Taken together, individual characteristics such as age, gender, origin, or impairments should be considered when studying MW in technology-related settings.

In this study we focus on age-related difference on MW while using different types of systems. Age should be investigated because perceptual (e.g., vision, auditory), cognitive (e.g., memory capacity, attentional control) and psychomotor (e.g., fine motoric, coordination) abilities decline with age [19]. Research has shown that these abilities influence the degree of MW (e.g., [9, 15]). Moreover, these abilities are powerful predictors of technology use [19]. Therefore, age-related changes in ability must be considered, e.g., when designing IS [19]. For example, as demographic change leads to an aging workforce, this critical aspect should be considered when introducing new IS in workplaces.

3   Research Model

According to literature, our research model distinguishes between hedonic and utilitarian systems [6]. Hedonic systems are systems that "aim to provide self-fulfilling rather than instrumental value to the user, are strongly connected to home and leisure activities, focus on the fun-aspect of using information systems and encourage prolonged rather than productive use" [7, p. 695]. Utilitarian systems "provide value that is external to the interaction between the user and system (e.g., improved performance)" [35, p. 445]. Based on this distinction, we propose a research model that investigates whether the relationship between the underlying system and the degree of MW is moderated by age (Fig. 1).

Research suggests that the use of hedonic systems differs from the use of utilitarian systems. For example, Lowry et al. [36] showed that cognitive absorption is stronger in a hedonistic context than in a utilitarian context. This may be explained by the fact that there are different motivational factors when it comes to hedonic (e.g., enjoyment) or utilitarian systems (e.g., job relevance). In line with [6], we argue that the use of hedonic systems leads to a higher degree of MW since users are primarily interested in enjoying a system instead of following instrumental goals. Hedonic usage is an effortless activity, which facilitates MW [6]. In this line, we propose our first hypothesis:

H1: The use of hedonic systems results in a higher degree of MW than utilitarian systems.

An important finding on cognitive aging is that older people have lower working memory capacity than younger people. (e.g., [19, 37]). Literature emphasized that older people have less capacity in working memory to attend to a task, leaving them with less residual capacity for MW [9, 15].

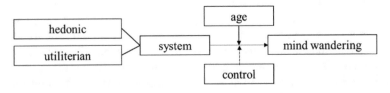

Fig. 1 Proposed research model

Utilitarian systems are mostly employed provide users value and improve productivity [7, 38]. In contrast, hedonic systems are mainly used in homes or leisure environments and are employed for pleasure and relaxation [7, 38, 39]. Thus, we argue that utilitarian systems require a higher working memory capacity than hedonic systems. Combining the above arguments, we propose our second hypothesis:

H2a: *The thoughts of older individuals wander less than those of younger individuals while using utilitarian systems.*
H2b: *The thoughts of older and younger individuals wander in the same degree while using hedonic systems.*

Moreover, we consider additional demographic variables (e.g., gender) to control for randomness or biases.

4 Methodology

Experimental design. Based on our research model (c.f. Fig. 1), we use a between-subject design to manipulate the system type (hedonic/utilitarian). Building up on the work of [15], who investigated the age-related differences between young and older adults on MW in a non-technology context, we acquire data from young–young adults (20–30 years old), young adults (31–64 years old), young–old adults (65–74 years old), and old–old adults (75–85 years old). Since we investigate MW in a technology context, we assume that the investigation of MW requires some degree of habitudinal use of technology because otherwise, when individuals use technology for the first time, the demands are too high to let the mind wander [40]. In other words, habitudinal use of technology was expected to lead to some degree of cognitive ease, which is a prerequisite for MW [41]. Consequently, we only collect data from individuals who indicate that they use their smartphones on a daily basis. Moreover, we ask the participants to use their own smartphones as users perceive their own devices as easier to use and more intuitive [42, 43].

Measurement Instruments. Since MW is an "internal mental experience" it can be measured by self-reports [11, p. 489]. We use established measurement scales for MW on seven-point Likert-Scales. To investigate the psychometric attributes of MW, we select four items from existing multi-measure scales [6, 32].

Experimental Procedure. The experimental procedure will be carried out in four phases: First, participants will be welcomed and informed about the general setting. Second, the participants will be asked to accomplish one of two tasks on their smartphone (approximately 5 min), which are briefly described below. Third, they will be asked to complete a questionnaire assessing their self-reported degree of MW, along with demographic questions. Fourth, they will be thanked and debriefed.

Task 1 ("Facebook"): A common type of hedonic systems relates to social media use. Therefore, we will ask the participants to do tasks on Facebook including navigate through commercials, comments, and postings.

Task 2 ("Email"): A common type of utilitarian technology is writing email. We will ask the participants to write an email to make a hotel reservation.

5   Outlook and Contribution

Our research will contribute to theory, practice, and design alike: From a theoretical perspective, our paper seeks to extent literature on the role of MW in technological settings with a particular emphasize on age-related differences. This goes in line with current literature on MW, emphasizing the relevance of age [9, 15]. Our paper contributes to a better understanding of how age influences individuals' MW while using different types of systems, i.e., hedonic and utilitarian systems. Therefore, research can benefit from this study as a point of departure for further research on how individual characteristics influence MW while using technology. For example, other individual differences (e.g., culture, gender) can be explored. Furthermore, in addition to the measurement scales we use, eye-tracking [44] or Electroencephalography (EEG) [45] could be integrated to provide not only a subjective but also an objective insight into individuals' MW. The investigation of MW as supplement to established concepts in IS, including mindfulness (e.g., [46]) and cognitive absorption (e.g., [47]), is an important step to a more holistic understanding of human cognition and behavior in technology-related settings.

From a design perspective, our research provides insights in how the design and the use experience of certain systems affect MW in light of age. We contribute to a better understanding of how IS should be designed by considering individual characteristics (e.g., age) to influence individuals' MW. This goes in line with literature on human computer interaction, emphasizing the importance of individual characteristics [48, 49].

Our research is also beneficial from a practical perspective. It contributes to a better understanding of the relationship between use behavior and MW. Therefore, it provides important insights to stimulate (e.g., creative jobs) and reduce individuals' MW (e.g., jobs that depend on productivity). Organizations should take MW in consideration when designing future workplaces since MW can provide unique benefits, including a positive influence on creativity, which can lead to performance increases in the long term [8]. Our paper contributes to a better understanding how to consider individual characteristics, such as age, to enhance individuals' creativity or productivity.

References

1. Mooneyham, B. W., & Schooler, J. W. (2013). The costs and benefits of mind-wandering: A review. *Canadian Journal of Experimental Psychology, 67*(1), 11–18.

2. Killingsworth, M. A., & Gilbert, D. T. (2010). A wandering mind is an unhappy mind. *Science, 330*(6006), 932.
3. Sullivan, Y. W., Davis, F., & Koh, C. (2015). Exploring mind wandering in a technological setting. In *Proceedings of the 36th International Conference on Information Systems.*
4. Drescher, L. H., van den Bussche, E., & Desender, K. (2018). Absence without leave or leave without absence: Examining the interrelations among mind wandering, metacognition and cognitive control. *PLoS ONE, 13*(2), 1–18.
5. Agnoli, S., Vanucci, M., Pelagatti, C., & Corazza, G. E. (2018). Exploring the link between mind wandering, mindfulness, and creativity: A multidimensional approach. *Creativity Research Journal, 30*(1), 41–53.
6. Oschinsky, F. M., Klesel, M., Ressel, N., & Niehaves, B. (2019). Where are your thoughts? On the relationship between technology use and mind wandering. In *Proceedings of the 52nd Hawaii International Conference on System Sciences.*
7. van der Heijden, H. (2004). User acceptance of hedonic information systems. *MIS Quarterly, 28*(4), 695–704.
8. Dane, E. (2018). Where is my mind? Theorizing mind wandering and its performance-related consequences in organizations. *Academy of Management Review, 43*(2), 179–197.
9. Maillet, D., Beaty, R. E., Jordano, M. L., Touron, D. R., Adnan, A., Silvia, P. J., Kwapil, T. R., Turner, G. R., Spreng, R. N., & Kane, M. J. (2018). Age-related differences in mind-wandering in daily life. *Psychology and Aging, 33*(4), 643–653.
10. Mowlem, F. D., Skirrow, C., Reid, P., Maltezos, S., Nijjar, S. K., Merwood, A., Barker, E., Cooper, R., Kuntsi, J., & Asherson, P. (2016). Validation of the mind excessively wandering scale and the relationship of mind wandering to impairment in adult ADHD. *Journal of Attention Disorders, 23*(6), 624–634.
11. Smallwood, J., & Schooler, J. W. (2015). The science of mind wandering: Empirically navigating the stream of consciousness. *Annual Review of Psychology, 66*, 487–518.
12. Elie-Dit-Cosaque, C., Pallud, J., & Kalika, M. (2011). The influence of individual, contextual, and social factors on perceived behavioral control of information technology: A field theory approach. *Journal of Management Information Systems, 28*(3), 201–234.
13. Kahneman, D. (2012). *Thinking, fast and slow.* Penguin Books.
14. Marchiori, D. M., Mainardes, E. W., & Rodrigues, R. G. (2019). Do individual characteristics influence the types of technostress reported by workers? *International Journal of Human-Computer Interaction, 35*(3), 218–230.
15. Zavagnin, M., Borella, E., & Beni, R. (2014). When the mind wanders: Age-related differences between young and older adults. *Acta Psychologica, 145*, 54–64.
16. Christoff, K., Irving, Z. C., Fox, K. C., Spreng, R. N., & Andrews-Hanna, J. R. (2016). Mind-wandering as spontaneous thought: A dynamic framework. *Nature Reviews Neuroscience, 17*(11), 718–731.
17. Tams, S. (2022). Helping older workers realize their full organizational potential: A moderated mediation model of age and IT-enabled task performance. *MIS Quarterly, 46*(1), 1–33.
18. Morris, M. G., & Venkatesh, V. (2000). Age differences in technology adoption decisions: Implications for a changing work force. *Personnel Psychology, 53*(2), 375–403.
19. Charness, N., & Boot, W. R. (2009). Aging and information technology use. *Current Directions in Psychological Science, 18*(5), 253–258.
20. Giambra, L. M. (1995). A laboratory method for investigating influences on switching attention to task-unrelated imagery and thought. *Consciousness and Cognition, 4*(1), 1–21.
21. Posner, M. I., & Petersen, S. E. (1990). The attention system of the human brain. *Annual Review of Neuroscience, 13*, 25–42.
22. Schooler, J. W., Smallwood, J., Christoff, K., Handy, T. C., Reichle, E. D., & Sayette, M. A. (2011). Meta-awareness, perceptual decoupling and the wandering mind. *Trends in Cognitive Sciences, 15*(7), 319–326.
23. Baldwin, C. L., Roberts, D. M., Barragan, D., Lee, J. D., Lerner, N., & Higgins, J. S. (2017). Detecting and quantifying mind wandering during simulated driving. *Frontiers in Human Neuroscience, 11*(406), 1–15.

24. Zhang, Y., Kumada, T., & Xu, J. (2017). Relationship between workload and mind-wandering in simulated driving. *PLoS ONE, 12*(5), 1–12.
25. Epel, E. S., Putermanet, E., Lin, J., Blackburn, E., Lazaro, A., & Mendes, W. B. (2013). Wandering minds and aging cells. *Clinical Psychological Science, 1*(1), 75–83.
26. Sayette, M. A., Dimoff, J. D., Levine, J. M., Moreland, R. L., & Votruba-Drzal, E. (2012). The effects of alcohol and dosage-set on risk-seeking behavior in groups and individuals. *Journal of the Society of Psychologists in Addictive Behaviors, 26*(2), 194–200.
27. Smallwood, J., O'Connor, R. C., Sudbery, M. V., & Obonsawin, M. (2007). Mind-wandering and dysphoria. *Cognition and Emotion, 21*(4), 816–842.
28. Seli, P., Risko, E. F., Smilek, D., & Schacter, D. L. (2016). Mind-wandering with and without intention. *Trends in Cognitive Sciences, 20*(8), 605–617.
29. Giambra, L. M. (1989). Task-unrelated thought frequency as a function of age: A laboratory study. *Psychology and Aging, 4*(2), 136–143.
30. Seli, P., Risko, E. F., & Smilek, D. (2016). Assessing the associations among trait and state levels of deliberate and spontaneous mind wandering. *Consciousness and Cognition, 41*, 50–56.
31. Conrad, C., & Newman, A. (2019). Measuring the impact of mind wandering in real time using an auditory evoked potential. In F. D. Davis et al. (Eds.), *Information systems and neuroscience* (Lecture notes in information systems and organisation) (Vol. 32, pp. 37–45).
32. Wati, Y., Koh, C., & Davis, F. (2014) Can you increase your performance in a technology-driven society full of distractions? In *Proceedings of the 35th International Conference on Information Systems*.
33. Franklin, M. S., Mrazek, M. D., Anderson, C. L., Johnston, C., Smallwood, J., Kingstone, A., & Schooler, J. W. (2017). Tracking distraction: The relationship between mind-wandering, meta-awareness, and ADHD symptomatology. *Journal of Attention Disorders, 21*(6), 475–486.
34. Christian, B. M., Miles, L. K., Parkinson, C., & Macrae, C. N. (2013). Visual perspective and the characteristics of mind wandering. *Frontiers in Psychology, 4*(699), 1–33.
35. Lin, H.-H., Wang, Y.-S., & Chou, C.-H. (2012). Hedonic and utilitarian motivations for physical game systems use behavior. *International Journal of Human-Computer Interaction, 28*(7), 445–455.
36. Lowry, P. B., Gaskin, J., Twyman, N., Hammer, B., & Roberts, T. (2013). Taking "fun and games" seriously: Proposing the hedonic-motivation system adoption model (HMSAM). *Journal of the Association for Information Systems, 14*(11), 617–671.
37. Salthouse, T. A., & Babcock, R. L. (1991). Decomposing adult age differences in working memory. *Developmental Psychology, 27*(5), 763–776.
38. Wu, J., & Lu, X. (2013). Effects of extrinsic and intrinsic motivators on using utilitarian, hedonic, and dual-purposed information systems: A meta-analysis. *Journal of the Association for Information Systems, 14*(3), 153–191.
39. Brown, S. A., & Venkatesh, V. (2005). Model of adoption of technology in households: A baseline model test and extension incorporating household life cycle. *MIS Quarterly, 29*(3), 399–426.
40. Ferratt, T. W., Prasad, J., & Dunne, E. J. (2018). Fast and slow processes underlying theories of information technology use. *Journal of the Association for Information Systems, 19*(1), 1–22.
41. Fox, K. C. R., & Roger, E. B. (2019). Mind-wandering as creative thinking: Neural, psychological, and theoretical considerations. *Current Opinion in Behavioral Sciences, 27*, 123–130.
42. Harris, J., Ives, B., & Junglas, I. (2012). IT consumerization: When gadgets turn into enterprise IT tools. *MIS Quarterly Executive, 11*, 99–112.
43. Niehaves, B., Köffer, S., & Ortbach, K. (2012). IT consumerization—A theory and practice review. In *Proceedings of the 18th Americas Conference on Information Systems*.
44. Klesel, M., Schlechtinger, M., Oschinsky, F. M., Conrad, C., & Niehaves, B (2020) Detecting mind wandering episodes in virtual realities using eye tracking. In F. D. Davis et al. (Eds.), *Information systems and neuroscience* (Lecture notes in information systems and organisation) (Vol. 43, pp. 163–171).

45. Klesel, M., Oschinsky, F. M., & Niehaves, B. (2019). Investigating the role of mind wandering in computer-supported collaborative work: A proposal for an EEG study. In F. D. Davis et al. (Eds.), *Information systems and neuroscience* (Lecture notes in information systems and organisation) (Vol. 32, pp. 53–62).
46. Thatcher, J. B., Wright, R. T., Sun, H., Zagenczyk, T. J., & Klein, R. (2018). Mindfulness in information technology use: Definitions, distinctions, and a new measure. *MIS Quarterly, 42*(3), 831–847.
47. Agarwal, R., & Karahanna, E. (2000). Time flies when you're having fun: Cognitive absorption and beliefs about information technology usage. *MIS Quarterly, 24*(4), 665–694.
48. Attig, C., Wessel, D., & Franke, T. (2017). Assessing personality differences in human-technology interaction: An overview of key self-report scales to predict successful interaction. In C. Stephanidis (Ed.), *HCI international 2017—Posters' extended abstracts.* (Communications in computer and information science) (pp. 19–29). Springer International Publishing.
49. Aykin, N. M., & Aykin, T. (1991). Individual differences in human-computer interaction. *Computers & Industrial Engineering, 20*(3), 373–379.

A Brief Review of Information Security and Privacy Risks of NeuroIS Tools

Rosemary Tufon and Adriane B. Randolph

Abstract The use of neurophysiological tools in information systems research has increased dramatically in the last decade as scholars employ these tools to better understand how humans think, feel, and behave while interacting with technology and develop innovative neuro-adaptive systems. These tools are expanding the realm of digital data capture and storage, fueling the ongoing debate on data security and privacy issues. This paper highlights the privacy and security risks posed by the use of neurophysiological tools in the domain of NeuroIS as fuel for further examination in the field.

Keywords NeuroIS · Neurophysiological tools · Information security · Privacy · Neuro-adaptive systems

1 Introduction

Data security and privacy concerns with the use of technology was a hot topic long before the emergence of research in the domain of neuro-information systems (NeuroIS) and now should be given more credence here. NeuroIS includes the use of neurophysiological tools in information systems research to better understand how humans think, feel, and behave while interacting with technology and develop innovative neuro-adaptive systems [1, 2]. Further, NeuroIS includes the process of capturing real-time data from the human body with the aid of neurophysiological tools [3]. The use of neurophysiological tools is gaining momentum in research due to their ability to complement, contradict, or supplement existing sources of information with data captured directly from the human body [4]. Most of these studies have focused on the benefits of these tools and there is a large body of literature showing

R. Tufon · A. B. Randolph (✉)
Kennesaw State University, Kennesaw, GA 30144, USA
e-mail: arandol3@kennesaw.edu

R. Tufon
e-mail: rtufon@students.kennesaw.edu

© The Author(s), under exclusive license to Springer Nature Switzerland AG 2022
F. D. Davis et al. (eds.), *Information Systems and Neuroscience*,
Lecture Notes in Information Systems and Organisation 58,
https://doi.org/10.1007/978-3-031-13064-9_33

329

their extensive use particularly in neuromarketing [5, 6]. Several scholars [4, 7–10] have identified and characterized methods and measurements for the application and use of these tools in information systems research but discussions on security and privacy issues in relation to the use of these tools in NeuroIS research is lacking.

Ienca and Haselager [11] raised concerns about the potential for these tools to become a target for cybercriminals through misuse in the brain-computer interface community. Li et al. [12] noted that manufacturers of some of these tools do not pay much attention to security and privacy related issues during development. The use of these tools in NeuroIS research involves the possibility of access to private and sensitive information from the brain of users thus presenting a significant threat to privacy and data protection. Like many ubiquitous technologies [13], neurophysiological tools have the potential to reveal private user attributes, personality traits, and even manipulate an individual's memory [14, 15] through the data collected or in the process of interpreting and communicating the information. For example, neuro-data from EEG signals is considered ideal for biometrics [16]. This is a method used to identify individuals for surveillance or security raising concerns about data gathering and sharing practices using this technology. Furthermore, when combined with artificial intelligence, neurophysiological tools offer the possibility to access and control our deepest thoughts raising novel legal and ethical questions on the privacy, confidentiality, and security of brain data in ways that are yet to be fully-explored.

Advances in information and communication technologies continue to pose significant threats to information security and privacy as the fundamental designs of these technologies rely on electronic data capture, transmission, and even interpretation of the data. This raises important questions on the concept of personal identity, privacy, and security even if the technology is still in its nascent stages of development. Yet the potential for these tools to impact data security and privacy remains relatively unexplored in the field. It is therefore important to understand how the use of these tools in NeuroIS research can impact the security, privacy, and integrity of data by asking important questions regarding storage and transmission of the data, data ownership, and security mechanisms provided in the design and implementation in order to limit access to an individual's personal thoughts.

Riedl and Léger [7] proposed a framework to categorize neurophysiological tools used in information systems research under three categories in terms of the type of measurement captured and mechanism of data acquisition and transmission. Positron Emission Tomography (PET), functional Magnetic Resonance Imaging (fMRI), functional Near-Infrared Spectroscopy (fNIRS), Electroencephalography (EEG), Transcranial Magnetic Stimulation (TMS), and Transcranial Direct-Current Stimulation (TDCS) are functional brain imaging methods that measure and manipulate neural activity in the central nervous system (CNS). The second category includes Electrocardiogram (EKG) and Electrodermal Activity (EDA) which deal with measurement of the Peripheral Nervous System (PNS). This category also includes Eye tracking and Electromyogram (EMG) technologies. The third class of neurophysiological tools is related to measurements of the hormonal system. A brief description of some commonly used neurophysiological tools and their application in industry and research is summarized in Table 1. This list is summarized from Riedl and Léger's classification [7] with a restatement of what they measure and some examples of their use in research and practice. This is not meant to be a definitive list of applications but just some examples of how these tools have been used in academia and practice.

2 Addressing Information Security and Privacy Concerns in NeuroIS Research

This work is an exploratory study that serves as a precursor to a larger study which intends to uncover how some of the commonly used neurophysiological tools interact with information, how the tools are being employed in research and practice, and how information can be compromised when using the tools. A brief review is presented here highlighting some of the risks associated with the use of NeuroIS tools. This proposal will be followed by later work with a more extensive review of the literature on privacy and data security of these tools in NeuroIS research.

Because the NeuroIS field is still developing, it is important for scholars not only to be familiar with the methods, tools, and measurements—which is the focus of most of the research in extant literature—but they also need to be aware of the data privacy and security issues that may arise with the use of these tools in their research. This problem of data privacy and security of NeuroIS tools is particularly significant for several reasons. The majority of neurophysiological tools have applications based in medicine [18, 21, 31] and are being used in many cases to assist patients whose neuromuscular functions have been impaired by a CNS disease or trauma, as with brain-computer interfaces. This exposes an already vulnerable population of users who may not have the ability to seclude confidential or inherently sensitive information about themselves, thus subjecting them to intrusion into their privacy [32].

In this paper, we first discuss how neurophysiological tools can be a target for computer criminals looking to compromise information security. Then we present an initial outline of the risks on data privacy and security issues associated with the neurophysiological tools listed in Table 1. This paper aims to alert researchers about privacy and security risks of neurophysiological tools in NeuroIS research and serves as a precursor to a more critical analysis of the literature. This more extensive review will provide several findings and contributions for the literature. First, we will identify privacy and security concerns of neurophysiological tools that have been used in NeuroIS research and initiate a discussion of the privacy implications of these tools. Then, we will identify research gaps and provide suggested directions and research questions that can be studied on a deeper level for each of the tools.

Research Objectives

1. Outline security risks associated with each neurophysiological tool listed in Table 1.

Table 1 Commonly used neurophysiological tools adapted from Riedl and Léger's classification [7]

Neurophysiological tools	Measure	Example application	Source
CNS: measurement and stimulation of the central nervous system (brain and spinal cord)			
Electroencephalogram tools (EEG)	Measures and records electrical activity of the brain	BCI-powered EEG headsets that record brain signals and spell out their messages for others to read [17]	Machado et al. (2010)
		BCI-actuated smart wheelchair system [18]	Tang et al. (2018)
		Brain oscillations control hand orthosis in a tetraplegic [19]	Pfurtscheller et al. (2000)
		At home with BCI allows patients with loss muscle tone to control systems in their environment, such as lights, stereo sets, television sets, telephones, and front doors [20, 21]	Nicolas-Alonso and Gomez-Gil (2012), Cincotti et al. (2008)
		Perform motor functions with artificial limbs, leading to functional recovery [22]	Kruse et al. (2020)
Functional magnetic resonance imaging (fMRI)	Measures neural activity by detecting changes in blood flow	Presurgical mapping of patients with brain tumors and epilepsy [23]	Vakamudi et al. (2020)
Positron emission tomography (PET)	Use of radioactive isotope to measure brain activity	Multitude of clinical applications in cardiology, neurology, and psychiatry as well [24]	Davis et al. (2020, December)
Functional near-infrared spectroscopy (fNIRS)	Uses near-infrared spectroscopy to measure brain activity from light sources applied on the scalp	Ambulatory brain activity assessment in real world environments [25, 26]	Pinti et al. (2015), Gefen et al. (2014)

(continued)

Table 1 (continued)

Neurophysiological tools	Measure	Example application	Source
Transcranial magnetic stimulation (TMS)	Use magnetic field to manipulate neural activity	Assess the functionality of a brain area [27]	Nardone et al. (2014)
Transcranial direct-current stimulation (tDCS)	Employs low-amplitude current that is applied directly to the scalp via electrodes to stimulate brain activity	Potentiate the effects of virtual reality training on static and functional balance among children with cerebral palsy [28]	Lazzari et al. (2017)
PNS: measurement and stimulation of the peripheral nervous system			
Electrocardiogram (EKG)	Measures electrical activity of the heart on the skin	Detect heart problems	
Electrodermal activity (EDA)	Measures conductance of the skin in a specific context, or in response to a particular stimulus	Assess physiological reactivity to trauma-related cues [2]	Riedl
Eye tracking	Measures conditions and movements of the eyes	Using the eyes for navigation and controls [29]	Duchowski (2002)
Facial electromyography (fEMG)	Measures electrical impulses caused by muscle fibers	Web usability [30]	Benedek and Hazlett (2005)
Measurement of the hormone system			
Salivary hormone measurement	Measures levels of the stress hormone cortisol	Technostress [2]	Riedl
Adrenaline hormone measurement	Adrenaline concentrations in urine samples	Technostress [2]	Riedl

2. Identify how NeuroIS researchers have addressed privacy and security concerns of the tools used. (This is distinct from NeuroIS researchers using neurophysiological tools to examine privacy and security questions.)
3. Identify research gaps and propose future research directions.

3 Information Security and Privacy Risks Associated with NeuroIS Tools

Neurophysiological tools capture and transmit data electronically and like any other technology that deals with electronic data capture, data ownership, security, privacy, and information sharing are key challenges that need to be addressed in this emerging environment. Security and privacy are two very closely related constructs. In this work, we define security as the protection of people, assets, and information from harm or abuse, and privacy as the freedom from being observed or disturbed. Security and privacy are interwoven as security is needed to safeguard privacy while at the same time privacy requires security. To identify the risks associated with these technologies, we adopted the classification offered by Riedl and Léger [7] and reviewed the potential risks under the three categories which include interactions with the CNS, interactions with the PNS, and interactions with the endocrine or hormonal system. Using the search terms security and privacy of neurophysiological tools, we performed an initial exploration on Google Scholar to derive a list of potential risks associated with neurophysiological tools. This information is summarized in Table 2.

We have seen that security and privacy threats from the use of neurophysiological tools can be malicious or unintentional, depending on the technology in question. The next objective in this study is to identify how and when these tools have been used in NeuroIS research in the last decade in the context of privacy and security. To accomplish this step, we plan to perform an extended, systematic literature review (SLR) related to the security and privacy of each class of neurophysiological tools to determine the security and privacy features of the technologies. We plan to carry out an in-depth analysis of issues related to the security and privacy features of each technology reported in published literature using the Preferred Reporting Items for Systematic reviews and Meta-Analysis (PRISMA) framework [40]. The search terms will be information security, data privacy, neurophysiological tools and NeuroIS in top outlets for NeuroIS research as identified in Riedl et al. [41].

Overall, this work aims to perform a systematic review to identify the work being done in NeuroIS on the security and privacy of neurophysiological tools. Concerns for information security and privacy have increased as innovative technologies flood the industry. This work will help further the discussion on security and neuro-adaptive tools and provide evidence from research to support the development of guidelines for using these tools in practice and academia.

Table 2 Summary of security and privacy risks with neurophysiological tools

Neurophysiological tool	Risk	Citation
CNS: measurement and stimulation of the central nervous system (brain and spinal cord)		
Electroencephalogram tools (EEG)	Input manipulation by altering the stimuli presented to the user leading to Exposure to sensitive information, intrusion, and decisional interference	Li et al. [12], Rosenfeld et al. [32], Bernal et al. [33]
	Manipulation of measurement by interfering with the signals to generate faulty output leading to data integrity and availability issues	Li et al. [12], Bernal et al. [33], Ienca and Haselager [11]
	Interference with data processing and conversion by disrupting the analog-to-digital conversion that occurs during neural data acquisition, as well as the translation of firing patterns to particular stimulation devices	Bernal et al. [33], Ienca and Haselager [11]
	Introduce malicious algorithms to alter the decoding and encoding of communication compromising data integrity and availability	Bonaci et al. [31]
	Feedback manipulation of user's intended action	Landau et al. [34], Ienca and Haselager [11]
PNS: measurement and stimulation of the peripheral nervous system		
Electrocardiogram (EKG)	Eavesdropping to record all encryption keys leading to data breach	Alsadhan and Khan [35]
	Flood network with illegitimate messages in Denial of Service attack compromising the data integrity	Alsadhan and Khan [35]
	Capture legitimate messages and replay them in the network compromising data freshness through replay	Alsadhan and Khan [35]
	Unauthorized access, false data injection attacks	Alsadhan and Khan [35]

(continued)

Table 2 (continued)

Neurophysiological tool	Risk	Citation
Eye tracking	Compromised passwords and unintended authentications; observation attacks, video attacks, device malfunction leading to information leakage about identity and interests	Liebling and Preibusch [13], Shen et al. [36], Katsini et al. [37], Katsini et al. [38]
Facial electromyography (fEMG)	Intentional malicious insider leading to Information theft, information fraud, information corruption and sabotage	Jouini et al. [39]
	Unintentional apathetic insider noncompliance with security requirement causing accidental information corruption or loss	Jouini et al. [39]

References

1. Fehrenbacher, D. D. (2017). Affect infusion and detection through faces in computer-mediated knowledge-sharing decisions. *Journal of the Association for Information Systems, 18*(10), 2.
2. Riedl, R., Davis, F. D., & Hevner, A. R. (2014). Towards a NeuroIS research methodology: Intensifying the discussion on methods, tools, and measurement. *Journal of the Association for Information Systems, 15*(10), 4.
3. Qian, J., & Law, R. (2016). *Vincenzo Morabito: The future of digital business innovation: Trends and practices* (pp. 459–461).
4. Dimoka, A. (2012). How to conduct a functional magnetic resonance (fMRI) study in social science research. *MIS Quarterly*, 811–840.
5. Alvino, L., Pavone, L., Abhishta, A., & Robben, H. (2020). Picking your brains: Where and how neuroscience tools can enhance marketing research. *Frontiers in Neuroscience, 14*, 1221.
6. Songsamoe, S., Saengwong-ngam, R., Koomhin, P., & Matan, N. (2019). Understanding consumer physiological and emotional responses to food products using electroencephalography (EEG). *Trends in Food Science & Technology, 93*, 167–173.
7. Riedl, R., & Léger, P.-M. (2016) Tools in NeuroIS research: An overview. *Fundamentals of NeuroIS*, 47–72.
8. Dimoka, A., Davis, F. D., Gupta, A., Pavlou, P. A., Banker, R. D., Dennis, A. R., Ischebeck, A., et al. (2012). On the use of neurophysiological tools in IS research: Developing a research agenda for NeuroIS. *MIS Quarterly*, 679–702.
9. Randolph, A. B., Petter, S. C., Storey, V. C., & Jackson, M. M. (2022). Context-aware user profiles to improve media synchronicity for individuals with severe motor disabilities. *Information Systems Journal, 32*(1), 130–163.
10. Wegrzyn, S. C., Hearrington, D., Martin, T., & Randolph, A. B. (2012). Brain games as a potential nonpharmaceutical alternative for the treatment of ADHD. *Journal of Research on Technology in Education, 45*(2), 107–130.
11. Ienca, M., & Haselager, P. (2016). Hacking the brain: Brain–Computer interfacing technology and the ethics of neurosecurity. *Ethics and Information Technology, 18*(2), 117–129.
12. Li, Q. Q., Ding, D., & Conti, M. (2015). Brain-computer interface applications: Security and privacy challenges. In *2015 IEEE Conference on Communications and Network Security (CNS)* (pp. 663–666). IEEE.

13. Liebling, D. J., & Preibusch, S. (2014). Privacy considerations for a pervasive eye tracking world. In *Proceedings of the 2014 ACM International Joint Conference on Pervasive and Ubiquitous Computing: Adjunct Publication* (pp. 1169–1177).

14. Davies, N., Friday, A., Clinch, S., Sas, C., Langheinrich, M., Ward, G., & Schmidt, A. (2015). Security and privacy implications of pervasive memory augmentation. *IEEE Pervasive Computing, 14*(1), 44–53.

15. Uzzaman, S., & Joordens, S. (2011). The eyes know what you are thinking: Eye movements as an objective measure of mind wandering. *Consciousness and Cognition, 20*(4), 1882–1886.

16. Ma, L., Minett, J. W., Blu, T., & Wang, W. S. Y. (2015). Resting state EEG-based biometrics for individual identification using convolutional neural networks. In *2015 37th Annual International Conference of the IEEE Engineering in Medicine and Biology Society (EMBC)* (pp. 2848–2851). IEEE.

17. Machado, S., Araújo, F., Paes, F., Velasques, B., Cunha, M., Budde, H., Basile, L. F., et al. (2010). EEG-based brain-computer interfaces: An overview of basic concepts and clinical applications in neurorehabilitation. *Reviews in the Neurosciences, 21*(6), 451–468.

18. Tang, J., Liu, Y., Dewen, H., & Zhou, Z. T. (2018). Towards BCI-actuated smart wheelchair system. *Biomedical Engineering Online, 17*(1), 1–22.

19. Pfurtscheller, G., Guger, C., Müller, G., Krausz, G., & Neuper, C. (2000). Brain oscillations control hand orthosis in a tetraplegic. *Neuroscience Letters, 292*(3), 211–214.

20. Nicolas-Alonso, L. F., & Gomez-Gil, J. (2012). Brain computer interfaces, a review. *Sensors, 12*(2), 1211–1279.

21. Cincotti, F., Mattia, D., Aloise, F., Bufalari, S., Schalk, G., Oriolo, G., Cherubini, A., Marciani, M. G., & Babiloni, F. (2008). Non-invasive brain–computer interface system: Towards its application as assistive technology. *Brain Research Bulletin, 75*(6), 796–803.

22. Kruse, A., Suica, Z., Taeymans, J., & Schuster-Amft, C. (2020). Effect of brain-computer interface training based on non-invasive electroencephalography using motor imagery on functional recovery after stroke-a systematic review and meta-analysis. *BMC Neurology, 20*(1), 1–14.

23. Vakamudi, K., Posse, S., Jung, R., Cushnyr, B., & Chohan, M. O. (2020). Real-time presurgical resting-state fMRI in patients with brain tumors: Quality control and comparison with task-fMRI and intraoperative mapping. *Human Brain Mapping, 41*(3), 797–814.

24. Davis, K. M., Ryan, J. L., Aaron, V. D., & Sims J. B. (2020). PET and SPECT imaging of the brain: History, technical considerations, applications, and radiotracers. In *Seminars in Ultrasound, CT and MRI* (Vol. 41, No. 6, pp. 521–529). WB Saunders.

25. Pinti, P., Aichelburg, C., Lind, F., Power, S., Swingler, E., Merla, A., Hamilton, A., Gilbert, S., Burgess, P., & Tachtsidis, I. (2015). Using fiberless, wearable fNIRS to monitor brain activity in real-world cognitive tasks. *JoVE (Journal of Visualized Experiments), 106*, e53336.

26. Gefen, D., Ayaz, H., & Onaral, B. (2014). Applying functional near infrared (fNIR) spectroscopy to enhance MIS research. *AIS Transactions on Human-Computer Interaction, 6*(3), 55–73.

27. Nardone, R., Tezzon, F., Höller, Y., Golaszewski, S., Trinka, E., & Brigo, F. (2014). Transcranial magnetic stimulation (TMS)/Repetitive TMS in mild cognitive impairment and Alzheimer's disease. *Acta Neurologica Scandinavica, 129*(6), 351–366.

28. Lazzari, R. D., Politti, F., Belina, S. F., Grecco, L. A. C., Santos, C. A., Dumont, A. J. L., Lopes, J. B. P., Cimolin, V., Galli, M., & Oliveira, C. S. (2017). Effect of transcranial direct current stimulation combined with virtual reality training on balance in children with cerebral palsy: A randomized, controlled, double-blind, clinical trial. *Journal of Motor Behavior, 49*(3), 329–336.

29. Duchowski, A. T. (2002). A breadth-first survey of eye-tracking applications. *Behavior Research Methods, Instruments, & Computers, 34*(4), 455–470.

30. Benedek, J., & Hazlett, R. (2005). Incorporating facial emg emotion measures as feedback in the software design process. In *Proceedings of Human Computer Interaction Consortium*.

31. Bonaci, T., Calo, R., & Chizeck, H. J. (2014). App stores for the brain: Privacy & security in brain-computer interfaces. In *2014 IEEE International Symposium on Ethics in Science, Technology and Engineering* (pp. 1–7). IEEE.

32. Rosenfeld, J. P., Biroschak, J. R., & Furedy, J. J. (2006). P300-based detection of concealed autobiographical versus incidentally acquired information in target and non-target paradigms. *International Journal of Psychophysiology, 60*(3), 251–259.

33. Bernal, S. L., Celdrán, A. H., Pérez, G. M., Barros, M. T., & Balasubramaniam, S. (2021). Security in brain-computer interfaces: State-of-the-art, opportunities, and future challenges. *ACM Computing Surveys (CSUR), 54*(1), 1–35.

34. Landau, O., Puzis, R., & Nissim, N. (2020). Mind your mind: EEG-based brain-computer interfaces and their security in cyber space. *ACM Computing Surveys (CSUR), 53*(1), 1–38.

35. Alsadhan, A., & Khan, N. (2013). An LBP based key management for secure wireless body area network (WBAN). In *2013 14th ACIS International Conference on Software Engineering, Artificial Intelligence, Networking and Parallel/Distributed Computing* (pp. 85–88). IEEE.

36. Shen, M., Liao, Z., Zhu, L., Mijumbi, R., Du, X., & Hu, J. (2018). *IriTrack: Liveness detection using irises tracking for preventing face spoofing attacks.* arXiv:1810.03323

37. Katsini, C., Fidas, C., Belk, M., Samaras, G., & Avouris, N. (2019). A human-cognitive perspective of users' password choices in recognition-based graphical authentication. *International Journal of Human-Computer Interaction, 35*(19), 1800–1812.

38. Katsini, C., Abdrabou, Y., Raptis, G. E., Khamis, M., & Alt, F. (2020). The role of eye gaze in security and privacy applications: Survey and future HCI research directions. In *Proceedings of the 2020 CHI Conference on Human Factors in Computing Systems* (pp. 1–21).

39. Jouini, M., Rabai, L. B. A., & Aissa, A. B. (2014). Classification of security threats in information systems. *Procedia Computer Science, 32*, 489–496.

40. Liberati, A., Altman, D. G., Tetzlaff, J., Mulrow, C., Gøtzsche, P. C., Ioannidis, J. P. A., Clarke, M., Devereaux, P. J., Kleijnen, J., & Moher, D. (2009). The PRISMA statement for reporting systematic reviews and meta-analyses of studies that evaluate health care interventions: Explanation and elaboration. *Journal of Clinical Epidemiology, 62*(10), e1–e34.

41. Riedl, R., Fischer, T., Léger, P.-M., & Davis, F. D. (2020). A decade of NeuroIS research: Progress, challenges, and future directions. *ACM SIGMIS Database: The DATABASE for Advances in Information Systems, 51*(3), 13–54.

Picture Classification into Different Levels of Narrativity Using Subconscious Processes and Behavioral Data: An EEG Study

Leonhard Schreiner, Hossein Dini, Harald Pretl, and Luis Emilio Bruni

Abstract In this study, the narrativity of pictures is evaluated using behavioral scales and subconscious processes. The narrative context of the stimulus pictures was classified into four different Levels. For eliciting evoked potentials (EPs), a P300-based picture ranking system was adopted. The EPs were analyzed on significant differences between seen/unseen and Levels of the pictures. In the first paradigm, pictures were continuously presented for 15 s, and the subjects were asked to focus on the picture's narrative. In the second paradigm, the pictures were randomly flashed, whereby one of the previously presented images was chosen as the target and unseen (non-target) pictures across Levels. The preliminary results from this Work in Progress (WIP) study show that seen images cause significantly different EPs compared to unseen images, especially in pictures with abstract and dramatic narratives. Therefore, target stimuli are ranked higher by the picture ranking system. In addition, the N600 potential is evident with abstract narrative stimuli, which have been previously reported to indicate memory function and post-perceptual processing. Further investigation will focus on differences in ERPs and ranking results across Levels and the extraction of possible EEG-biomarkers for narrative Levels in visual stimuli.

Keywords Narrativity · EEG · EP · P300 · Pictures

L. Schreiner (✉)
G.Tec Medical Engineering GmbH, Sierningstraße 14, 4521 Schiedlberg, Austria
e-mail: schreiner@gtec.at

H. Dini · L. E. Bruni
Aalborg University Copenhagen, A. C. Meyers Vænge 15, 2450 Copenhagen, SV, Denmark

L. Schreiner · H. Pretl
Johannes Kepler University, Altenberger Straße 69, 4040 Linz, Austria

© The Author(s), under exclusive license to Springer Nature Switzerland AG 2022   339
F. D. Davis et al. (eds.), *Information Systems and Neuroscience*,
Lecture Notes in Information Systems and Organisation 58,
https://doi.org/10.1007/978-3-031-13064-9_34

1   Introduction

To understand the influence a story has on people, research has focused on the narrative experience to which the audience is exposed [1]. Therefore, liking behavioral data with subconscious information from narratives and narrative engagement is of particular interest [2]. When it comes to narrative stimuli, as presented by Naany in 2009, it does not need movies or a series of connected events to create narratives. Because a picture can represent more than what it depicts, it is possible to interpret a picture as a narrative [3]. Ryan et al. categorizes the conditions of narrativity into spatial, temporal, mental, formal, and pragmatic dimensions [4]. Based on these narrative theories, the picture stimuli chosen in this study are expected to hold narrative information that can be divided into several Levels according to their complexity and semantic content. Human Electroencephalographic (EEG) biomarkers have been utilized to understand the neural processing of narratives [5]. Thus, written narratives in textual form reduced motor activity for the second language [6], engaging movie clips cause higher inter-subject correlation in EEG than non-engaging ones [7], and N400 Evoked Potentials (EPs) are subconscious measures for incongruent brand logos [8]. Moreover, P300 EPs indicate visual attention and focus [9], and N600 are prominent during visual memory tasks [10]. Yet, it is not unclear whether or not visual attention tasks can be utilized to draw inferences from known narratives to unseen visual stimuli. Hence, in this study, an EP-based picture ranking system, initially introduced by Sutaj et al., was adopted to find differences between seen/unseen stimuli and Levels of Narrativity [11]. The paper at hand explores narrative experience and engagement using EEG, which is an important and major tool used in NeuroIS research [12–16].

2   Materials and Methods

2.1   Subjects

We gathered data from 27 healthy young adults (18 male, 9 female) mean age 27.1 ± 3.3 years. No subject that participated in any of the two paradigms took any psychoactive medication or suffered from psychiatric or neurological disease. The subjects provided written informed consent and were approved by the Aalborg University Copenhagen (AAU).

2.2   Experimental Design

In the first part of the experiment, 40 pictures (10 of each Level of Narrativity—see Sect. 2.4) were presented pseudo-randomly for 15 s on a computer screen. The

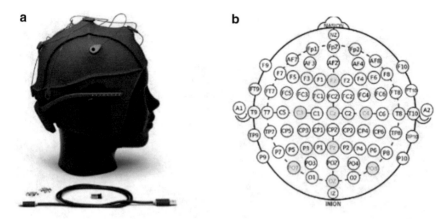

Fig. 1 **a** EEG system used in the study; **b** 8 electrode positions used for signal acquisition

subjects were sitting at approximately 1 m distance to the screen. The participants were asked to focus on the narrative and semantic content. After each picture presentation, the subjects filled out a questionnaire on the PC screen within 40 s. Subsequently, a fixation cross presented for 5 s indicated the appearance of the next stimulus. As a next step, a picture flashing software that generates EPs for image ranking purposes was adopted for stimulus presentation [11]. The EEG system used for analysis in this paper is the Unicorn Hybrid Black system (G.Tec Medical Engineering GmbH) (see Fig. 1a). The 8-channel EEG amplifier was connected to a computer using a Bluetooth interface. The channel positions used in this study are Fz, C3, Cz, C4, Pz, PO7, Oz, PO8 (see Fig. 1b).

Four subsets of pictures were created to elicit target EPs. Each subset represents a narrative Level and contains one target picture that was shown in the first paradigm, representing the respective Level (1–4) (see Fig. 2) and ten non-target pictures (randomized unseen pictures from all Levels). The classification into Levels was done according to Sects. 2.4 and 2.5. Finally, this selection results in 4 picture sets of each 11 pictures (1 target, 10 non-target). Each image was shown for 150 ms directly, followed by the next image without any dark screen. The application randomly chooses the order of the images.

2.3 Stimuli

In Fig. 2, the target stimuli from all 4 Levels (classification according to Sect. 2.5) can be seen. Level 1 represents abstract pictures (Fig. 2a), Level 2 low-Level narratives (Fig. 2b), Level 3 includes mid to high-Level narrativity (Fig. 2c), and Level 4 shows non-habitual dramatic narratives (Fig. 2d).

Fig. 2 The four target pictures were chosen according to the pre-test ranking from narrative Level 1 (**a**), Level 2 (**b**), Level 3 (**c**), Level 4 (**d**)

2.4 Behavioral Parameters

The pictures used in this study were classified into 4 different Levels of Narrativity according to the Questions Q1.1–Q1.4 as seen in the flow chart in Fig. 3. Additionally, the narrative engagement of every picture used was estimated with the Questions Q2.1–Q2.4. In an online pre-test survey, 50 participants (17 male, 33 female) and an average age of 26.44 ± 4.19 years were asked to answer the Questions Q1.1–Q1.4 and Q2.1–Q2.4. The survey was created via the platform Survey Monkey (https://www.surveymonkey.com) and participants were recruited using the platform Prolific (www.prolific.co). The participants were asked to inspect the pictures for approx. 10 s and answer the 8 questions.

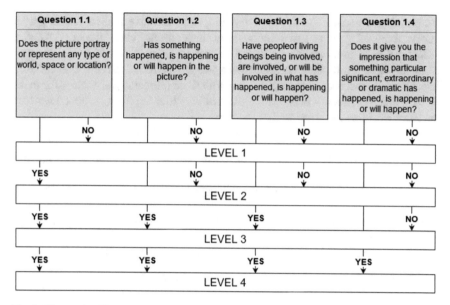

Fig. 3 Picture classification according questions Q1.1–Q1.4

2.5 Level of Narrativity

The flow chart of the division of the images depending on the Yes or No answer can be seen in Fig. 3. If all of the 4 questions were answered with No, the picture is classified to have no narrativity (Level 1—Abstract). Answering just Q1.1 with yes gives the figure a sensuous meaning with at least a setting of narrativity (Level 2). Additionally, answering Q1.2 with Yes links the stimulus with an event. This event-related narrative provides a medium Level of narrativity (Level 2–3). Further classification parameters are humans or living beings appearing in the picture, allowing us to assign it to Level 3. Level 4 pictures have dramatic non-habitual or character involvement in events.

An overall percentage of 100% for No, and 0% of Yes answers indicated that the stimulus is Level 1 and the picture has the lowest possible narrativity. On the contrary, 100% of the questions answered with Yes and 0% answered with No show that the images have the highest possible narrativity. Therefore, the mean Yes answer was taken as a score for the Level of narrativity.

2.6 Narrative Engagement

Questions to estimate the Narrative engagement in the picture itself were done according to Busselle and Bilandzic [17]. Thereby it is discriminated in Narrative Understanding, Emotional Engagement and Attentional Focus. The first factor, labeled narrative understanding, comprises narrative realism and cognitive perspective-taking items, and describes how viewers comprehend the story. Another aspect, referred to as attentional focus, relates to viewers' focus on, or attention from, the figure. A third factor, called emotional engagement, refers to viewers' feelings about characters, either empathizing with them (empathy), or feeling for them. Finally, attentional focus deals with a sense of transitioning from the actual world to the story world and is composed of the items.

The questions used in the study were formulated as follows:

- **Q2.1 Narrative presence**: It took me a long time to interpret what was happening in the picture.
- **Q2.2 Emotional Engagement**: What happens in the picture moved me emotionally.
- **Q2.3 Narrative understanding**: It was Easy to understand what was going on in the picture
- **Q2.4 Attentional Focus**: I had a hard time keeping my mind on the picture.

For quantification, the 7 point Likert scale was adopted [18].

2.7 EEG and Picture Ranking

Data were acquired and digitized at 250 Hz and further preprocessed with a 2nd order Butterworth Notch filter at 50 Hz and a 2nd order Butterworth bandpass filter with a band from 0.5 to 30 Hz. Following, data is epoched into 1.1 s trials with 100 ms pre- and 1000 ms post-visual stimulus. After data screening, six datasets had to be removed due to artifacts and bad signal quality. Evoked responses were averaged across the remaining 21 subjects of previously seen visual stimuli and unseen ones (target and non-target). Both seen and unseen visual stimuli, were taken from the same narrativity Level and yielded similar scores during the questionnaire. Hence, they were expected to elicit similar attention and mental workload.

Triggered by the visual stimulus, EEG features are extracted and passed to train the machine learning algorithm using time-variant Linear Discrimination Analysis (LDA). The LDA distance (score value) was used to rank the pictures and correlate them to the behavioral measures. Cai et al. (2013) showed that a linear regression in the LDA subspace is mathematically equivalent to a low-rank linear regression [19]. Accordingly, the features generated by subsequent images during the ranking process are projected into the LDA subspace and the distance (score value) resulting from that projection is used to rank the new images.

Table 1 Picture ranking order and Behavioural Question answers of the online survey of 50 participants from the picture set used for Level 1 picture ranking

Ranking	1 (target)	2	3	4	5	6	7	8	9	10	11
Picture level	1	1	3	4	2	1	4	3	1	2	2
Level score	0.20	0.08	0.66	0.92	0.38	0.18	0.96	0.72	0.12	0.74	0.40
Narr. presence	4.67	3.93	2.38	3.65	1.74	4.63	2.67	2.22	5	2.16	2.48
Emotional eng.	2.48	2.52	2.23	4.80	3.85	2.26	3.26	2.96	2.15	2.69	3.26
Narr. Underst.	2.56	3.11	5.04	4.77	6.33	2.52	5.41	5.70	2.15	5.5	4.85
Attentional focus	3.41	3.07	2.43	2.73	2.11	2.59	2.78	2.89	4.15	2.73	2.67

Columns are ordered according to the ranking of the system stated in the first row. Bold marked pictures from Level 1

3   Results

3.1   Behavioral and Ranking Results

In Table 1, the averaged score results from the online survey are listed. The Level Score was calculated as mentioned in Sect. 2.4. The scores from Q2.1–Q2.4 are the average from the 7-point Likert scale. Marked in bold are the columns for all Level 1 pictures. Ranked first was the target picture from Level 1. All Level 1 pictures are marked bold. Levels from all pictures are stated in the Picture Level row.

3.2   Evoked Potentials

Figure 4 depicts the grand mean from the evoked potentials averaged from 21 subjects, target and non-target trials of channel 7 (Oz). Every Level contains ten unseen non-target (mixed Levels) and one target picture within the Level that was shown in the first part of the experiment. The difference between the narrativity paradigm presented target stimulus and the unseen non-target stimuli are significant for Levels 1 and 4, with a p-value of 0.01, 0.4 s after stimulus onset. Clearly recognizable is also a negative peak at around 0.25 s after the trigger for Levels 1 and 4. Especially important to highlight is the significant difference around 600 ms after the stimulus (N600 peak) for target pictures of Level one. For Levels 2 and 3, minor differences between target and non-target stimuli are found. The baseline oscillation occurs due to the fast flashing of 150 ms and the overlap of the evoked potentials.

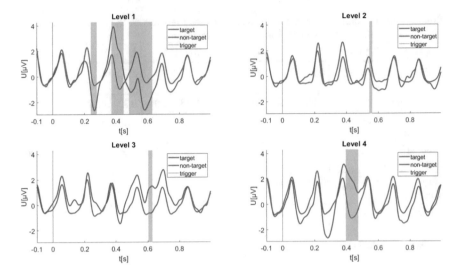

Fig. 4 Evoked potentials recorded from channel 7 (Oz), averaged over 21 subjects and all trials. Trials from one target (blue) and 10 non-target (red) pictures averaged with a significant difference of p < 0.05 marked in dark green and p < 0.01 marked in light green

4   Discussion and Conclusion

Preliminary results indicate that EPs obtained from seen or familiar visual stimuli of narrativity Level 1 and Level 4 are significantly different from unseen ones across all Levels. Although narrativity Levels and related scores are similar for the seen and unseen stimuli, a potential difference is mainly visible at a latency of 250, 400 ms for Level 1 and Level 4 target pictures. This implies that not only the narrativity Level may affect the EP, but also a known narrative may alter the EP waveform. Additionally, a more negative deflection of the already seen narrative around 600 ms after the stimulus is evident for the Level one target picture. The N600 potential has been previously reported to indicate memory function and post-perceptual and post-decision information processing [10, 20]. Hence, a more negative N600 for known stimuli may cause further post-processing of the narrative, while for unknown stimuli, the narrative may not be fully recognized during the short stimulation period of 150 ms. In contrast, subjects could perceive the narrative of the known stimulus over 15 s before the EEG attention task. The current analysis includes a P300-based ranking system trained using pictures of the 4 Levels. The Level 1 target picture could successfully be ranked highest based on the differences in the EPs. Such a ranking could be used to determine the narrative Level score of each picture. However, given the preliminary results that known pictures elicit different EP waveforms, the ranking scores may be different for known and unknown target training data sets.

Acknowledgements This work was partially funded via the European Commission project RHUMBO—H2020-MSCA-ITN-2018-813234.

References

1. Escalas, J. E. (2007). Self-referencing and persuasion: Narrative transportation versus analytical elaboration. *Journal of Consumer Research, 33*, 421–429. https://doi.org/10.1086/510216
2. Kalaganis, F. P., Georgiadis, K., Oikonomou, V. P., et al. (2021). Unlocking the subconscious consumer bias: A survey on the past, present, and future of hybrid EEG schemes in neuromarketing. *Frontiers in Neuroergonomics, 2.*
3. Nanay, B. (2009). Narrative pictures. *Journal of Aesthetics and Art Criticism, 67*, 119–129. https://doi.org/10.1111/j.1540-6245.2008.01340.x
4. Ryan, M.-L. (2007). Toward a definition of narrative. In D. Herman (Ed.), *The Cambridge companion to narrative* (1st ed., pp. 22–36). Cambridge University Press.
5. Jääskeläinen, I. P., Klucharev, V., Panidi, K., & Shestakova, A. N. (2020). Neural Processing of Narratives: From Individual Processing to Viral Propagation. *Frontiers in Human Neuroscience, 14*, 253. https://doi.org/10.3389/fnhum.2020.00253
6. Birba, A., Beltrán, D., Martorell Caro, M., et al. (2020). Motor-system dynamics during naturalistic reading of action narratives in first and second language. *NeuroImage, 216*, 116820. https://doi.org/10.1016/j.neuroimage.2020.116820
7. Cohen, M. X. (2008). Assessing transient cross-frequency coupling in EEG data. *Journal of Neuroscience Methods, 168*, 494–499. https://doi.org/10.1016/j.jneumeth.2007.10.012
8. Dini, H., Simonetti, A., Bigne, E., & Bruni, L. E. (2022). EEG theta and N400 responses to congruent versus incongruent brand logos. *Science and Reports, 12*, 4490. https://doi.org/10.1038/s41598-022-08363-1
9. Pritchard, W. S. (1981). Psychophysiology of P300. *Psychological Bulletin, 89*, 506–540. https://doi.org/10.1037/0033-2909.89.3.506
10. Harauzov, A. K., Shelepin, Y. E., Noskov, Y. A., et al. (2016). The time course of pattern discrimination in the human brain. *Vision Research, 125*, 55–63. https://doi.org/10.1016/j.visres.2016.05.005
11. Sutaj, N., Walchshofer, M., & Schreiner, L., et al. (2021). Evaluating a novel P300-based real-time image ranking BCI. *Frontiers in Computer Science, 3.*
12. Dimoka, A., Davis, F. D., Gupta, A., et al. (2012). On the use of neurophysiological tools in IS research: Developing a research agenda for NeuroIS. *MIS Quarterly, 36*, 679–702. https://doi.org/10.2307/41703475
13. Müller-Putz, G., Riedl, R., & Wriessnegger, S. (2015). Electroencephalography (EEG) as a research tool in the information systems discipline: Foundations, measurement, and applications. *Communications of the Association for Information Systems, 37*, 911–948. https://doi.org/10.17705/1CAIS.03746
14. Riedl, R., Fischer, T., Léger, P.-M., & Davis, F. D. (2020). A decade of NeuroIS research: Progress, challenges, and future directions. *SIGMIS Database, 51*, 13–54. https://doi.org/10.1145/3410977.3410980
15. Müller-Putz, G. R., Tunkowitsch, U., Minas, R. K., et al. (2021). On electrode layout in EEG studies: 13th annual information systems and neuroscience, NeuroIS 2021. *Information Systems and Neuroscience—NeuroIS Retreat, 2021*, 90–95. https://doi.org/10.1007/978-3-030-88900-5_10
16. Riedl, R., Minas, R., Dennis, A., & Müller-Putz, G. (2020). *Consumer-grade EEG instruments: Insights on the measurement quality based on a literature review and implications for NeuroIS research* (pp. 350–361).
17. Busselle, R., & Bilandzic, H. (2009). Measuring narrative engagement. *Media Psychology, 12*, 321–347. https://doi.org/10.1080/15213260903287259
18. Joshi, A., Kale, S., Chandel, S., & Pal, D. (2015). Likert scale: Explored and explained. *BJAST, 7*, 396–403. https://doi.org/10.9734/BJAST/2015/14975
19. Cai, X., Ding, C., Nie, F., & Huang, H. (2013). On the equivalent of low-rank linear regressions and linear discriminant analysis based regressions.

20. Althen, H., Banaschewski, T., Brandeis, D., & Bender, S. (2020). Stimulus probability affects the visual N700 component of the event-related potential. *Clinical Neurophysiology, 131*, 655–664. https://doi.org/10.1016/j.clinph.2019.11.059

Usability Evaluation of Assistive Technology for ICT Accessibility: Lessons Learned with Stroke Patients and Able-Bodied Participants Experiencing a Motor Dysfunction Simulation

Félix Giroux, Loic Couture, Camille Lasbareille, Jared Boasen, Charlotte J. Stagg, Melanie K. Fleming, Sylvain Sénécal, and Pierre-Majorique Léger

Abstract The recruitment of disabled participants for conducting usability evaluation of accessible information and communication technologies (ICT) is a challenge that current research faces. To overcome these challenges, researchers have been calling upon able-bodied participants to undergo disability simulations. However, this practice has been criticized due to the different experiences and expectations that

F. Giroux (✉) · L. Couture · J. Boasen · S. Sénécal · P.-M. Léger
Tech3Lab, HEC Montréal, Montréal, Québec, Canada
e-mail: felix.giroux@hec.ca

L. Couture
e-mail: loic.couture@hec.ca

J. Boasen
e-mail: jared.boasen@hec.ca

S. Sénécal
e-mail: sylvain.senecal@hec.ca

P.-M. Léger
e-mail: pierre-majorique.leger@hec.ca

P.-M. Léger
Department of Information Technologies, HEC Montréal, Montréal, Québec, Canada

S. Sénécal
Department of Marketing, HEC Montréal, Montréal, Québec, Canada

J. Boasen
Faculty of Health Sciences, Hokkaido University, Sapporo, Japan

C. Lasbareille · C. J. Stagg · M. K. Fleming
Wellcome Centre for Integrative Neuroimaging, Department of Clinical Neuroscience, University of Oxford, Nuffield, England
e-mail: camille.lasbareilles@ndcn.ox.ac.uk

C. J. Stagg
e-mail: charlotte.stagg@ndcn.ox.ac.uk

© The Author(s), under exclusive license to Springer Nature Switzerland AG 2022
F. D. Davis et al. (eds.), *Information Systems and Neuroscience*,
Lecture Notes in Information Systems and Organisation 58,
https://doi.org/10.1007/978-3-031-13064-9_35

disabled and able-bodied participants may have with ICT. This paper presents the methodology and lessons learned from ongoing mixed method-based usability evaluation of a suboptimal conventional computer mouse and an assistive gesture-based interface (i.e., the Leap Motion Controller) by stroke patients with upper-limb impairment and able-bodied participants experiencing a motor dysfunction simulation. The paper concludes with recommendations for future multidisciplinary research on ICT accessibility by people with disabilities.

Keywords Accessibility · Usability Evaluation · Disability Simulation ·
Gesture-Based Interface · Assistive Technology

1 Introduction

With the growing development of assistive technologies for Information and Communication Technology (ICT) accessibility, there is a need for evaluating their usability to encourage their adoption and continued use by disabled people [1–3]. However, research faces the challenge of recruiting disabled participants to perform usability evaluations [4]. Therefore, able-bodied participants experiencing disability simulations are often used for identifying preliminary usability issues and design recommendations with technologies targeting people with disabilities [5–8].

However, literature has criticized the relevance of simulating disabilities in able-bodied participants in research [9–11]. Since disability simulations are only temporary experiences of a newly acquired disability [12], it has been argued that they do not reflect the true experience of disabled individuals who have developed compensatory techniques to cope with their disability [13, 14]. Furthermore, when experiencing a disability simulation, able-bodied individuals may focus on what they cannot do rather than focusing on issues that are relevant to the targeted users [9]. Therefore, able-bodied participants have experiences with and expectations about technologies that may influence their perceptions and consequently the relevance of their usability evaluation.

Usability evaluations of assistive technologies are often conducted according to international standards (ISO-9241) [15] by assessing the effectiveness, efficiency, and satisfaction with the technology. Typically, mixed-method approaches using a combination of objective measures (e.g., task performance) and subjective measures (e.g., Likert scales and post-task interviews) [8, 10, 16–21] are used. The advantages of mixed-method approaches are that they allow the simultaneous investigation of confirmatory and exploratory research questions, providing stronger inferences for complementary findings, as well as allowing the investigation of divergent findings [22–24]. Therefore, mixed method-based usability evaluations may allow the investigation of potential perception bias in able-bodied participants' subjective assessment of assistive technologies.

M. K. Fleming
e-mail: melanie.fleming@ndcn.ox.ac.uk

Here, we build upon the findings of a previous research proposal [25] by using a mixed method-based usability evaluation of a conventional technology (i.e., computer mouse) and a novel assistive technology (i.e., gesture interface) in stroke patients suffering from upper-limb spasticity and able-bodied participants experiencing a motor dysfunction simulation. Based on task-technology fit literature [26–28] suggesting that alignment between one's characteristics and a technology's characteristics can predict performance, we expected that the assistive technology supporting the motor dysfunction would lead to better performance and perceived usability than the suboptimal conventional technology. This single-session study is not trying to measure participants' intention to use the technology, but rather how the alignment between one's abilities and a technology's functionalities can predict performance at a task. Therefore, task technology fit literature and more specifically individual technology fit framework [26–28] is used to manipulate the expected performance of both technologies, which is expected to influence one's perceptions and attitudes towards the technologies.

To assess the usability of the two input devices, we used: (a) objective measures of task performance and workload via electroencephalography (EEG) and electromyography (EMG), (b) subjective measures of workload, usability, and self-efficacy via Likert scales, and (c) qualitative insights on usability issues and design recommendations via semi-structured interviews. This paper presents the methodology and lessons learned from the mixed method-based usability evaluation conducted with able-bodied participants and stroke survivors. This study contributes to the field of NeuroIS which has been studying digital technologies with patient groups ranging from cardiovascular conditions to Parkinson Disease patients [29–32]. Moreover, NeuroIS researchers have studied the influence of input devices like mice or touchscreen on user's memory retrieval in online product selection task [33, 34]. Therefore, this study also contributes to NeuroIS research by using neurophysiological measurements to assess the influence of two input devices of users' performance, as well as perceptions and attitudes towards the technologies.

2 Methods

2.1 Participants

In a first study, we recruited 22 right-handed able-bodied participants with no neurological disorders. For these participants, a hand spasticity disability simulation splint was designed out of 3D printed plastic and rubber bands to replicate the clenched hand and wrist position of people suffering from post-stroke upper limb spasticity (Fig. 1) [35]. In the second ongoing study, we recruited five chronic stroke survivors via online forums and platforms in the UK including *My stroke guide* and *Different Strokes* [36, 37] and from a database of participants who had taken part in previous research at the Wellcome Centre for Integrative Neuroimaging.

Fig. 1 The 3D printed hand
spasticity disability
simulation splint with an
elastic band to replicate the
clenched hand and wrist
position of people with
post-stroke upper limb
spasticity

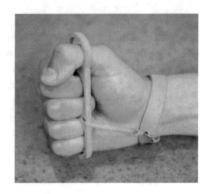

2.2 Input Devices

A mouse is a computer input device with one or more buttons, capable of a two-
dimensional rolling motion that can control a cursor on display [38]. Conventional
mice are designed to be gripped and manipulated using fingers and wrist movement.
They are thus reliant on relatively intact upper motor control and therefore can pose
problems to people with motor dysfunction such as hand spasticity. As alternatives
to conventional mice, assistive technologies such as joysticks or trackballs have been
developed to assist people with physical disabilities in controlling a computer cursor
[17]. One promising, yet unexplored, assistive technology for controlling a cursor
is a gesture interface [39] enabled by devices such as the Leap Motion Controller
(LMC) (Fig. 2). The LMC is a USB device that has three infrared LED lights and
two infrared cameras that track upper limb position and allow moving a cursor on a
display with upper limb movements [40]. The LMC is an accessible device that has
been used extensively in rehabilitation therapy for stroke patients suffering with upper
limb motor deficits [41–44]. Yet, to our knowledge, no research has investigated the
potential of this input device as an alternative to a mouse for computer use by people
with motor dysfunctions.

Fig. 2 The Leap Motion
Controller (LMC) sensing a
participant's fist position to
control the cursor in a
computer interface

Fig. 3 From left to right, the one-dimension Fitts task performed by a participant using a conventional mouse, the two-dimension Fitts task and the spiral tracing task using the LMC

2.3 Usability Tasks

The usability of pointing devices such as a mouse, touchpad, trackball, or the LMC have been typically evaluated using the ISO 9241–9 serial and multidimensional point-selection tasks, also known as the one-dimension and two-dimension Fitts tasks [40, 45–48]. These tasks, which consist in moving a cursor as accurately and rapidly as possible from a fixed target to another and selecting them by hovering over the targets for specified dwell time, were used as the first and second tasks of the study (Fig. 3). While the Fitts tasks assessed the speed and accuracy of reaching movements, a third spiral tracing task was used to assess fine motor control with the cursor. Spiral tracing tasks have been used in past research across a myriad of neurological disorders including Multiple Sclerosis and Parkinson's Disease [49, 50].

2.4 Usability Evaluation

Our mixed method-based usability evaluation considers measures of effectiveness via task performance, efficiency via subjective and neurophysiological workload, and satisfaction via self-efficacy with and usability of the input devices. Performance for the Fitts tasks was determined by Fitts' throughput) [51], a metric that assesses accuracy of rapid aimed movements at targets, and thus is independent from the speed-accuracy tradeoff. As for the spiral tracing task, we designed a metric that combines the mean error of tracing (i.e., distance between spiral tracing and spiral template) adjusted for tracing time, such that faster spiral tracing can increase the accuracy score and overall tracing performance.

We assessed subjective workload, which corresponds to the physical and cognitive demands placed on the user [52], via the NASA-TLX scale [53]. We measured cortico-muscular coherence using a 32-channel EEG system and EMG electrodes positioned on participants' extensor carpi. Cortico-muscular coherence has been found to increase with motor learning in healthy humans [54–56] and reflects motor function recovery after a stroke [57, 58], and can be used as an index of neuromotor efficiency [59].

The satisfaction dimension of usability was measured via a 5-item self-efficacy scale [60, 61], which assesses one's perception about his or her abilities to perform a task well, and 13-item device assessment questionnaire (DAQ) that includes aspects of usability (e.g., pointing accuracy, operation speed), physical and mental effort, ease of use, comfort, and fatigue (i.e., finger, wrist, arm, shoulder, and neck fatigue), and which is typically used to assess the usability of new pointing devices [47, 62]. Finally, we investigated usability issues and design recommendations via semi-structured open-question interviews. We asked participants about their overall experience with the input device(s) and the tasks, the main issues that they have encountered, the consequent design recommendations that they would suggest for improving the usability of the device(s).

2.5 Procedure

Study 1 was conducted in a controlled usability lab in North America during July 2021. We adopted a between-group design by manipulating the input device. All 22 healthy participants performed 10 trials of the one-dimension Fitts task, followed by 10 trials of the two-dimension Fitts task, and 15 trials of the spiral task, with either the LMC or the mouse according to their randomized group. After each trial, participants were presented performance feedback which they were instructed to try increasing over the trials. After each task, participants answered the SES. At the end of the experiment, participants completed the DAQ to assess the device over all three tasks. The experiment lasted for about 60 min and concluded with a 15-min interview.

Study 2 is ongoing but so far, the data of 5 participants have been collected at a Clinical Neuroscience lab (N = 1) or at the patient's home (N = 4) in the UK from January to March 2022. Unlike Study 1, we used a within-subject design to evaluate the two input devices in a random order. We assessed patients' level of motor impairment and cognitive functions using the Fugl-Meyer upper-limb assessment and the Montreal Cognitive Assessment (MoCA). The 5 patients performed six trials of the one-dimension Fitts task, followed by 6 trials of the two-dimension Fitts task, and 12 trials of the spiral task, with both devices in a random order. After each task, participants were administered the NASA-TLX, the DAQ, and the SES. The experiment lasted for about 90 min and was concluded with a 15-min interview. Finally, as suggested by literature on conducting usability testing with disabled participants [63], we carefully monitored levels of fatigue and offered frequent breaks whenever needed as well as reinforcing the fact that the study could be stopped at any time.

3 Lessons Learned

Study 1 allowed us to identify preliminary usability issues (e.g., excessive range of motion with the LMC) and design recommendations (e.g., increasing the sensitivity of the LMC to allow cursor control with smaller movements) that were validated in Study 2. These insights will be used to improve the design of the assistive technology, which will be evaluated with additional stroke patients.

In Study 1, we designed a disability simulation splint replicating early stages of hand spasticity (i.e., clenched fist) such that able-bodied participants could not put their fingers around the mouse to stabilize it. However, the motor dysfunction of interest (i.e., poststroke upper limb hemiparesis) varies significantly on an individual basis in a clinical population in terms of severity and affected segment of the upper limb. This was clearly highlighted by the wide range of upper limb impairments in our stroke patients' Fugl-Meyer assessment scores in Study 2. For example, some patients had limited forearm and wrist movement but retained good use of their shoulder, while other patients had good function of their hand and wrist but had a lot of difficulty moving their affected bicep and shoulder. These insights highlight that our fixed disability simulation did not consider arm and shoulder motion limitations and the impact that this would have on the usability of and perceptions about the LMC. Therefore, as suggested by previous research [5], future studies should consider simulating different levels of impairments to account for a more representative range of disabilities.

Apart from motor function disabilities like limb spasticity, stroke patients may have other conditions like mental and physical fatigue, speech or language disorders, visual impairments, and cognitive function disorders like memory or attention span limitations. In Study 2, all five stroke patients reported that the experiment was very tiring, both physically and mentally. This was corroborated by subjective reports of fatigue and high workload, with three patients requiring an early termination of the experimental tasks. Moreover, during the final interview, one patient could not remember about her experience with the device tested in the first part of the experiment, which could be explained by her low score in the delayed recall section of the MoCA. From this we learned that future research with clinical populations should be shorter in time, and that interviews should be conducted immediately following the use of the devices to avoid reliance on intact memory recall.

Finally, in Study 1, able-bodied participants seem to have underestimated their ability to effectively use a familiar mouse, as contrasted with the novel LMC, in tasks where they clearly performed better with the former input device. This could be explained by the fact that able-bodied participants realized that they would have performed the tasks better with the familiar mouse without the disability simulation, thereby negatively impacting their perception about its usability. Although more analysis is required to investigate the relation between devices' performance and users' perceptions among able-bodied and disabled participants' groups, future research should be cautious about the reliability of subjective measures assessing able-bodied participants' perceptions and attitudes about a technology with which they may have experiences and expectations.

4 Conclusion

This paper supports previous literature on the benefits of working with healthy participants with disability simulations to identify preliminary usability issues in assistive technologies' evaluation but stresses the reliability of assessing subjective measures by this group. Future analysis will include neurophysiological data to contrast objective and subjective workload. Finally, it is hoped that this study will encourage the field of Information System (IS) to investigate the adoption and continued use of assistive technologies for disabled people [29–34, 64–66].

Acknowledgements This study was funded in part by the Wellcome Trust and the Royal Society (102584/Z/13/Z) and supported by the National Institute for Health Research (NIHR) Oxford Biomedical Research Centre. MKF is supported by Guarantors of Brain and the Wellcome Trust. The Wellcome Centre for Integrative Neuroimaging is supported by core funding from the Wellcome Trust (203139/Z/16/Z). Finally, this study was also funded by M. Jean Chagon via the HEC Montréal Foundation, as well as NSERC (IRCPJ/514835-16) and Prompt (61_Léger-Deloitte 2016.12).

References

1. Hoppestad, B. S. (2007). Inadequacies in computer access using assistive technology devices in profoundly disabled individuals: An overview of the current literature. *Disability and Rehabilitation: Assistive Technology, 2*(4), 189–199.
2. Millán, J. D. R., Rupp, R., Mueller-Putz, G., Murray-Smith, R., Giugliemma, C., Tangermann, M., et al. (2010). Combining brain–computer interfaces and assistive technologies: state-of-the-art and challenges. Frontiers in neuroscience *161*
3. Cook, A. M., & Polgar, J. M. (2014). Assistive technologies-e-book: principles and practice. *Elsevier Health Sciences.*
4. Sears, A., & Hanson, V. (2011). Representing users in accessibility research. In: *Proceedings of the SIGCHI conference on Human factors in computing systems,* (pp. 2235–2238).
5. Chen, H. C., Chen, C. L., Lu, C. C., & Wu, C. Y. (2009). Pointing device usage guidelines for people with quadriplegia: A simulation and validation study utilizing an integrated pointing device apparatus. *IEEE Transactions on neural systems and rehabilitation engineering, 17*(3), 279–286.
6. Palani, H. P., & Giudice, N. A. (2017). Principles for designing large-format refreshable haptic graphics using touchscreen devices: An evaluation of nonvisual panning methods. *ACM Transactions on Accessible Computing, 9*(3), 1–25.
7. El Lahib, M., Tekli, J., & Issa, Y. B. (2018). Evaluating Fitts' law on vibrating touch-screen to improve visual data accessibility for blind users. *International Journal of Human-Computer Studies, 112*, 16–27.
8. Manresa-Yee, C., Roig-Maimó, M. F., & Varona, J. (2019). Mobile accessibility: Natural user interface for motion-impaired users. *Universal Access in the Information Society, 18*(1), 63–75.

9. Bennett, C. L., & Rosner, D. K. (2019). The promise of empathy: Design, disability, and knowing the "Other". In: *Proceedings of the 2019 CHI conference on human factors in computing systems*, (pp. 1–13).
10. Bajcar, B., Borkowska, A., & Jach, K. (2020). Asymmetry in usability evaluation of the assistive technology among users with and without disabilities. *International Journal of Human-Computer Interaction, 36*(19), 1849–1866.
11. Tigwell, G. W. (2021). Nuanced Perspectives Toward Disability Simulations from Digital Designers, Blind, Low Vision, and Color-Blind People. In: *Proceedings of the 2021 CHI Conference on Human Factors in Computing Systems (CHI '21)*. Association for Computing Machinery, New York, NY, USA, Article 378, 1–15.
12. French, S. (1996). Simulations exercises in disability awareness training: A critique. In: G. Hales (Ed.), *Beyond disability: Towards an enabling society*, (pp. 114 –123).
13. Burgstahler, S., & Doe, T. (2004). Disability-related simulations: If, when and how to use them in professional development. *Review of Disability Studies, 1*(2), 8–18.
14. Nario-Redmond, M. R., Gospodinov, D., & Cobb, A. (2017). Crip for a day: The unintended negative consequences of disability simulations. *Rehabilitation psychology, 62*(3), 324.
15. ISO 9241–11 (2018). Ergonomics of human-system interaction—Usability: definitions and concepts. *International Organization for Standardization, 9241*(11).
16. Kübler, A., Holz, E. M., Riccio, A., Zickler, C., Kaufmann, T., Kleih, S. C., & Mattia, D. (2014). The user-centred design as novel perspective for evaluating the usability of BCI-controlled applications. *PLoS One, 9*(12), e112392
17. Pousada, T., Pareira, J., Groba, B., Nieto, L., & Pazos, A. (2014). Assessing mouse alternatives to access to computer: A case study of a user with cerebral palsy. *Assistive Technology, 26*(1), 33–44.
18. Choi, I., Rhiu, I., Lee, Y., Yun, M. H., & Nam, C. S. (2017). A systematic review of hybrid brain-computer interfaces: Taxonomy and usability perspectives. *PLoS ONE, 12*(4), e0176674.
19. Menges, R., Kumar, C., & Staab, S. (2019). Improving user experience of eye tracking-based interaction: Introspecting and adapting interfaces. *ACM Transactions on Computer-Human Interaction, 26*(6), 1–46.
20. Šumak, B., Špindler, M., Debeljak, M., Heričko, M., & Pušnik, M. (2019). An empirical evaluation of a hands-free computer interaction for users with motor disabilities. *Journal of biomedical informatics, 96*, 103249.
21. Bogza, L. M., Patry-Lebeau, C., Farmanova, E., Witteman, H. O., Elliott, J., Stolee, P., & Giguere, A. M. (2020). User-centered design and evaluation of a web-based decision aid for older adults living with mild cognitive impairment and their health care providers: mixed methods study. *Journal of Medical Internet Research, 22*(8), e17406.
22. Creswell, J. W., & Clark, V. L. P. (2007). *Designing and conducting mixed methods research.* Sage Publications.
23. Tashakkori, A., & Teddlie, C. (2009). Integrating qualitative and quantitative approaches to research. *The SAGE Handbook of Applied Social Research Methods, 2*, 283–317.
24. Venkatesh, V., Brown, S. A., & Bala, H. (2013). Bridging the qualitative-quantitative divide: Guidelines for conducting mixed methods research in information systems. *MIS Quarterly.* 21–54.
25. Giroux, F., Boasen, J., Stagg, C. J., Sénécal, S., Coursaris, C., & Léger, P. M., (2021). Motor dysfunction simulation in able-bodied participants for usability evaluation of assistive technology: a research proposal. In: *NeuroIS Retreat*, (pp. 30–37). Springer, Cham.
26. Goodhue, D. L., & Thompson, R. L. (1995). Task-technology fit and individual performance. *MIS Quarterly.* 213–236.
27. Dishaw, M. T., & Strong, D. M. (1999). Extending the technology acceptance model with task–technology fit constructs. *Information & management, 36*(1), 9–21.
28. Randolph, A. B., & Moore Jackson, M. M. (2010). Assessing fit of nontraditional assistive technologies. *ACM Transactions on Accessible Computing 2*(4), 1–31.
29. Moore, M., Storey, V., & Randolph, A. (2005). User profiles for facilitating conversations with locked-in users. In: *ICIS 2005 Proceedings*, 73.

30. Randolph, A., Karmakar, S., & Jackson, M. (2006). Towards predicting control of a Brain-computer interface. In: *ICIS 2006 Proceedings*. 53.
31. Javor, A., Ransmayr, G., Struhal, W., & Riedl, R. (2016). Parkinson patients' initial trust in avatars: Theory and evidence. *PLoS ONE, 11*(11), e0165998.
32. Vogel, J., Auinger, A., Riedl, R., Kindermann, H., Helfert, M., & Ocenasek, H. (2017). Digitally enhanced recovery: Investigating the use of digital self-tracking for monitoring leisure time physical activity of cardiovascular disease (CVD) patients undergoing cardiac rehabilitation. *PLoS ONE, 12*(10), e0186261.
33. Mirhoseini, S., Leger, P. M., Senecal, S., Fredette, M., Cameron, A. F., & Riedl, R. (2013). Investigating the effect of input device on memory retrieval: Evidence from Theta and Alpha Band Oscillations. In: *SIGHCI 2013 Proceedings*. 15
34. Sénécal, S., Léger, P. -M., Fredette, M., Courtemanche, F., Cameron, A. -F., Mirhoseini, S. M. M., Paquet, A., & Riedl, R. (2013). Mouse vs. touch screen as input device: does it influence memory retrieval? In: *International Conference on Information Systems (ICIS 2013)*.
35. Bhakta, B. B. (2000). Management of spasticity in stroke. *British medical bulletin, 56*(2), 476–485.
36. https://www.stroke.org.uk/
37. https://differentstrokes.co.uk/
38. ISO 9241–400 (2007). Ergonomics of human-system interaction—Principles and requirements for physical input devices. *International Organization for Standardization*, 9241(400).
39. ISO 9241–171 (2007). Ergonomics of human-system interaction—Principles and requirements for physical input devices. *International Organization for Standardization*. 9241(400).
40. Bachmann, D., Weichert, F., & Rinkenauer, G. (2015). Evaluation of the leap motion controller as a new contact-free pointing device. *Sensors, 15*(1), 214–233.
41. Callejas-Cuervo, M., Díaz, G. M., & Ruíz-Olaya, A. F. (2015). Integration of emerging motion capture technologies and videogames for human upper-limb telerehabilitation: A systematic review. *Dyna., 82*(189), 68–75.
42. Iosa, M., Morone, G., Fusco, A., Castagnoli, M., Fusco, F. R., Pratesi, L., & Paolucci, S. (2015). Leap motion controlled videogame-based therapy for rehabilitation of elderly patients with subacute stroke: A feasibility pilot study. *Topics in stroke rehabilitation, 22*(4), 306–316.
43. Barrett, N., Swain, I., Gatzidis, C., & Mecheraoui, C. (2016). The use and effect of video game design theory in the creation of game-based systems for upper limb stroke rehabilitation. *Journal of Rehabilitation and Assistive Technologies Engineering, 3*, 2055668316643644.
44. Tarakci, E., Arman, N., Tarakci, D., & Kasapcopur, O. (2020). Leap Motion Controller–based training for upper extremity rehabilitation in children and adolescents with physical disabilities: A randomized controlled trial. *Journal of Hand Therapy, 33*(2), 220–228.
45. MacKenzie, I. S. (1992). Fitts' law as a research and design tool in human-computer interaction. *Human-computer interaction, 7*(1), 91–139.
46. MacKenzie, I.S., Buxton, W. (1992). Extending Fitts' law to two-dimensional tasks. In: *Proceedings of the CHI'92: ACM Conference on Human Factors in Computing Systems*, (pp. 219–226).
47. Soukoreff, R. W., & MacKenzie, I. S. (2004). Towards a standard for pointing device evaluation, perspectives on 27 years of Fitts' law research in HCI. *International Journal of Human-Computer Studies, 61*(6), 751–789.
48. Jones, K. S., McIntyre, T. J., & Harris, D. J. (2020). Leap motion-and mouse-based target selection: Productivity, perceived comfort and fatigue, user preference, and perceived usability. *International Journal of Human-Computer Interaction, 36*(7), 621–630.
49. Longstaff, M. G., & Heath, R. A. (2006). Spiral drawing performance as an indicator of fine motor function in people with multiple sclerosis. *Human movement science, 25*(4–5), 474–491.
50. Danna, J., Velay, J. L., Eusebio, A., Véron-Delor, L., Witjas, T., Azulay, J. P., & Pinto, S. (2019). Digitalized spiral drawing in Parkinson's disease: A tool for evaluating beyond the written trace. *Human movement science, 65*, 80–88.
51. MacKenzie, I.S. and Isokoski, P. (2008). Fitts' throughput and the speed-accuracy tradeoff. In: *Proceedings CHI '08*, 1633–1636.

52. Rubio, S., Díaz, E., Martín, J., & Puente, J. M. (2004). Evaluation of subjective mental work-load: A comparison of SWAT, NASA-TLX, and workload profile methods. *Applied psychology, 53*(1), 61–86.

53. Hart, S. G., & Staveland, L. E. (1988). Development of NASA-TLX (Task Load Index): Results of empirical and theoretical research. *In Advances in psychology, 52*, 139–183.

54. Kristeva-Feige, R., Fritsch, C., Timmer, J., & Lücking, C. H. (2002). Effects of attention and precision of exerted force on beta range EEG-EMG synchronization during a maintained motor contraction task. *Clinical Neurophysiology, 113*(1), 124–131.

55. Kristeva, R., Patino, L., & Omlor, W. (2007). Beta-range cortical motor spectral power and corticomuscular coherence as a mechanism for effective corticospinal interaction during steady-state motor output. *NeuroImage, 36*(3), 785–792.

56. Mendez-Balbuena, et al. (2012). Corticomuscular coherence reflects interindividual differences in the state of the corticomuscular network during low-level static and dynamic forces. *Cerebral Cortex, 22*(3), 628–638.

57. Fang, Y., Daly, J. J., Sun, J., Hvorat, K., Fredrickson, E., Pundik, S., & Yue, G. H. (2009). Functional corticomuscular connection during reaching is weakened following stroke. *Clinical Neurophysiology, 120*(5), 994–1002.

58. Zheng, Y., Peng, Y., Xu, G., Li, L., & Wang, J. (2018). Using corticomuscular coherence to reflect function recovery of paretic upper limb after stroke: A case study. *Frontiers in neurology, 8*, 728.

59. Franco-Alvarenga, P. E., Brietzke, C., Canestri, R., Goethel, M. F., Viana, B. F., & Pires, F. O. (2019). Caffeine increased muscle endurance performance despite reduced cortical activation and unchanged neuromuscular efficiency and corticomuscular coherence. *Nutrients, 11*(10), 2471.

60. Maddux, J. E. (1995). *Self-efficacy Theory*. Springer.

61. Compeau, D. R., & Higgins, C. A. (1995). Application of social cognitive theory to training for computer skills. *Information systems research, 6*(2), 118–143.

62. Douglas, S. A., Kirkpatrick, A. E., & MacKenzie, I. S. (1999). Testing pointing device perfor-mance and user assessment with the ISO 9241, Part 9 standard. In: *Proceedings of the SIGCHI conference on Human Factors in Computing Systems*, (pp. 215–222).

63. Pernice, K., & Nielsen, J. (2001). *How to conduct usability studies for accessibility*. Technical Report. Nielsen Norman Group.

64. Dimoka, A., Davis, F. D., Gupta, A., Pavlou, P. A., Banker, R. D., Dennis, A. R., ... & Weber, B. (2012). On the use of neurophysiological tools in IS research: Developing a research agenda for NeuroIS. *MIS Quarterly*. 679–702.

65. Olbrich, S., Trauth, E. M., Niedermann, F., & Gregor, S. (2015). Inclusive design in IS: Why diversity matters. *Communications of the Association for Information Systems, 37*(1), 37.

66. Pethig, F., & Kroenung, J. (2019). Specialized information systems for the digitally disadvan-taged. *Journal of the Association for Information Systems, 20*(10), 5.

Correction to: Information Systems and Neuroscience

Fred D. Davis, René Riedl⊕, Jan vom Brocke, Pierre-Majorique Léger, Adriane B. Randolph, and Gernot R. Müller-Putz

Correction to:
F. D. Davis et al. (eds.), *Information Systems*
and Neuroscience, **Lecture Notes in Information Systems**
and Organisation 58,
https://doi.org/10.1007/978-3-031-13064-9

Chapter 12 includes a small error in the Chapter Title. It has been updated as "Resolving the Paradoxical Effect of Human-Like Typing Errors by Conversational Agents".

The following belated corrections have been incorporated in Chapter 28 "Measurement of Heart Rate and Heart Rate Variability: A Review of NeuroIS Research with a Focus on Applied Methods":

The numbering of the first six references as well as an citations in the text has now been updated in sequential order. The correction chapter and the book have been updated with the changes.

The updated original version of these chapters can be found at
https://doi.org/10.1007/978-3-031-13064-9_12
https://doi.org/10.1007/978-3-031-13064-9_28

Author Index

A
Addas, Shamel, 201
Aguila Del, Laurène, 229

B
Bartholomeyczik, Karen, 23
Barth, Simon, 89
Bieber, Gerald, 155
Blankenship, Joe, 53
Blicher, Andreas, 61
Boasen, Jared, 81, 349
Bouvier, Frédérique, 179
Brendel, Alfred Benedikt, 113
Brieugne, David, 179
Bruni, Luis Emilio, 339
Büttner, Lea, 155

C
Chakravarty, Sumit, 53
Chandler, Chantel, 229
Chan, Yao-Cheng, 187
Chen, Shang-Lin, 179
Clement, Jesper, 61
Conrad, Colin, 105, 261
Constantiou, Ioanna, 61
Côté, Myriam, 229
Courtemanche, Francois, 229
Couture, Loic, 349

D
Dini, Hossein, 339
Dorner, Verena, 219

Dupuis, Mariko, 229

E
Early, Spencer, 73

F
Fanfan, Ernst R., 53
Fellmann, Michael, 155
Fernández-Shaw, Juan, 229
Fleming, Melanie K., 349
Flöck, Alessandra, 133
Fredette, Marc, 229

G
Giroux, Felix, 81, 179, 349
Gleasure, Rob, 61
Gordon, Nicolette, 311
Greif-Winzrieth, Anke, 123
Greulich, R. Stefan, 113
Gwizdka, Jacek, 187

H
Hamza, Zeyad, 229
Hassanein, Khaled, 73
Henry, Maya L., 187

K
Karran, Alexander J., 229
Klesel, Michael, 105, 301, 319
Knierim, Michael Thomas, 23, 147

Korosec-Serfaty, Marion, 163
Kraft, Dimitri, 155

L
Laerhoven Van, Kristof, 155
Lambusch, Fabienne, 155
Langner, Moritz, 89
Lasbareille, Camille, 349
Laufer, Ilan, 97
Léger, Pierre-Majorique, 81, 163, 179, 229,
    349
Leybourne, Robin, 1
Lutz, Bernhard, 9

M
Maedche, Alexander, 35, 89
Mannina, Sophia, 201
Mayhew, Kydra, 105
Mehu, Marc, 133
Mirhoseini, Mahdi, 73
Mizrahi, Dor, 97
Moore, Kimberly Weston, 311
Müller-Putz, Gernot R., 301

N
Neumann, Dirk, 9
Newman, Aaron J., 261
Niehaves, Björn, 301, 319
Nieken, Petra, 23
Nikolajevic, Nevena, 147

O
Öksüz, Nurten, 171
O'Neil, Kiera, 105
Ortiz, Cesar Enrique Uribe, 219
Oschinsky, Frederike Marie, 105, 155, 301,
    319

P
Paquette, Michel, 81
Pavlevchev, Samuil, 1
Peukert, Christian, 123, 251
Poirier, Sara-Maude, 179
Popp, Florian, 9
Pretl, Harald, 339

R
Radhakrishnan, Kavita, 187
Randolph, Adriane B., 53, 329
Reßing, Caroline, 301, 319
Riedl, René, 211, 269, 285, 301

S
Saigot, Maylis, 237
Schenkluhn, Marius, 251
Schmidt, Angelina, 155
Schreiner, Leonhard, 339
Seitz, Julia, 23, 35
Sénécal, Sylvain, 81, 163, 179, 229, 349
Snow, Pascal, 229
Stagg, Charlotte J., 349
Stangl, Fabian J., 269, 285
Stano, Fabio, 23
Subramaniam, Kajamathy, 81
Suwandjieff, Patrick, 301

T
Tazi, Salima, 229
Tessmer, Rachel, 187
Toreini, Peyman, 89, 123
Tufon, Rosemary, 329

U
Usai, Francesco, 105

V
Vilone, Domenico, 229

W
Walla, Peter, 1
Weinhardt, Christof, 23, 123, 147, 251
Wolfartsberger, Josef, 211
Wriessnegger, Selina C., 301

Z
Zeuge, Anna, 319
Zuckerman, Inon, 97

Printed in the United States
by Baker & Taylor Publisher Services